Lecture Notes in Mathematics 1677

Editors:
A. Dold, Heidelberg
F. Takens, Groningen

Springer
Berlin
Heidelberg
New York
Barcelona
Budapest
Hong Kong
London
Milan
Paris
Santa Clara
Singapore
Tokyo

Nikolai Proskurin

Cubic Metaplectic Forms and Theta Functions

Springer

Author

Nikolai Proskurin
St. Petersburg Branch of the Mathematical Institute
Russian Academy of Sciences
Fontanka 27
St. Petersburg 191011, Russia
e-mail: proskn@pdmi.ras.ru

Library of Congress Cataloging-in-Publication Data

Proskurin, Nikolai, 1953-
Cubic metaplectic forms and theta functions / Nikolai Proskurin.
p. cm. -- (Lecture notes in mathematics ; 1677)
Includes bibliographical references and index.
ISBN 3-540-63751-6 (softcover : alk. paper)
1. Automorphic forms. 2. Discontinuous groups. 3. Functions, Theta. I. Title. II. Series: Lecture notes in mathematics (Springer-Verlag) ; 1677.
QA3.L28 no. 1677
[QA243]
510 s--dc21
[512'.7] 97-45869
CIP

Mathematics Subject Classification (1991): Primary: 11F55, 11F60, 11F30
Secondary: 33C10

ISSN 0075-8434
ISBN 3-540-63751-6 Springer-Verlag Berlin Heidelberg New York

Typesetting: Camera-ready TEX output by the author
SPIN: 10553411 46/3143-543210 - Printed on acid-free paper

Preface

The subject of these notes does not have a long history. It takes its origin from the very short paper published by Kubota in 1965. Let F be a totally imaginary algebraic number field containing the full group of mth roots of 1, denoted by $\mu_m(F)$, and let $\Gamma^{(n)}_{\text{princ}}(q)$ be the principal congruence subgroup module ideal q in $\text{SL}(n, \mathcal{O}_F)$, $\mathcal{O}_F$ being the integers ring of F. Kubota showed [51] that, under some conditions on q, the reciprocity low yields

$$\begin{pmatrix} a & b \\ c & d \end{pmatrix} \mapsto \begin{cases} \left(\frac{c}{d}\right)_m & \text{if } c \neq 0 \\ 1 & \text{if } c = 0 \end{cases}$$

is a group homomorphism $\Gamma^{(2)}_{\text{princ}}(q) \to \mu_m(F)$; here we write $\left(\frac{\cdot}{\cdot}\right)_m$ for the mth degree residue symbol. This theorem has very far-reaching consequences.

In a series of papers [52], [53], ..., [57] Kubota studied automorphic forms under the group $\Gamma^{(2)}_{\text{princ}}(q)$ with the homomorphism above as a multiplier system (= a factor of automorphy), the so called metaplectic forms of degree m. That are real analytic forms, in the sence of Maaß and Selberg, defined on $\mathbf{H}^r$, where $\mathbf{H} \simeq \text{SL}(2, \mathbf{C})/\text{SU}(2)$ is the 3-dimentional hyperbolic space, and r is the number of complex places of F. The metaplectic Eisenstein series are of particular interest. Their Fourier coefficients are the Dirichlet series whose coefficients are the Gauß sums of degree m. The general principles yield then that these Dirichlet series have meromorphic continuations and satisfy some functional equations. This remarkable observation gives the key to solve Kummer problem [37].

Taking the residues of the metaplectic Eisenstein series at some 'exceptional' poles Kubota constructed metaplectic forms which it is reasonable to consider as m^{th} degree analoques of the classical quadratic theta function. In particular, taking $F = \mathbf{Q}(\sqrt{-3})$, $q = (3)$ and $m = 3$, we get the cubic theta function $\Theta_{\text{K-P}}: \mathbf{H} \to \mathbf{C}$. Patterson [69] could find an explicit form for its Fourier expansion. We refer $\Theta_{\text{K-P}}$ as the Kubota–Patterson cubic theta function, it playes a crucial role in these notes. For $m \geq 4$, Fourier coefficients of the mth degree theta functions are not known yet. Some particular results were obtained by

Suzuki [94] and by Eckhardt and Patterson [20] for the biquadratic theta series, i.e., for the case $m = 4$. Patterson considered also the case $m = 6$. It seems reasonable to think that unknown Fourier coefficients of the mth degree theta functions, $m \geq 4$, could be treated as Gauß sums in some extended sence, in this connection see [72], [73].

One more line of thought relates the Kubota homomorphism with the congruence subgroup problem. The relation becames clear if we notice that the kernel of the Kubota homomorphism is the subgroup of finite index in $\mathrm{SL}(2, \mathcal{O}_F)$, and that it does not contain any principle congruence subgroup. To deal with the congruence subgroup problem in more general context, Bass, Milnor and Serre [5] constructed homomorphisms extending Kubota's one.

Due to Bass, Milnor and Serre, we have homomorphisms $\Gamma^{(n)}_{\text{princ}}(q) \to \mu_m(F)$, $n \geq 2$, and we can treat them as multiplier systems to define metaplectic forms on spaces others than that considered originally by Kubota. This is just the point of view accepted by the author in the series of papers [78], ... , [84] and in the present notes.

In other words, the metaplectic forms we define and deal with are 'classical' but not 'adelic' ones. The adelic point of view was accepted by Kubota in [54] and then by Deligne [18], Flicker [27], Kazhdan and Patterson [46], [47], Patterson and Piatetski-Shapiro [74], Flicker and Kazhdan [28]. In this framework metaplectic forms are treated as adelic automorphic forms defined on metaplectic groups.

These notes are organized into three parts.

Part 0 contains essentially known material (except that in subsections 0.3.14 and 0.4.3). Writing Part 0 we had in mind to prepare the necessary background for our research in the forthcoming two parts, and also to gather together the main results concerning the cubic metaplectic forms on $\mathbf{H}$ given by Kubota and by Patterson, which one can find yet in original papers only. The contents of Part 0 is described in details in subsection 0.1.1.

In Part 1 and Part 2 we study cubic metaplectic forms on the symmetric space $\mathbf{X} \simeq \mathrm{SL}(3, \mathbf{C})/\mathrm{SU}(3)$ and, respectively, $\mathbf{X} \simeq \mathrm{Sp}(4, \mathbf{C})/\mathrm{Sp}(4)$. These two parts are independent one from another. For $\mathrm{SL}(3, \mathbf{C})$-case our results are more complete and our exposition is more detailed rather than that for $\mathrm{Sp}(4, \mathbf{C})$-case. In the meantime, Part 1 and Part 2 have entirely similar structure, and this should emphasize the similarity of the methods. In both cases, given a cubic metaplectic form $f\colon \mathbf{H} \to \mathbf{C}$, we have the Eisenstein series $\mathrm{E}(\cdot, s; f)\colon \mathbf{X} \to \mathbf{C}$, $s \in \mathbf{C}$, attached to f in accordance with the general Eisenstein series theory. Our primary goal is to find their Fourier coefficients. For this we apply specific '$sl(2)$-triples technique' developed in [81], [82] and [84]. Then we consider one particular case $f = \Theta_{\text{K-P}}$. The series $\mathrm{E}(\cdot, s; \Theta_{\text{K-P}})$ are very interesting metaplectic forms. We find they have some 'exceptional' poles, and, taking residues, we get cubic theta function Θ on $\mathrm{SL}(3, \mathbf{C})/\mathrm{SU}(3)$ and two cubic theta functions, Θ_{I} and Θ_{II}, on $\mathrm{Sp}(4, \mathbf{C})/\mathrm{Sp}(4)$.

The function Θ on $\mathrm{SL}(3, \mathbf{C})/\mathrm{SU}(3)$ has been constructed by the author [78] and by Kazhdan and Patterson [46]. This theta function, as well as $\Theta_{\text{K-P}}$, occurs in [46] as particular representative of a wide class of theta functions defined on metaplectic coverings of the general linear groups. The Fourier coefficients of Θ were evaluated in [78]. The technique used in [78] is not perfect, and this is one reason to review [78] again, in order to simplify and clarify it.

Two cubic theta functions on $\mathrm{Sp}(4,\mathbf{C})/\mathrm{Sp}(4)$, $\Theta_\natural$ and $\Theta_\sharp$ in the notations of these notes, were constructed in [80], [81]. One of them is the residue of the Eisenstein series $\mathrm{E}(\cdot, s; \Theta_{\text{K-P}})$ at the maximal pole, and it looks like symplectic analogue of the theta functions described in Kazhdan–Patterson theory. The second one, being the residue of $\mathrm{E}(\cdot, s; \Theta_{\text{K-P}})$ at the second pole, has slightly different origin. For better understanding it would be pleasureable to involve both $\Theta_\natural$ and $\Theta_\sharp$ into general symplectic metaplectic forms theory. We hope our observations might by useful to develope such theory.

It should be pointed out that a lot of things we deal with in first four sections of Part 1 as well as of Part 2 are not related to cubic metaplectic forms and theta functions only. In particular, the basic theorems which give us expressions for the Eisenstein series $\mathrm{E}(\cdot, s; f)$ Fourier coefficients in subsections 1.4.1 and 2.4.1 are valid for the Eisenstein series with almost arbitrary multiplier systems. Among other such things, there is our treatment of Whittaker functions on the group $\mathrm{SL}(3,\mathbf{C})$ in subsections 1.3.10, 1.4.7 and on the group $\mathrm{Sp}(4,\mathbf{C})$ in subsections 2.3.7, 2.4.7. We show that the integrals defining Whittaker functions can be evaluated, that gives rise to simple and useful expressions.

In [84] we studied cubic metaplectic forms on the Lie group $\mathrm{G}_2(\mathbf{C})$ of type G_2. There were found two 'exceptional' poles of the Eisenstein series $\mathrm{E}(\cdot, s; \Theta_{\text{K-P}})$ on $\mathrm{G}_2(\mathbf{C})$. Unfortunatly there are too many open questions concerning the theta functions associated with these poles. For this reason, we do not include this part of our research into these notes having in mind first to resolve at least some of them and only then to overview the subject.

I would like to thank U. Christian, H. Helling, S. J. Patterson and D. Hejhal for invitations to Göttingen, Bielefeld and Uppsala where I have worked supported by SFB-170, SFB-343 and Uppsala University. This work was supported also by RFFI (grant 96-01-00663). I especially wish to thank S. J. Patterson for his interest to my work and for stimulating encouragement over many years.

Contents

Part 0

0.1 Preliminaries

0.1.1 Suggestions to the reader. We would like to give some commentary on the present section and the whole Part 0, to save the time and effort of the reader.

In subsection 0.1.2 we have collected some basic notations which will be used throughout these notes. In subsection 0.1.3 we state the properties of the cubic residue symbol. The next two subsections — 0.1.4 and 0.1.5, — contain elementary facts on Gauß and Ramanujan sums, some arithmetic functions, the Dedekind zeta function $\zeta_{\mathbf{Q}(\sqrt{-3})}$ and some cubic Hecke series. We only deal with the field $\mathbf{Q}(\sqrt{-3})$, so prerequisited knowlege of the algebraic number theory is not significant. We hope however the reader feels free with the computations like that in 0.1.4. The most convenient source is the book of Ireland and Rosen [41], where we find detailed description of the arithmetic of the field $\mathbf{Q}(\sqrt{-3})$, including the proof of the cubic reciprocity low. That is all we need except some facts on the Dedekind zeta function and Hecke series for which we can refer to Weil [102]. For reference convenience, we give in subsection 0.1.6 the definitions and some useful facts on special functions — the Euler gamma-function Γ, the Bessel–MacDonald function K_m, the Airy function Ai and the hypergeometric functions ${}_2F_1$ and ${}_3F_2$.

Section 0.2 begins, subsection 0.2.1, with the definitions of some congruence subgroups of $\mathrm{SL}(n, \mathcal{O})$. In the next subsections — 0.2.2,..., 0.2.5 — we present the remarkable Kubota's theorem and its generalization given by Bass, Milnor and Serre. These theorems are of the fundamental importance for us. As to the explicit formulae for the Bass–Milnor–Serre homomorphisms, given in subsection 0.2.5, they are not so important, and we shall not use them.

Section 0.3 is written as a short review of the theory of cubic metaplectic forms on 3-dimensional hyperbolic space $\mathbf{H} = \mathbf{C} \times \mathbf{R}_+^* \simeq \mathrm{SL}(2, \mathbf{C})/\mathrm{SU}(2)$. By the cubic metaplectic forms are understood the automorphic functions with specific multiplier system discovered by Kubota. The material is taken in the main from the works of Kubota [53], [54] and Patterson [69], [70], and includes some new things, particularly in subsection 0.3.14. We give all necessary definitions and statements, but only a few proofs. There is no book on this subject yet, and our operating assumption is that our reader knows somethat the basic concepts of the theory of automorphic functions on complex upper half-plane, including Maaß theory of real analytic automorphic functions. This knowlege would be very helpful for a complete understanding of our exposition. (We can recommend Koblitz [50], Shimura [91], Kubota [58], Venkov [99].) The classical complex upper half-plane is nothing but 2-dimensional hyperbolic space and it can be understood also as $\mathrm{SL}(2,\mathbf{R})/\mathrm{SO}(2)$. It is not a wonder that a lot of things go on $\mathrm{SL}(2,\mathbf{C})/\mathrm{SU}(2)$ quite similar to that on $\mathrm{SL}(2,\mathbf{R})/\mathrm{SO}(2)$. The main difference is that $\mathrm{SL}(2,\mathbf{C})/\mathrm{SU}(2)$ is not a complex analytic manifold, but only a real analytic one. For this reason the automorphic functions one can define on $\mathrm{SL}(2,\mathbf{C})/\mathrm{SU}(2)$ are analogues of the Maaß wave forms, but not of the classical analytic automorphic forms. Certainly, the theory of the automorphic functions on $\mathrm{SL}(2,\mathbf{C})/\mathrm{SU}(2)$ can be viewed as a part of the general automorphic functions theory developed by Selberg [87], [88], Harish-Chandra [32], Langlands [59], Jacquet [43] and others, and such viewpoint leads to better understanding. (We can recommend Baily [2].)

To demonstrate the importance of the cubic metaplectic forms theory for the number theory, we collected in Section 0.4 some of its consequences concerninig the Dirichlet series whose coefficients are the cubic Gauß sums and the squares of the cubic Gauß sums.

0.1.2 Notations. $\mathbf{Z}$, $\mathbf{Q}$, $\mathbf{R}$, $\mathbf{C}$ are the ring of rational integers, the field of rational, the field of real and the field of complex numbers; $\mathbf{C}^*$ is the multiplicative group of $\mathbf{C}$; $\mathbf{R}_+^*$ is the multiplicative group of real positive numbers.

$\Re(z)$, $\Im(z)$, $|z|$ are the real part, the imaginary part and the absolute value of $z \in \mathbf{C}$; $\bar{z}$ is the complex conjugate of z and $e(z) = \exp\big(2\pi i(z+\bar{z})\big)$.

For $z \in \mathbf{C} \setminus (-\infty, 0]$ and $s \in \mathbf{C}$ we write z^s for $\exp(s \log z)$, log being the principle logarithm, i. e., $\log z \in \mathbf{R}$ for $z \in \mathbf{R}_+^*$.

${}^t\gamma$ is the matrix transpose to γ, e_n is the identity matrix $n \times n$, and $\mathrm{diag}(c_1, c_2, \ldots, c_n)$ is $n \times n$ diagonal matrix with c_i at the intersection of ith row and i^{th} column.

All integrals we shall deal with are intergals over the standard Lebesgue measures on $\mathbf{R}$ or on $\mathbf{C}$, and it will be easy to distinguish one case from another just from context.

We shall use standard notation $\mathrm{SL}(n,\mathbf{C})$ for the special linear group of order n over $\mathbf{C}$, and $\mathrm{SU}(n)$ for the special unitary groups of order n,

$$\mathrm{SU}(n) = \big\{k \in \mathrm{SL}(n,\mathbf{C}) \,\big|\, k\,{}^t\bar{k} = e_n\big\}.$$

$\mathrm{Sp}(4,\mathbf{C})$ will denote the symplectic group of rank 2 over $\mathbf{C}$, see 2.1.1 for the definition.

$\mathcal{O} = \mathbf{Z}[\omega]$ is the ring of integers of the field $\mathbf{Q}(\sqrt{-3}\,)$,

$$\omega = \exp(2\pi i/3) = (-1+\sqrt{-3}\,)/2.$$

$\|\cdot\|: \mathbf{Q}(\sqrt{-3}) \to \mathbf{Q}$ is the norm, $\|z\| = z\bar{z}$ for all $z \in \mathbf{Q}(\sqrt{-3})$.

By q we mean an ideal of $\mathcal{O}$, also considered as a lattice in $\mathbf{C}$. Then, $\mathbf{C}/q$ is a fundamental domain of the lattice q in $\mathbf{C}$ and $\mathrm{vol}(\mathbf{C}/q)$ is its volume (with respect to the Lebesgue measure on $\mathbf{C}$). We shall assume that $q \subset (3)$ and $q \neq 0$. The fractional ideal dual to q is

$$q^* = \{c \in \mathbf{Q}(\sqrt{-3}) \mid (cd + \overline{cd}) \in \mathbf{Z} \text{ for all } d \in q\}.$$

One can describe q^* also as the set of all $c \in \mathbf{Q}(\sqrt{-3})$ such that $e(cd) = 1$ for all $d \in q$. If $q = (r)$, i.e., q is generated by $r \in \mathcal{O}$, then $\mathrm{vol}(\mathbf{C}/q) = \sqrt{3}\|r\|/2$, and $q^* = (\sqrt{-3}\,r)^{-1}\mathcal{O}$.

Given $l \in \mathcal{O}$ and $A, B \subset \mathcal{O}$, we write lA for the set $\{la \mid a \in A\}$, $l + A$ for the set $\{l + a \mid a \in A\}$, and $A + B$ for the set $\{a + b \mid a \in A,\ b \in B\}$.

We recall [11], [41] that $\mathcal{O}$ is a ring of principal ideals with the Euclidean algorithm and with a unique factorization of the elements into prime factors. Its group of units is $\mathcal{O}^* = \{\zeta \in \mathbf{C}^* \mid \zeta^6 = 1\} = \{\pm 1, \pm\omega, \pm\omega^2\}$. One can represent each element $k \in \mathcal{O}$, $k \neq 0$, uniquely as the product

$$k = \zeta(\sqrt{-3})^m c \tag{0.1.1}$$

with $\zeta \in \mathcal{O}^*$, $m \in \mathbf{Z}$, $m \geq 0$, $c \in \mathcal{O}$, $c \equiv 1 \pmod 3$.

$\mathcal{O}_{\mathrm{ass}}$ is the subset of $\mathcal{O}$ consisting of 0 and of the numbers $k \in \mathcal{O}$ with $\zeta = 1$ in factorization (0.1.1). Sometimes we shall not make a distinction between an ideal of $\mathcal{O}$ and its generator in $\mathcal{O}_{\mathrm{ass}}$. For each prime $p \in \mathcal{O}$, there exists a unique prime $p' \in \mathcal{O}_{\mathrm{ass}}$ which is associated with p. Certainly, $p' \equiv 1 \pmod 3$ or $p' = \sqrt{-3}$. Throughout these notes, by primes in $\mathcal{O}$ we shall mean primes in $\mathcal{O}_{\mathrm{ass}}$ only. With this agreement, for each prime p we have either $p \equiv 1 \pmod 3$ or $p = \sqrt{-3}$, and sometimes we write $p \equiv 1 \pmod 3$ only to exclude from the consideration $\sqrt{-3}$.

As usual, if $a, b \in \mathcal{O}$, $a \neq 0$, then $a \mid b$ denotes that a divides b; $a \nmid b$ denotes that a does not divide b; $a \mid b^\infty$ denotes that $a \mid b^r$ for some rational integer $r \geq 1$.

For $c \in \mathbf{Q}(\sqrt{-3}) \setminus \{0\}$ and prime $p \in \mathcal{O}$ we denote by $\mathrm{ord}_p\, c$ the rational integer t such that $c = p^t ab^{-1}$ with some $a, b \in \mathcal{O}$, $p \nmid ab$. Sometimes we write $p \mid c$ instead of $\mathrm{ord}_p\, c \geq 1$, and $p^t \parallel c$ instead of $\mathrm{ord}_p\, c = t$. We set also $\mathrm{ord}_p\, 0 = \infty$.

We shall use $\mid$ and $\nmid$ also in a little more general context than it is described above. For $a \in \mathcal{O} \setminus \{0\}$ and $b \in \mathbf{Q}(\sqrt{-3})$:

$a \mid b$ means that, for each prime p, if $\mathrm{ord}_p\, a \geq 1$, then $\mathrm{ord}_p\, b \geq \mathrm{ord}_p\, a$;

$a \nmid b$ means that there exists prime p so that $\mathrm{ord}_p\, a \geq 1$, $\mathrm{ord}_p\, b < \mathrm{ord}_p\, a$;

$a \mid b^\infty$ means that, for each prime p, if $\mathrm{ord}_p\, a \geq 1$, then $\mathrm{ord}_p\, b \geq 1$.

Certainly, for $b \in \mathcal{O}$ these definitions coincide with those given before.

For $c_1, \ldots, c_n \in \mathcal{O}$ let $k \in \mathcal{O}_{\mathrm{ass}}$ be so that, for each prime p, $\mathrm{ord}_p\, k = \min\{\mathrm{ord}_p\, c_j \mid j = 1, \ldots, n\}$. We say that k is the greatest common divisor of $c_1, \ldots, c_n$ and denote it as $\gcd(c_1, \ldots, c_n)$. In some cases we have $a, b \in \mathbf{Q}(\sqrt{-3})$ and we have to say that there is no prime $p \in \mathcal{O}$ such that both

$\text{ord}_p a > 0$ and $\text{ord}_p b > 0$. For this we shall write $\gcd(a,b) = 1$. Certainly, this coincides with usual $\gcd(a,b) = 1$ (as defined above), if it so happens that $a, b \in \mathcal{O}$, and thus this will not lead to misunderstanding.

We say that $k \in \mathcal{O} \setminus \{0\}$ is square-free if k is of the form (0.1.1) with $\zeta = \pm 1$, $m = 0, 1$ and $\text{ord}_p c \leq 1$ for all prime $p \equiv 1 \pmod 3$. Eeach $l \in \mathcal{O} \setminus \{0\}$ can be factored as $l = kr^2$ with square-free k, uniquely determined by l, and with $r \in \mathcal{O} \setminus \{0\}$.

We say that $k \in \mathcal{O} \setminus \{0\}$ is cube-free if k is of the form (0.1.1) with $\zeta = 1, \omega, \omega^2$, $m = 0, 1, 2$ and $\text{ord}_p c \leq 2$ for all prime $p \equiv 1 \pmod 3$. Each $l \in \mathcal{O} \setminus \{0\}$ can be factored as $l = kr^3$ with cube-free k, uniquely determined by l, and with $r \in \mathcal{O} \setminus \{0\}$.

We say that $l \in \mathbf{Q}(\sqrt{-3})$ is a cube, or is a cube in $\mathbf{Q}(\sqrt{-3})$, if $l = r^3$ with some $r \in \mathbf{Q}(\sqrt{-3})$. Clearly, if $l \in \mathcal{O}$ is a cube in $\mathbf{Q}(\sqrt{-3})$, then $l = r^3$ with $r \in \mathcal{O}$, and this case we can say l is a cube in $\mathcal{O}$.

Let $c \in \mathcal{O} \setminus \{0\}$. A set $\mathfrak{c}(c) \subset \mathcal{O}$ is said to be a complete residue system $\bmod c$ if it contains a unique representative of each residue class $\bmod c$. It is said to be a reduced residue system $\bmod c$ if it contains a unique representative of each residue class $\bmod c$ coprime with c. We shall use script gothic letters to denote residue systems.

For short, we shall write sometimes $a \equiv b\,(c)$ instead of $a \equiv b \pmod c$.

0.1.3 Cubic residue symbol. If $c, d \in \mathcal{O}$, $\gcd(3c, d) = 1$ and $d = \varepsilon p_1^{\omega_1} \cdots p_n^{\omega_n}$ is the canonical factorization into prime factors, $\varepsilon \in \mathcal{O}^*$, then we set

$$\left(\frac{c}{d}\right) = \prod_{1 \leq j \leq n} \left(\frac{c}{p_j}\right)^{\omega_j},$$

where, for prime $p \equiv 1 \pmod 3$, the symbol $\left(\frac{c}{p}\right)$ is equal to $\zeta \in \mathbf{C}^*$ uniquely determined by the conditions

$$\zeta^3 = 1 \quad \text{and} \quad c^{(\tau-1)/3} \equiv \zeta \pmod p \quad \text{with} \quad \tau = \|p\|.$$

(The multiplicative group of the ring $\mathcal{O}/(p)$ has order $\tau - 1$, and it contains the cyclic subgroup of order 3 generated by ω. This yields $c^{\tau-1} \equiv 1 \pmod p$ and $\tau - 1 \equiv 0 \pmod 3$, and so, the ζ required exists.)

The cubic residue symbol $\left(\frac{\cdot}{\cdot}\right)$ has the following properties.

(a) If $a \equiv b\,(d)$, then $\left(\frac{a}{d}\right) = \left(\frac{b}{d}\right)$;

(b) $\left(\frac{c}{ab}\right) = \left(\frac{c}{a}\right)\left(\frac{c}{b}\right)$;

(c) $\left(\frac{ab}{d}\right) = \left(\frac{a}{d}\right)\left(\frac{b}{d}\right)$;

(d) If $c, d \equiv \pm 1\,(3)$, then $\left(\frac{c}{d}\right) = \left(\frac{d}{c}\right)$ (the cubic reciprocity law);

(e) If $\omega = \exp(2\pi i/3)$ and $d = \zeta(1 + 3(m + n\omega))$, $\zeta \in \mathcal{O}^*$, $m, n \in \mathbf{Z}$, then

$$\left(\frac{\omega}{d}\right) = \omega^{-m-n}, \quad \left(\frac{\sqrt{-3}}{d}\right) = \omega^{-n}$$

(the supplement to the reciprocity law);

(f) If $a \equiv b\,(c)$ and $a \equiv b\,(9)$, then $\left(\frac{c}{a}\right) = \left(\frac{c}{b}\right)$;

(g) $\left(\frac{\pm 1}{d}\right) = 1$;

(h) If $c, d \in \mathbf{Z}$, then $\left(\frac{c}{d}\right) = 1$.

The cubic resiprocity law (d) and the supplement (e) are known due to Eisenstein [21], [22], [23]. Other points are rather simple. See [41], [34], [17].

Throughout these notes we shall assume that

$$\left(\frac{0}{d}\right) = 1 \qquad \text{for} \quad d \in \mathcal{O}, \quad \gcd(d, 3) = 1,$$

and sometimes we shall write for simplicity

$$\left(\frac{c}{d}\right) \quad \text{instead of} \quad \left(\frac{a}{d}\right)\left(\frac{b}{d}\right)^{-1}$$

if $c = a/b$ with $a, b \in \mathcal{O}$ and $\gcd(ab, d) = 1$. We hope, there will not be misunderstandings, though these agreements are not commonly used.

We shall use the residue symbol very often. At first we shall try to point out explicitly which of formulae (a),..., (g) we need, but then, we hope, this will not be necessary.

0.1.4 Arithmetic functions. For $c \in \mathcal{O} \setminus \{0\}$ we define

$$\tilde{\mu}(c) = \begin{cases} (-1)^l & \text{if } c \text{ is square-free with } l \text{ prime divisors,} \\ 0 & \text{if } c \text{ is not square-free} \end{cases} \qquad \text{(Möbius function),}$$

$$\tilde{\varphi}(c) = \sum_{k \in \mathfrak{c}(c)} 1 \qquad \text{(Euler's totient function),} \tag{0.1.2}$$

where $\mathfrak{c}(c)$ is a reduced residue system mod c. We have used ~ in the notations to emphasize the functions above are attached to the ring $\mathcal{O}$ (but not to $\mathbf{Z}$), and to save the letters μ, φ for other purposes.

One says that a function $f: \Omega \to \mathbf{C}$, Ω being a subset of $\mathcal{O}$, is multiplicative if $ab \in \Omega$ and $f(ab) = f(a)f(b)$ for all $a, b \in \Omega$ with $\gcd(a,b) = 1$. One can easily find the functions defined in (0.1.2) are multiplicative.

For $\lambda \in q^*$, $q \subset (3)$, $c \in 1+q$ let us write

$$C(\lambda, c) = \sum_{k \in \mathfrak{c}(c)} e(\lambda k/c) \quad \text{(Ramanujan sum)},$$
$$S(\lambda, c) = \sum_{k \in \mathfrak{c}(c)} \left(\frac{k}{c}\right) e(\lambda k/c) \quad \text{(Gauß sum)}, \tag{0.1.3}$$

where $\mathfrak{c}(c)$ is a reduced residue system $\bmod c$, which we can, have and do assume to be a subset of q. Next, if $c' \in \mathcal{O}$ can be represented in the form ld^3 with $l \,|\, c^\infty$, $\gcd(d,3) = 1$, $l, d \in \mathcal{O}$, then we set

$$S(\lambda, c, c') = \sum_{k \in \mathfrak{c}(c,c')} \left(\frac{k}{c}\right)\left(\frac{k}{c'}\right) e(\lambda k/c) \quad \text{(Gauß sum)}, \tag{0.1.4}$$

where $\mathfrak{c}(c, c') \subset q$ is a reduced residue system $\bmod c$ composed of numbers coprime with c'. Notice that $S(\lambda, 1, c') = 1$ if $c' = d^3$ with some $d \in \mathcal{O}$, $\gcd(d,3) = 1$, and that $S(\lambda, 1, c')$ is not defined for other c'. For λ, q as above and $c \in q$ we set

$$S_*(\lambda, c) = \sum_{\substack{k \in \mathfrak{c}(cq) \\ k \equiv 1(3)}} \left(\frac{c}{k}\right) e(\lambda k/c) \quad \text{(Gauß sum)}, \tag{0.1.5}$$

where $\mathfrak{c}(cq)$ is a reduced residue system $\bmod cq$.

Notice that the terms in all the sums in (0.1.3), (0.1.4), (0.1.5) do not depend on the choice of reduced residue systems involved in their definitions. The sums $S_*(\lambda, c)$ depend essentially on the ideal q involved. The sums in (0.1.3), (0.1.4) do not depend on q in a sense that whichever the ideal q with the properties $q \subset (3)$, $\lambda \in q^*$, $c \in 1+q$ is considered in their definitions, for given λ and c these sums are the same. One says λ to be a parameter and c to be a module of the sums (0.1.3), (0.1.4), (0.1.5), while c' should be considered as a supplementary module of the sum (0.1.4).

All the sums in (0.1.2), (0.1.3) are the sums of type (0.1.4). In fact we have

$$S(\lambda, c, c') = \begin{cases} S(\lambda, c) & \text{if } c' \text{ is a cube}, \\ C(\lambda, c) & \text{if } cc' \text{ is a cube}, \\ \tilde{\varphi}(c) & \text{if } cc' \text{ is a cube and } \lambda = 0. \end{cases} \tag{0.1.6}$$

Here are some elementary properties of the sums above:

(a) $\tilde{\varphi}(c) = \|c\| \prod_{\substack{p|c \\ p \text{ is prime}}} \left(1 - \frac{1}{\|p\|}\right)$;

(b) $S(0,c) = \begin{cases} \tilde{\varphi}(c) & \text{if } c \text{ is a cube,} \\ 0 & \text{otherwise;} \end{cases}$

(c) $C(\lambda, c_1c_2) = C(\lambda, c_1)C(\lambda, c_2)$ if $\gcd(c_1, c_2) = 1$;

(d) $C(\delta\lambda, c) = C(\lambda, c)$ if $\delta \in \mathcal{O}$, $\gcd(\delta, c) = 1$;

(e) For prime $p \equiv 1\,(3)$ one has $C(p^\varepsilon, p^\alpha) = \begin{cases} \tilde{\varphi}(p^\alpha) & \text{if } \alpha \le \varepsilon, \\ -\|p\|^{\alpha-1} & \text{if } \alpha = \varepsilon + 1, \\ 0 & \text{if } \alpha \ge \varepsilon + 2; \end{cases}$

(f) $|S(1,c)|^2 = \begin{cases} \|c\| & \text{if } c \text{ is square-free,} \\ 0 & \text{otherwise;} \end{cases}$

(g) $S(\lambda, c_1c_2, c_1'c_2') = \left(\frac{c_1}{c_2c_2'}\right)\left(\frac{c_2}{c_1c_1'}\right) S(\lambda, c_1, c_1') S(\lambda, c_2, c_2')$ if $\gcd(c_1c_1', c_2c_2') = 1$;

(h) For prime $p \equiv 1\,(3)$, if $\delta = \operatorname{ord}_p \lambda$ and $\lambda' = \lambda p^{-\delta}$, then

$$S(\lambda, p^\alpha, p^\beta) = \begin{cases} 1 & \text{if } \alpha = 0,\ \beta \equiv 0(3), \\ \tilde{\varphi}(p^\alpha) & \text{if } 1 \le \alpha \le \delta,\ \alpha + \beta \equiv 0(3), \\ -\|p\|^\delta & \text{if } \alpha = \delta + 1,\ \alpha + \beta \equiv 0(3), \\ \|p\|^\delta S(\lambda', p) & \text{if } \alpha = \delta + 1,\ \alpha + \beta \equiv 1(3), \\ \|p\|^\delta \overline{S(\lambda', p)} & \text{if } \alpha = \delta + 1,\ \alpha + \beta \equiv 2(3), \\ 0 & \text{in all other cases;}^\dagger \end{cases}$$

(i) $S(\lambda, c, c') \ne 0$ if and only if for each prime $p \equiv 1\,(3)$ one has either $\operatorname{ord}_p c = \operatorname{ord}_p \lambda + 1$ or $0 \le \operatorname{ord}_p c \le \operatorname{ord}_p \lambda$, $\operatorname{ord}_p c + \operatorname{ord}_p c' \equiv 0\,(3)$;

(j) $S(\delta\lambda, c, c') = \left(\frac{\delta}{cc'}\right)^{-1} S(\lambda, c, c')$ if $\delta \in \mathcal{O}$, $\gcd(\delta, cc') = 1$;

(k) Let $q = (3)$ and $\lambda = \xi(\sqrt{-3})^m\delta$, $c = \varepsilon(\sqrt{-3})^l$ with $\xi, \varepsilon \in \mathcal{O}^*$, $m, l \in \mathbf{Z}$, $m \ge -3$, $l \ge 2$, $\delta \in \mathcal{O}$, $\delta \equiv 1\,(3)$. Then, with $j, k \in \mathbf{Z}$ defined mod 6 by $\varepsilon\xi^{-1} = \exp(\pi i j/3)$ and $\varepsilon = \exp(-\pi i k/3)$, one has

$$S_*(\lambda, c) = 3^l e\left(\xi\varepsilon^{-1}(\sqrt{-3})^{m-l}\right)\left(\frac{\varepsilon(\sqrt{-3})^l}{\delta}\right)^{-1}$$

if $l - m \le 3$, $k \equiv l \equiv 0\,(3)$, or
$l - m = 4$, $l \equiv k + (-1)^j \equiv 0\,(3)$, or
$l - m = 5$, $l \equiv (-1)^j\,(3)$, $k \equiv (-1)^j j\,(3)$,
and $S_*(\lambda, c) = 0$ for all other j, k, l, m.

---

† Notice however that $S(\lambda, p^\alpha, p^\beta)$ is not defined for $\alpha = 0$, $\beta \not\equiv 0$ (mod 3).

All these facts are known, perhaps in slightly different form, and all of them can be proved by standard techniques which we find is reasonable to demonstrate in details to simplify the reader problems in what follows.

Part (a) follows by multiplicativity from obvious $\tilde{\varphi}(p^\alpha) = \|p\|^\alpha - \|p\|^{\alpha-1}$, where p is prime and $\alpha \geq 1$ is a rational integer.

For (b), let us take $t \in \mathcal{O}$, $\gcd(t,c) = 1$. If k runs over a reduced residue system $\bmod c$, then tk runs over a reduced residue system $\bmod c$ too, i.e., the set $t\mathfrak{c}(c)$ is, just like $\mathfrak{c}(c)$ in (0.1.3), a reduced residue system $\bmod c$. So we have

$$S(0,c) = \sum_{k \in \mathfrak{c}(c)} \left(\frac{k}{c}\right) \quad \text{and} \quad S(0,c) = \sum_{k \in \mathfrak{c}(c)} \left(\frac{tk}{c}\right).$$

This yields $S(0,c) = \left(\frac{t}{c}\right) S(0,c)$, because of (c) in 0.1.3, and thus

$$\text{either} \quad S(0,c) = 0 \quad \text{or} \quad \left(\frac{t}{c}\right) = 1 \quad \text{for all} \quad t \in \mathcal{O}, \quad \gcd(t,c) = 1.$$

The latter case takes place only if c is a cube in $\mathcal{O}$, that is (b).

We can proceed further in a similar manner, but it seems more convenient to take into account (0.1.6) and to notice that one can deduce (c), (d), (e) specifying parameters in (g), (h), (j). To be more precise: (c) follows from (g) if we choose c_l', $l = 1, 2$, in (g) in such a way that $c_l c_l'$ are cubes; (d) follows from (j) in a similar way; (e) follows from (h) with $\lambda = p^\varepsilon$ and $\alpha + \beta \equiv 0 \pmod 3$. Thus, accepting (g), (h), (j) we have (c), (d), (e).

For (f) we can take $\lambda = c_1' = c_2' = 1$ in (g) to get that the function $c \mapsto |S(1,c)|^2$ is multiplicative. This reduces the proving of (f) to the case $c = p^\alpha$, $\alpha \geq 1$ being a rational integer, p being a prime $\equiv 1 \pmod 3$. For $\alpha > 1$ we have $S(1,p^\alpha) = 0$, by (h) with $\lambda = 1$, $\beta = 0$. Thus, accepting (g), (h) it remains to evaluate $|S(1,p)|^2$ for prime $p \equiv 1 \pmod 3$. Just by definition,

$$S(1,p) = \sum_{k \in \mathfrak{c}(p)} \left(\frac{k}{p}\right) e(k/p), \tag{0.1.7}$$

where $\mathfrak{c}(p) \subset q$ is a reduced residue system $\bmod p$. If $k \in \mathfrak{c}(p)$ and t runs over some reduced residue system $\bmod p$, say $\mathfrak{c}'(p)$, then kt runs over a reduced residue system $\bmod p$ also. Hence

$$S(1,p) = \sum_{t \in \mathfrak{c}'(p)} \left(\frac{kt}{p}\right) e(kt/p), \quad \text{for any} \quad k \in \mathfrak{c}(p). \tag{0.1.8}$$

Combining (0.1.7) and (0.1.8) we get

$$\begin{aligned} |S(1,p)|^2 = S(1,p)\overline{S(1,p)} &= \sum_{\substack{t \in \mathfrak{c}'(p) \\ k \in \mathfrak{c}(p)}} \left(\frac{k}{p}\right) \overline{\left(\frac{kt}{p}\right)} e((1-t)k/p) \\ &= \sum_{t \in \mathfrak{c}'(p)} \overline{\left(\frac{t}{p}\right)} \sum_{k \in \mathfrak{c}(p)} e((1-t)k/p), \quad \text{because of (c) in 0.1.3.} \end{aligned} \tag{0.1.9}$$

To evaluate the interior sum in the right-hand side of (0.1.9) notice that

$$\sum_{k\in\mathfrak{c}''(p)} e(lk/p) = \begin{cases} \|p\| & \text{if } l \equiv 0(p),\ l \in \mathcal{O}, \\ 0 & \text{for other } l \in \mathcal{O}, \end{cases}$$

where $\mathfrak{c}''(p) = \mathfrak{c}(p) \cup \{0\}$ is a complete residue system $\bmod p$. Thus we have

$$\sum_{k\in\mathfrak{c}(p)} e((1-t)k/p) = \begin{cases} \|p\| - 1 & \text{if } t \equiv 1(p), \\ -1 & \text{if } t \not\equiv 1(p). \end{cases} \qquad (0.1.10)$$

Substituting (0.1.10) into (0.1.9) we get

$$|S(1,p)|^2 = \|p\|\left(\frac{1}{p}\right) - \sum_{t\in\mathfrak{c}'(p)} \overline{\left(\frac{t}{p}\right)} = \|p\| - \overline{S(0,p)} = \|p\|, \quad \text{as claimed.}$$

To prove (g) let us choose $\mathfrak{c}(c_1, c_1') \subset q$ and $\mathfrak{c}(c_2, c_2') \subset q$ so that

$$\begin{aligned} k_1 &\equiv 0 \pmod{c_2'} \quad \text{for each} \quad k_1 \in \mathfrak{c}(c_1, c_1'), \\ k_2 &\equiv 0 \pmod{c_1'} \quad \text{for each} \quad k_2 \in \mathfrak{c}(c_2, c_2'). \end{aligned} \qquad (0.1.11)$$

(see (0.1.4)). This choice is possible because of $\gcd(c_1c_1', c_2c_2') = 1$. Then we find that $(k_1, k_2) \mapsto c_1k_2 + c_2k_1$ maps injectively $\mathfrak{c}(c_1, c_1') \times \mathfrak{c}(c_2, c_2')$ onto some set $\subset q$ which is a reduced residue system $\bmod c_1c_2$ and consists of numbers coprime with $c_1'c_2'$, and which, so, we can take as $\mathfrak{c}(c_1c_2, c_1'c_2')$. Just the same one can describe as follows:

if $k_1 \in q$ runs over a reduced residue system $\bmod c_1$ consisting of numbers coprime with c_1' and divisable by c_2',
and $k_2 \in q$ runs over a reduced residue system $\bmod c_2$ consisting of numbers coprime with c_2' and divisable by c_1',
then $c_1k_2 + c_2k_1 \in q$ runs over a reduced residue system $\bmod c_1c_2$ consisting of numbers coprime with $c_1'c_2'$.

By definition (0.1.4) we have

$$S(\lambda, c_1c_2, c_1'c_2') = \sum_{k\in\mathfrak{c}(c_1c_2, c_1'c_2')} \left(\frac{k}{c_1c_2}\right)\left(\frac{k}{c_1'c_2'}\right) e(\lambda k/(c_1c_2)). \qquad (0.1.12)$$

The preceding considerations allow us to express here k as $c_1k_2 + c_2k_1$ and to sum over $k_1 \in \mathfrak{c}(c_1, c_1')$ and $k_2 \in \mathfrak{c}(c_2, c_2')$. In the meantime, we have

$$\begin{aligned} \left(\frac{k}{c_1c_2}\right) &= \left(\frac{k}{c_1}\right)\left(\frac{k}{c_2}\right) = \left(\frac{c_2k_1}{c_1}\right)\left(\frac{c_1k_2}{c_2}\right) \\ &= \left(\frac{c_2}{c_1}\right)\left(\frac{c_1}{c_2}\right)\left(\frac{k_1}{c_1}\right)\left(\frac{k_2}{c_2}\right), \\ \left(\frac{k}{c_1'c_2'}\right) &= \left(\frac{k}{c_1'}\right)\left(\frac{k}{c_2'}\right) = \left(\frac{c_2k_1}{c_1'}\right)\left(\frac{c_1k_2}{c_2'}\right) \\ &= \left(\frac{c_2}{c_1'}\right)\left(\frac{c_1}{c_2'}\right)\left(\frac{k_1}{c_1'}\right)\left(\frac{k_2}{c_2'}\right), \end{aligned} \qquad (0.1.13)$$

where we have used (0.1.11) and the formulae (a), (b), (c) in 0.1.3. Substituting the right-hand sides of (0.1.13) into (0.1.12), and taking into account that $e(\lambda k/(c_1c_2)) = e(\lambda k_1/c_1)e(\lambda k_2/c_2)$, we get

$$S(\lambda, c_1c_2, c_1'c_2')$$

$$= \left(\frac{c_2}{c_1c_1'}\right)\left(\frac{c_1}{c_2c_2'}\right) \sum_{\substack{k_1\in\mathfrak{c}(c_1,c_1')\\ k_2\in\mathfrak{c}(c_2,c_2')}} \left(\frac{k_1}{c_1}\right)\left(\frac{k_1}{c_1'}\right)\left(\frac{k_2}{c_2}\right)\left(\frac{k_2}{c_2'}\right) e(\lambda k_1/c_1)e(\lambda k_2/c_2)$$

$$= \left(\frac{c_2}{c_1c_1'}\right)\left(\frac{c_1}{c_2c_2'}\right) S(\lambda, c_1, c_1')S(\lambda, c_2, c_2'), \quad \text{as claimed.}$$

To prove (h) it is convenient to consider separately the cases $\alpha = 0$, $\alpha = 1$ and $\alpha \geq 2$. Let us deal with the case $\alpha \geq 2$. Just by definition (0.1.4) and by (b) in 0.1.3 we have

$$S(\lambda, p^\alpha, p^\beta) = \sum_{k\in\mathfrak{c}(p^\alpha)} \left(\frac{k}{p}\right)^{\alpha+\beta} e(\lambda k/p^\alpha), \tag{0.1.14}$$

where $\mathfrak{c}(p^\alpha)$ is a reduced residue system $\bmod p^\alpha$. Let $\mathfrak{c}(p)$ be a reduced residue system $\bmod p$ and $\mathfrak{c}'(p^{\alpha-1})$ be a complete residue system $\bmod p^{\alpha-1}$. We can choose $\mathfrak{c}(p^\alpha)$ in (0.1.14) to be $\mathfrak{c}(p) + p\mathfrak{c}'(p^{\alpha-1})$, because if l runs over $\mathfrak{c}(p)$ and t runs over $\mathfrak{c}'(p^{\alpha-1})$, then $l + pt$ runs over a reduced residue system $\bmod p^\alpha$. This allows us to replace k in (0.1.14) by $l + pt$ and to sum over $l \in \mathfrak{c}(p)$ and $t \in \mathfrak{c}'(p^{\alpha-1})$ to get

$$S(\lambda, p^\alpha, p^\beta) = \sum_{\substack{l\in\mathfrak{c}(p)\\ t\in\mathfrak{c}'(p^{\alpha-1})}} \left(\frac{l+pt}{p}\right)^{\alpha+\beta} e(\lambda(l+pt)/p^\alpha)$$

$$= \sum_{l\in\mathfrak{c}(p)} \left(\frac{l}{p}\right)^{\alpha+\beta} e(\lambda l/p^\alpha) \sum_{t\in\mathfrak{c}'(p^{\alpha-1})} e(\lambda t/p^{\alpha-1}),$$

where

the sum over t equals $\|p\|^{\alpha-1}$ if $\alpha \leq \delta + 1$;

the sum over t equals 0 if $\alpha \geq \delta + 2$;

$e(\lambda l/p^\alpha) = 1$ if $\alpha \leq \delta$;

$e(\lambda l/p^\alpha) = e(\lambda' l/p) = \overline{e(-\lambda' l/p)}$ if $\alpha = \delta + 1$;

$\left(\frac{l}{p}\right)^{\alpha+\beta}$ equals 1, $\left(\frac{l}{p}\right)$ or $\overline{\left(\frac{l}{p}\right)}$,

according to $\alpha + \beta \equiv 0$, 1 or 2(3).

This yields

$$\frac{S(\lambda,p^\alpha,p^\beta)}{\|p\|^{\alpha-1}} = \begin{cases} \sum_{l\in\mathfrak{c}(p)} 1 = \|p\|-1 & \text{if } \alpha\le\delta,\ \alpha+\beta\equiv 0(3),\\ \sum_{l\in\mathfrak{c}(p)} \left(\frac{l}{p}\right) = 0 & \text{if } \alpha\le\delta,\ \alpha+\beta\equiv 1(3),\\ \sum_{l\in\mathfrak{c}(p)} \overline{\left(\frac{l}{p}\right)} = 0 & \text{if } \alpha\le\delta,\ \alpha+\beta\equiv 2(3),\\ \sum_{l\in\mathfrak{c}(p)} e(\lambda' l/p) = -1 & \text{if } \alpha=\delta+1,\ \alpha+\beta\equiv 0(3),\\ \sum_{l\in\mathfrak{c}(p)} \left(\frac{l}{p}\right) e(\lambda' l/p) = S(\lambda',p) & \text{if } \alpha=\delta+1,\ \alpha+\beta\equiv 1(3),\\ \sum_{l\in\mathfrak{c}(p)} \overline{\left(\frac{l}{p}\right) e(-\lambda' l/p)} = \overline{S(\lambda',p)} & \text{if } \alpha=\delta+1,\ \alpha+\beta\equiv 2(3),\\ 0 & \text{if } \alpha\ge\delta+2=0, \end{cases}$$

that is nothing but (h) for $\alpha\ge 2$. The case $\alpha=1$ is rather simple, and the case $\alpha=0$ is trivial, let us omit the details.

Part (i) follows easily from (g) and (h). Indeed, applying (g) one can find that, up to a non-zero factor, $S(\lambda,c,c')$ is a product of the factors of type $S(\lambda,p^\alpha,p^\beta)$ (with $\alpha=\operatorname{ord}_p c$, $\beta=\operatorname{ord}_p c'$) taken over prime $p\,|\,c$. Then the result needed follows from (h).

Let us prove (j). By definition (0.1.4) and (c) of 0.1.3 we have

$$\left(\frac{\delta}{cc'}\right) S(\delta\lambda,c,c') = \sum_{k\in\mathfrak{c}(c,c')} \left(\frac{\delta k}{c}\right)\left(\frac{\delta k}{c'}\right) e(\lambda\delta k/c). \tag{0.1.15}$$

The set $\delta\mathfrak{c}(c,c')\subset q$ is, just like $\mathfrak{c}(c,c')$ itself, a reduced residue system mod c composed of numbers coprime with c'. Hence the sum in the right-hand side of (0.1.15) is nothing but $S(\lambda,c,c')$, and (j) is proved.

Let us refer to [69] (proposition 5.1†) for the proof of (k).

0.1.5 Dirichlet series. Let $\zeta_{\mathbf{Q}(\sqrt{-3})}$ be the Dedekind zeta function of the field $\mathbf{Q}(\sqrt{-3})$, and let

$$\zeta_*(s) = (1-3^{-s})\zeta_{\mathbf{Q}(\sqrt{-3})}(s), \qquad s\in\mathbf{C}. \tag{0.1.16}$$

So, the function ζ_* is obtained by excluding the factor attached to the prime $p=\sqrt{-3}$ from the Euler product for $\zeta_{\mathbf{Q}(\sqrt{-3})}$ (which we belive to be known).

† If $\lambda\in(\sqrt{-3})^{-3}\mathcal{O}$ and $c\equiv 1 \pmod 3$, then Patterson's $g(\lambda,c)$ in [69] is the same as our $S(\lambda,c)$.

We find it is more convenient for our purposes to deal with ζ_* rather than with original $\zeta_{\mathbf{Q}(\sqrt{-3})}$. We have

$$\zeta_*(s) = \sum_{n\equiv 1(3)} \frac{1}{\|n\|^s} = \prod_{\substack{p\equiv 1(3)\\ p \text{ is prime}}} \left(1-\frac{1}{\|p\|^s}\right)^{-1}, \qquad \Re(s) > 1;$$

$$\frac{\zeta_*(s-1)}{\zeta_*(s)} = \sum_{n\equiv 1(3)} \frac{\tilde{\varphi}(n)}{\|n\|^s}, \qquad \Re(s) > 2; \tag{0.1.17}$$

$$\sum_{n\equiv 1(3)} \frac{S(0,n)}{\|n\|^s} = \sum_{n\equiv 1(3)} \frac{\tilde{\varphi}(n)}{\|n\|^{3s-2}} = \frac{\zeta_*(3s-3)}{\zeta_*(3s-2)}, \qquad \Re(s) > 4/3;$$

where $p, n \in \mathcal{O}$, and all the series and the product converge absolutely. These formulae, and others of the similar type, follow easily from the simple principle: if f is a multiplicative function defined on numbers $\equiv 1 \pmod 3$, then

$$\sum_{n\equiv 1(3)} f(n) = \prod_{\substack{p\equiv 1(3)\\ p \text{ is prime}}} \{1+f(p)+f(p^2)+f(p^3)+f(p^4)+\ldots\}, \tag{0.1.18}$$

whenever the series and the product are absolutely convergent. Essentially, this principle is equivalent to the fact that each $n \equiv 1 \pmod 3$ can be written uniquely as a product of primes $p \equiv 1 \pmod 3$. For (0.1.17) we choose appropriately $f(n)$ and apply (a), (b) in 0.1.4 to find $f(p^\alpha)$ for $\alpha = 1, 2, \ldots$, and then we find $1+f(p)+f(p^2)+\ldots$ by means of the well-known formula for geometric series.

The Dedekind zeta function is regular except for a simple pole with residue $\pi/3^{3/2}$ at the point 1. It satisfies the functional equation

$$(2\pi)^{-s}\Gamma(s)\zeta_{\mathbf{Q}(\sqrt{-3})}(s) = 3^{1/2-s}(2\pi)^{s-1}\Gamma(1-s)\zeta_{\mathbf{Q}(\sqrt{-3})}(1-s),$$

Γ being Euler's gamma-function (see 0.1.6). One has $\zeta_{\mathbf{Q}(\sqrt{-3})}(0) = -1/6$ and $\zeta_{\mathbf{Q}(\sqrt{-3})}(-n) = 0$ for $n \in \mathbf{Z}$, $n \geq 1$. All other zeros of the Dedekind zeta function lie in $\{ s \in \mathbf{C} \mid 0 < \Re(s) < 1 \}$. As a consequence, we have:

> ζ_* is a regular function on $\mathbf{C}$ except for a simple pole at the point 1 with residue $2\pi/3^{5/2}$;
> $\zeta_*(-n) = 0$ for $n \in \mathbf{Z}$, $n \geq 1$;
> ζ_* has first order zeros at $2\pi i n/\log 3$, $n \in \mathbf{Z}$ (in particular, at 0);
> all other zeros of ζ_* lie in $\{ s \in \mathbf{C} \mid 0 < \Re(s) < 1 \}$.

For $\mu, \nu \in (\sqrt{-3})^{-3}\mathcal{O} \setminus \{0\}$ we set

$$L_{\mu,\nu}(s) = \sum_{n\equiv 1(3)} \frac{\chi(n)}{\|n\|^s}, \tag{0.1.19}$$

where $s \in \mathbf{C}$, $\Re(s) > 1$, and[†]

$$\chi(n) = \begin{cases} \overline{\left(\frac{\mu}{n}\right)\left(\frac{\nu}{n}\right)} & \text{if } \gcd(\mu\nu, n) = 1, \\ 0 & \text{otherwise.} \end{cases}$$

The Dirichlet series (0.1.19) are very particular examples of Hecke series. By the Hecke theory, the series (0.1.19) have meromorphic continuations to all of $\mathbf{C}$. If μ/ν is a cube in $\mathbf{Q}(\sqrt{-3})$, then $\chi(n) = 1$ for all $n \equiv 1 \pmod 3$ with $\gcd(\mu\nu, n) = 1$, and we have

$$L_{\mu,\nu}(s) = \zeta_*(s) \prod_{\substack{p \mid \mu\nu \\ p \text{ is prime, } p \equiv 1(3)}} \left(1 - \frac{1}{\|p\|^s}\right).$$

In this case $L_{\mu,\nu}$ has a simple pole at the point 1 and has no other singularities,

$$\operatorname*{Res}_{s=1} L_{\mu,\nu}(s) = 2\pi/3^{5/2} \prod_{\substack{p \mid \mu\nu \\ p \text{ is prime, } p \equiv 1(3)}} \left(1 - \frac{1}{\|p\|}\right).$$

If μ/ν is not a cube in $\mathbf{Q}(\sqrt{-3})$, then $L_{\mu,\nu}$ is an entire function. See [17], [102].

0.1.6 Special functions. Euler's gamma-function can be defined on the half-plane $s \in \mathbf{C}$, $\Re(s) > 0$, by the integral

$$\Gamma(s) = \int_0^\infty \exp(-x) x^{s-1}\, dx,$$

and it can be continued meromorphically to all of $\mathbf{C}$ by means, for example, of the functional equation or the reflection formula —

$$\Gamma(s+1) = s\Gamma(s), \qquad \Gamma(1-s)\Gamma(s) = \frac{\pi}{\sin \pi s}, \qquad s \in \mathbf{C}.$$

It is easy to see that $\Gamma(1) = \Gamma(2) = 1$ and $\Gamma(1/2) = \sqrt{\pi}$. For any positive integer n one has the Legendre–Gauß multiplication formula

$$\prod_{k=0}^{n-1} \Gamma(s + k/n) = (2\pi)^{(n-1)/2} n^{1/2 - ns} \Gamma(ns), \qquad s \in \mathbf{C},$$

† In the formulae below we first time use our agreements 0.1.2 on the symbols gcd and $|$. Recall, $\gcd(\mu\nu, n) = 1$ means that there is no prime p such that both $\operatorname{ord}_p(\mu\nu) \geq 1$ and $\operatorname{ord}_p n \geq 1$. Also, $p \mid \mu\nu$ means $\operatorname{ord}_p(\mu\nu) \geq 1$.

known also as the duplication formula if $n = 2$ and as the triplication formula if $n = 3$. The gamma-function has no zeros and has no singularities except simple poles at non-positive integers. One has

$$\Gamma(s)\Gamma(t) = B(s,t)\Gamma(s+t), \qquad s,t \in \mathbf{C},$$

where B denotes Euler's beta-function. By definition,

$$B(s,t) = \int_0^1 (1-x)^{t-1} x^{s-1}\, dx \quad \text{if } s,t \in \mathbf{C}, \ \Re(s) > 0, \ \Re(t) > 0.$$

As an obvious consequence we have

$$\int_0^\infty \frac{x^{s-1}\, dx}{(1+x)^{s+t}} = \frac{\Gamma(s)\Gamma(t)}{\Gamma(s+t)} \quad \text{if } s,t \in \mathbf{C}, \ \Re(s) > 0, \ \Re(t) > 0. \tag{0.1.20}$$

The Bessel–MacDonald function is defined for $z \in \mathbf{C}$, $\Re(z) > 0$, by the integral

$$K_m(z) = \frac{1}{2}\int_0^\infty \exp\Big(-(t+t^{-1})\frac{z}{2}\Big)\frac{dt}{t^{m+1}}, \qquad m \in \mathbf{C}.$$

It satisfies the Bessel differential equation

$$z^2 K_m''(z) + zK_m'(z) - (z^2+m^2)K_m(z) = 0$$

and can be characterized, up to a constant factor, as the solution which decays exponentially as $z \to \infty$, $z \in \mathbf{R}$. One can continue K_m to a regular function on $\mathbf{C} \setminus (-\infty, 0]$. Given any $z \in \mathbf{C} \setminus (-\infty, 0]$, we have $m \mapsto K_m(z)$ is an entire function on $\mathbf{C}$ satisfying the functional equation $K_m(z) = K_{-m}(z)$. As $K_m(z)$ as its derivatives over z or m are majorized locally uniformly on m by $z^{-1/2}\exp(-z)$ as $z \to \infty$, $z \in \mathbf{R}$. One has

$$\int_0^\infty K_m(cx)x^{s-1}\, dx = 2^{s-2}c^{-s}\Gamma\Big(\frac{s+m}{2}\Big)\Gamma\Big(\frac{s-m}{2}\Big), \tag{0.1.21}$$

$$\int_0^\infty K_m(c\sqrt{1+x})(1+x)^{-m/2}x^{s-1}\, dx = 2^s\Gamma(s)c^{-s}K_{m-s}(c), \tag{0.1.22}$$

where $m, s \in \mathbf{C}$, $c \in \mathbf{R}_+^*$, and $\Re(s) > |\Re(m)|$ in (0.1.21), $\Re(s) > 0$ in (0.1.22).

The Airy function Ai is an entire function on $\mathbf{C}$ defined for $z \in \mathbf{R}$ by the Airy integral

$$Ai(z) = \frac{1}{\pi}\int_0^\infty \cos(t^3/3 + zt)\, dt,$$

which converges due to increasingly rapid oscillations of the integrand. One has also contour integral expression

$$Ai(z) = \frac{1}{2\pi i}\int_{\nabla} \exp(t^3/3 - zt)\,dt, \tag{0.1.23}$$

valid for all $z \in \mathbf{C}$ if the contour ∇ runs from ∞ to 0 inside the sector $-\pi/2 + \varepsilon < \arg t < -\pi/6 - \varepsilon$ and then from 0 to ∞ inside the sector $\pi/6 + \varepsilon < \arg t < \pi/2 - \varepsilon$, ε being any real positive number. The Airy function has very pleasant properties, in particular,

$$\begin{gathered} Ai''(z) = zAi(z), \qquad (Ai^2)'''(z) - 4z(Ai^2)'(z) - 2Ai^2(z) = 0, \\ Ai(z) + \omega Ai(\omega z) + \omega^2 Ai(\omega^2 z) = 0, \end{gathered} \tag{0.1.24}$$

where $\omega = \exp(2\pi i/3)$, $z \in \mathbf{C}$. Also, one has

$$\begin{aligned} Ai(0) &= 2^{-1}3^{-1/6}\pi^{-1}\Gamma(1/3) = 3^{-2/3}\Gamma(2/3)^{-1}, \\ Ai'(0) &= -2^{-1}3^{1/6}\pi^{-1}\Gamma(2/3) = -3^{-1/3}\Gamma(1/3)^{-1}, \end{aligned} \tag{0.1.25}$$

and the Wirtinger formulae

$$\begin{aligned} K_{1/3}(x) &= 2^{1/3}3^{1/6}\pi x^{-1/3} Ai\big((3x/2)^{2/3}\big), \\ K_{2/3}(x) &= -2^{2/3}3^{-1/6}\pi x^{-2/3} Ai'\big((3x/2)^{2/3}\big), \qquad \text{for} \quad x \in \mathbf{R}_+^*. \end{aligned} \tag{0.1.26}$$

Given real $\varepsilon > 0$, one has the asymptotics

$$\begin{aligned} Ai(z) &\sim \frac{1}{2\pi^{1/2}z^{1/4}} \exp\Big(-\frac{2}{3}z^{3/2}\Big), \\ Ai'(z) &\sim \frac{1}{2\pi^{1/2}} z^{1/4} \exp\Big(-\frac{2}{3}z^{3/2}\Big) \end{aligned}$$

as $z \to \infty$, $|\arg z| \leq \pi - \varepsilon$, $z \in \mathbf{C}$.

To introduce the hypergeometric functions ${}_2F_1$ and ${}_3F_2$, let us[†] first consider the series

$${}_2\widetilde{F}_1(a, b; h; z) = \sum_{n=0}^{\infty} \frac{(a)_n (b)_n}{\Gamma(h+n)} \frac{z^n}{n!},$$

$${}_3\widetilde{F}_2(a, b, c; g, h; z) = \sum_{n=0}^{\infty} \frac{(a)_n (b)_n (c)_n}{\Gamma(g+n)\Gamma(h+n)} \frac{z^n}{n!},$$

where $a, b, c, g, h, z \in \mathbf{C}$ and, for any $m \in \mathbf{C}$ and integer $n \geq 2$,

$$(m)_0 = 1, \quad (m)_1 = m, \quad (m)_n = m(m+1)\ldots(m+n-1).$$

[†] Here we do make like Olver [68].

If $n+h$ or $n+g$ is a pole of the gamma-function (that is possible only if h or g is a negative integer or zero) we believe n^{th} term of the series to be zero. These series converge absolutely and define the regular functions ${}_2\widetilde{F}_1(a,b;h;\cdot)$ and ${}_3\widetilde{F}_2(a,b,c;g,h;\cdot)$ in $|z|<1$ which can be continued to regular functions on $\mathbf{C}\setminus[1,\infty)$. For any $z\in\mathbf{C}\setminus[1,\infty)$, ${}_2\widetilde{F}_1(\ldots;z)$ and ${}_3\widetilde{F}_2(\ldots;z)$ are regular functions of parameters $(a,b,c,g,h\in\mathbf{C})$. Under the restriction $\Re(g+h)>\Re(a+b+c)$ the series ${}_3\widetilde{F}_2(a,b,c;g,h;z)$ converges absolutely in $|z|\le 1$ and ${}_3\widetilde{F}_2(\ldots;1)$ is a regular function of parameters.

By definition, for $z\in\mathbf{C}\setminus[1,\infty)$ one has

$$ {}_2F_1(a,b;h;z)=\Gamma(h)\,{}_2\widetilde{F}_1(a,b;h;z), $$

where $a,b,h\in\mathbf{C}$ and h is not a negative integer or zero. Actually we are only need in ${}_2F_1(a,b;h;z)$ with real $z\in(-\infty,1)$.

Next, let $a,b,c,g,h\in\mathbf{C}$ and g,h be not negative integers or zero. By definition,

$$ {}_3F_2(a,b,c;g,h;z)=\Gamma(g)\Gamma(h)\,{}_3\widetilde{F}_2(a,b,c;g,h;z), $$

where either $z\in\mathbf{C}\setminus[1,\infty)$ or $\Re(g+h)>\Re(a+b+c)$, $z\in\mathbf{C}\setminus(1,\infty)$. Actually we are interested only in special values at $z=1$.

The functions ${}_3\widetilde{F}_2$ and ${}_2\widetilde{F}_1$ are more convenient than the original hypergeometric functions ${}_3F_2$ and ${}_2F_1$ if we do not want to exclude negative integer and zero values of the denominator parameters h,g. For example, termwise integration gives

$$ \int_0^\infty {}_2\widetilde{F}_1\Big(a,b;h;\frac{x}{1+x}\Big)\frac{x^{c-1}\,dx}{(1+x)^g}=\Gamma(c)\Gamma(g-c)\,{}_3\widetilde{F}_2(a,b,c;g,h;1) \qquad (0.1.27) $$

under the only conditions $\Re(g)>\Re(c)>0$, $\Re(g+h)>\Re(a+b+c)$, but

$$ \int_0^\infty {}_2F_1\Big(a,b;h;\frac{x}{1+x}\Big)\frac{x^{c-1}\,dx}{(1+x)^g}=\frac{\Gamma(c)\Gamma(g-c)}{\Gamma(g)}\,{}_3F_2(a,b,c;g,h;1), \qquad (0.1.28) $$

subject to the supplementary condition that neither g nor h is a negative integer or zero. If $\Re(h)>\Re(b)>0$, one has Euler's identity

$$ \begin{aligned} \int_0^\infty x^{b-1}(1+x)^{a-h}(z+x)^{-a}\,dx &= \Gamma(b)\Gamma(h-b)\,{}_2\widetilde{F}_1(a,b;h;1-1/z)z^{-a} \\ &= \frac{\Gamma(b)\Gamma(h-b)}{\Gamma(h)}\,{}_2F_1(a,b;h;1-1/z)z^{-a}, \qquad z\in\mathbf{C}\setminus(-\infty,0], \end{aligned} \qquad (0.1.29) $$

where the last expression is valid if h is not a negative integer or zero.

Euler's transformation

$$ {}_2F_1(a,b;h;z) = (1-z)^{-b}\,{}_2F_1\Big(h-a,b;h;\frac{z}{z-1}\Big) \tag{0.1.30} $$

has place for all $a, b, h \in \mathbf{C}$, $z \in \mathbf{C} \setminus [1, \infty)$, unless h is a negative integer or zero, and it can be rewritten as

$$ {}_2\widetilde{F}_1(a,b;h;z) = (1-z)^{-b}\,{}_2\widetilde{F}_1\Big(h-a,b;h;\frac{z}{z-1}\Big) \tag{0.1.31} $$

to eliminate the restriction on h.

We conclude this subsection by two formulae for the Fourier transform. Let $\lambda, r \in \mathbf{C}$, $h \in \mathbf{R}_+^*$ and $\Re(r) > 1$. Then[†]

$$ \int\limits_{\mathbf{C}} \exp(-h|z|^2)\,e(\lambda z)\,dz = \pi h^{-1}\exp(-4\pi^2|\lambda|^2 h^{-1}), \tag{0.1.32} $$

$$ \int\limits_{\mathbf{C}} (h^2+|z|^2)^{-r} e(\lambda z)\,dz = \begin{cases} \dfrac{\pi}{r-1} h^{2-2r} & \text{if } \lambda = 0, \\ \dfrac{(2\pi)^r |\lambda|^{r-1} h^{1-r}}{\Gamma(r)} K_{r-1}(4\pi|\lambda|h) & \text{if } \lambda \neq 0. \end{cases} \tag{0.1.33} $$

Let us prove only the first of them, leaving the other one for the reader (see [54] if necessary). Let $x = \Re(z)$, $y = \Im(z)$ and $a = \Re(\lambda)$, $b = \Im(\lambda)$. One can rewrite the left-hand side part of (0.1.32) as

$$ \iint\limits_{\mathbf{R}\,\mathbf{R}} \exp(-h(x^2+y^2) + 4\pi i(ax - by))\,dx\,dy = \nabla(a)\nabla(-b) \tag{0.1.34} $$

with

$$ \begin{aligned} \nabla(t) &= \int\limits_{\mathbf{R}} \exp(-hu^2 + 4\pi itu)\,du \\ &= \exp(-4\pi^2 t^2 h^{-1}) \int\limits_{\mathbf{R}} \exp(-(h^{1/2}u - 2\pi ith^{-1/2})^2)\,du. \end{aligned} $$

It is easy to see the last integral does not depend on t. So, we have

$$ \nabla(t) = \exp(-4\pi^2 t^2 h^{-1}) \int\limits_{\mathbf{R}} \exp(-hu^2)\,du = h^{-1/2}\Gamma(1/2) = (\pi/h)^{1/2}. $$

Substituting this into (0.1.34) we get (0.1.32).

For the special functions one can recommend the classical course of Whittaker and Watson [103]. Also: Euler's gamma-function and related topics are treated in Bourbaki [12]; Watson [101] gives the most complete treatement of

---

[†] Hereafter we write $\ldots dz$ having in mind an integration over the Lebesgue measure on $\mathbf{C}$.

the Bessel functions theory; Olver [68], Luke [60] and Bailey [3] give detailed exposition of the hypergeometric functions; for the Airy function see Olver [68].

0.2 Kubota and Bass–Milnor–Serre homomorphisms

0.2.1 Congruence subgroups. We shall deal with several congruence subgroups mod q in $\mathrm{SL}(n, \mathcal{O})$. The first one is the principal congruence subgroup $\Gamma^{(n)}_{\mathrm{princ}}(q)$. By definition, $\Gamma^{(n)}_{\mathrm{princ}}(q)$ consists of all matrices $(a_{ij}) \in \mathrm{SL}(n, \mathcal{O})$ which satisfy the conditions

$$\begin{aligned} &a_{ii} \equiv 1 \ (\mathrm{mod}\ q), \quad i = 1, \ldots, n; \\ &a_{ij} \in q, \quad i, j = 1, \ldots, n, \quad i \neq j. \end{aligned}$$

The second one, denoted by $\Gamma_n(q)$, is the group of all matrices $(a_{ij}) \in \mathrm{SL}(n, \mathcal{O})$ which satisfy the conditions

$$\begin{aligned} &a_{ii} \equiv 1 \ (\mathrm{mod}\ 3), \quad i = 1, \ldots, n; \\ &a_{ij} \in q, \quad i, j = 1, \ldots, n, \quad i \neq j. \end{aligned}$$

We shall write

$\widetilde{\Gamma}^{(n)}_{\mathrm{princ}}(q)$ for the subgroup of $\mathrm{SL}(n, \mathcal{O})$ generated by $\Gamma^{(n)}_{\mathrm{princ}}(q)$ and $\mathrm{SL}(n, \mathbf{Z})$,

$\widetilde{\Gamma}_n(q)$ for the subgroup of $\mathrm{SL}(n, \mathcal{O})$ generated by $\Gamma_n(q)$ and $\mathrm{SL}(n, \mathbf{Z})$.

Generally, a group $\Gamma \subset \mathrm{SL}(n, \mathcal{O})$ is said to be a congruence subgroup if Γ contains some principal congruence subgroup. In this sense all above defined groups are congruence subgroups —

$$\Gamma^{(n)}_{\mathrm{princ}}(q) \subset \Gamma_n(q) \subset \widetilde{\Gamma}_n(q), \qquad \Gamma^{(n)}_{\mathrm{princ}}(q) \subset \widetilde{\Gamma}^{(n)}_{\mathrm{princ}}(q).$$

Sometimes we find it is convenient to replace in the notations the ideal q by its generator, and to write, for example, $\Gamma^{(n)}_{\mathrm{princ}}(r)$ for $\Gamma^{(n)}_{\mathrm{princ}}(q)$ with $q = (r)$. We have $\Gamma_n(3) = \Gamma^{(n)}_{\mathrm{princ}}(3)$, $\widetilde{\Gamma}_n(3) = \widetilde{\Gamma}^{(n)}_{\mathrm{princ}}(3)$, and

$$\gamma \in \widetilde{\Gamma}^{(n)}_{\mathrm{princ}}(3) \quad \text{if and only if} \quad \gamma \equiv \delta \ (\mathrm{mod}\ 3) \quad \text{with some} \quad \delta \in \mathrm{SL}(n, \mathbf{Z}).$$

The notations introduced here will be saved throughout these notes.

0.2.2 Kubota homomorphism.

Theorem 0.2.1 *The mapping* $\kappa\colon \Gamma^{(2)}_{\mathrm{princ}}(3) \to \mathbf{C}^*$ *defined by the equality*

$$\kappa(\gamma) = \left(\frac{c}{d}\right)^{-1} \quad \textit{for} \quad \gamma = \begin{pmatrix} * & * \\ c & d \end{pmatrix} \tag{0.2.1}$$

is a group homomorphism. □

Proof. Let $\gamma = \begin{pmatrix} a & b \\ c & d \end{pmatrix}$, $\gamma' = \begin{pmatrix} a' & b' \\ c' & d' \end{pmatrix}$. Then $\gamma\gamma' = \begin{pmatrix} * & * \\ ca' + dc' & cb' + dd' \end{pmatrix}$, and we have only to prove

$$\left(\frac{ca' + dc'}{cb' + dd'}\right) = \left(\frac{c}{d}\right)\left(\frac{c'}{d'}\right), \tag{0.2.2}$$

assuming $\gamma, \gamma' \in \Gamma^{(2)}_{\text{princ}}(3)$. First, let us exclude the trivial case $c = 0$ (where $d = 1$ and nothing to prove). Then, we shall prove (0.2.2) assuming

$$\gcd(c, d') = 1, \tag{0.2.3}$$

and then we shall show how to reduce the general case to the case (0.2.3).

Assuming (0.2.3), we have $\gcd(cb' + dd', d') = \gcd(b', d') = 1$, and we can write the left-hand side of (0.2.2) as

$$\left(\frac{(ca' + dc')d'}{cb' + dd'}\right)\left(\frac{d'}{cb' + dd'}\right)^{-1}. \tag{0.2.4}$$

Applying the cubic reciprosity low 0.1.2 (d) and then 0.1.2 (a), (c), (g) we find†

$$\left(\frac{d'}{cb' + dd'}\right)^{-1} = \left(\frac{cb' + dd'}{d'}\right)^{-1} = \left(\frac{cb'}{d'}\right)^{-1} = \left(\frac{c'}{d'}\right)\left(\frac{c}{d'}\right). \tag{0.2.5}$$

We have† $(ca' + dc')d' = c + (cb' + dd')c'$ and so $cb' + dd' \equiv dd' \pmod{c}$. We have also $cb' \equiv 0 \pmod{9}$. Thus, by 0.1.2 (a), (b), (f) we have

$$\left(\frac{(ca' + dc')d'}{cb' + dd'}\right) = \left(\frac{c}{cb' + dd'}\right) = \left(\frac{c}{dd'}\right) = \left(\frac{c}{d'}\right)\left(\frac{c}{d}\right). \tag{0.2.6}$$

Substituting (0.2.5) and (0.2.6) into (0.2.4) we get (0.2.2).

Thus we have proved $\kappa(\gamma\gamma') = \kappa(\gamma)\kappa(\gamma')$ for all pairs of matrices $\gamma, \gamma' \in \Gamma^{(2)}_{\text{princ}}(3)$ satisfying (0.2.3), and it remains only to avoid (0.2.3). For this, assuming the pair γ, γ' does not satisfy (0.2.3), we set

$$\delta = \begin{pmatrix} 1 & 0 \\ t & 1 \end{pmatrix}, \quad \text{where} \quad t \in q \quad \text{is so that}$$
$$\gcd(at + c, d') = \gcd(t, d) = \gcd(t, b'c + d'd) = 1$$

(use Chines remainder theorem, [41]). Certainly, $\kappa(\delta) = 1$. With such t, each pair of matrices — $\delta\gamma$ and γ', δ and γ, δ and $\gamma\gamma'$, — satisfies the condition of type (0.2.3) and thus we have

$$\kappa(\delta\gamma\gamma') = \kappa(\delta\gamma)\kappa(\gamma') = \kappa(\delta)\kappa(\gamma)\kappa(\gamma') = \kappa(\gamma)\kappa(\gamma'),$$
$$\kappa(\delta\gamma\gamma') = \kappa(\delta)\kappa(\gamma\gamma') = \kappa(\gamma\gamma').$$

This yields immediately (0.2.2) for all $\gamma, \gamma' \in \Gamma^{(2)}_{\text{princ}}(3)$. ■

---

† Notice that $a'd' - b'c' = 1$, $b'c' \equiv -1 \pmod{d'}$.

0.2.3 Brief on more general setting. The theorem above gives us only one example of a wide class of homomorphisms known due to Kubota [51] and Bass, Milnor, Serre [5]. In spite of our principle to deal only with the field $\mathbf{Q}(\sqrt{-3})$ and with the cubic residue symbol, we would like to outline here the subject from a more general point of view. First we should introduce some notations. For a Dedekind ring A and for any ideal q of A, let:

$\mathrm{GL}_n(A)$ be the general linear group of order n over A, i. e., the group of all $n \times n$ matrices (a_{ij}) with entries in A, such that $\det(a_{ij})$ is an invertable element of A;

$\mathrm{SL}_n(A)$ be the special linear group of order n over A, i. e., the subgroup of $\mathrm{GL}_n(A)$ consisting of matrices (a_{ij}) with $\det(a_{ij}) = 1$;

$\mathrm{GL}_n(A, q)$ and $\mathrm{SL}_n(A, q)$ be the subgroups of $\mathrm{GL}_n(A)$ and, respectivly, of $\mathrm{SL}_n(A)$ consisting of matrices (a_{ij}), such that $a_{ii} \in 1 + q$, $a_{ij} \in q$, for $i, j = 1, \dots, n$, $i \neq j$;

$W_q = \{(a, b) \mid a \in 1 + q,\ b \in q,\ aA + bA = A\}$.

These are the notations used in [5], and we find them most convenient for our present discussion.

According to Kubota [51], [54], if

A is the ring of intergers of the totally imaginary global field,

A contains the full group, say μ_m, of mth roots of 1, and $q \subset (m^2)$,

then the mapping $\kappa_2 \colon \mathrm{SL}_2(A, q) \to \mu_m$, defined by

$$\begin{pmatrix} a & b \\ c & d \end{pmatrix} \mapsto \left(\frac{b}{a}\right)_m, \quad \text{where} \quad \left(\frac{\cdot}{\cdot}\right)_m \quad \text{is the } m^{\text{th}} \text{ power residue symbol,}$$

is a group homomorphism. Kubota noticed also that in some cases one can replace $q \subset (m^2)$ by weaken condition. We have seen already one example in the preceding subsection.

Bass, Milnor and Serre generalized this dramatically. To explain, let

$$W_q \to C, \quad \text{say} \quad (a, b) \mapsto \left[\frac{b}{a}\right],$$

be a Mennicke symbol with values in any group C. By definition, this means that

(a) $\left[\dfrac{0}{1}\right] = 1;$

(b) $\left[\dfrac{b_1 b_2}{a}\right] = \left[\dfrac{b_1}{a}\right]\left[\dfrac{b_2}{a}\right]$ for all $(a, b_1), (a, b_2) \in W_q$;

(c) $\left[\dfrac{b + ka}{a}\right] = \left[\dfrac{b}{a}\right]$ for all $k \in q$, $(a, b) \in W_q$;

(d) $\left[\dfrac{b}{a + kb}\right] = \left[\dfrac{b}{a}\right]$ for all $k \in A$, $(a, b) \in W_q$.

As a consequence of these Mennicke symbol axioms we have

(e) $\left[\frac{b}{a_1 a_2}\right] = \left[\frac{b}{a_1}\right]\left[\frac{b}{a_2}\right]$ for all $(a_1, b), (a_2, b) \in W_q$

(Lam, Mennicke and Newman); and also

(f) $\left[\frac{-b}{a}\right] = \left[\frac{b}{a}\right]$ for all $(a, b) \in W_q$;

(g) $\left[\frac{b}{1+kb}\right] = 1$ for all $k, b \in q$;

(h) if $\begin{pmatrix} a & b \\ c & d \end{pmatrix} \in \mathrm{GL}_2(A, q)$, then $\left[\frac{c}{d}\right] = \left[\frac{b}{a}\right]$ and $\left[\frac{b}{a}\right]\left[\frac{b}{d}\right] = 1$.

It is shown in [5] that the mapping $\kappa_2 \colon \mathrm{GL}_2(A, q) \to C$, defined by

$$\begin{pmatrix} a & b \\ c & d \end{pmatrix} \mapsto \left[\frac{c}{d}\right],$$

is a group homomorphism. This statement implies Kubota's original one as a particular case, because under Kubota's circumstances the mth power residue symbol, considered as a mapping $W_q \to C$ with $C = \mu_m$, is a Mennicke symbol. We shall refer to κ_2 above as the Kubota homomorphism attached to Mennicke symbol.

To describe Bass–Milnor–Serre inductive procedure, let us introduce two conditions on a mapping $\kappa_n \colon \mathrm{GL}_n(A, q) \to C$. These are

$(1)_n$ $\kappa_n(\varepsilon) = 1$ for all elementary matrices[†] $\varepsilon \in \mathrm{SL}_n(A, q)$;

$(2)_n$ if $\kappa_n(\gamma) = 1$, then $\kappa_n({}^t\gamma) = 1$.

Assuming, as above, A to be a Dedekind ring, q to be an ideal of A, C to be any group, let us further assume $\kappa_n \colon \mathrm{GL}_n(A, q) \to C$ to be a group homomorphism satisfying $(1)_n$, $(2)_n$, and let $n \geq 2$. According to Bass, Milnor and Serre, under these assumptions we have the following.

(1) Any $\gamma \in \mathrm{GL}_{n+1}(A, q)$ can be factored in $\mathrm{GL}_{n+1}(A, q)$ as

$$\gamma = \begin{pmatrix} & & & * \\ & \alpha & & \vdots \\ & & & * \\ 0 & \cdots\cdots & 0 & 1 \end{pmatrix} \begin{pmatrix} 1 & 0 & \cdots\cdots & 0 \\ 0 & 1 & 0 \cdots & 0 \\ & & \ddots & \\ 0 \cdots & 0 & 1 & 0 \\ * & 0 & \cdots 0 & 1 \end{pmatrix} \begin{pmatrix} 1 & * & \cdots\cdots & * \\ 0 & & & \\ \vdots & & \beta & \\ 0 & & & \end{pmatrix}, \quad \alpha, \beta \in \mathrm{GL}_n(A, q).$$

(2) With γ, α, β as in (1), the product $\kappa_n(\alpha)\kappa_n(\beta)$ depends on γ only, and not on the particular choice of the factors in (1). Thus we get a mapping $\kappa_{n+1} \colon \mathrm{GL}_{n+1}(A, q) \to C$ by setting $\kappa_{n+1}(\gamma) = \kappa_n(\alpha)\kappa_n(\beta)$.

[†] A matrix (a_{ij}) is said to be elementary if and only if $a_{ii} = 1$ and there exists at most one pair i, j with $i \neq j$ such that $a_{ij} \neq 0$; $i, j = 1, \ldots, n$.

(3) If $n \geq 3$, the mapping $\kappa_{n+1}\colon \mathrm{GL}_{n+1}(A,q) \to C$, defined in (2), is a group homomorphism which extends κ_n in a sense that

$$\kappa_{n+1}\left(\begin{pmatrix} & & & 0 \\ & \delta & & \vdots \\ & & & 0 \\ 0 & \cdots & 0 & 1 \end{pmatrix}\right) = \kappa_n(\delta) \quad \text{for all} \quad \delta \in \mathrm{GL}_n(A,q),$$

and which satisfies $(1)_{n+1}$, $(2)_{n+1}$. Moreover, this κ_{n+1} is the only homomorphism $\mathrm{GL}_{n+1}(A,q) \to C$ which extends κ_n and satisfies $(1)_{n+1}$.

(4) If $n = 2$ and the mapping $\kappa_3\colon \mathrm{GL}_3(A,q) \to C$, defined in (2), satisfies $\kappa_3(\lambda\sigma\lambda^{-1}) = \kappa_3(\sigma)$ with

$$\lambda = \begin{pmatrix} 1 & 0 & 0 \\ 0 & 0 & 1 \\ 0 & 1 & 0 \end{pmatrix} \quad \text{and with any} \quad \sigma = \begin{pmatrix} * & * & * \\ * & * & * \\ 0 & 0 & 1 \end{pmatrix}\begin{pmatrix} 1 & 0 & 0 \\ 0 & 1 & 0 \\ * & 0 & 1 \end{pmatrix} \in \mathrm{GL}_3(A,q),$$

then κ_3 is a group homomorphism which extends κ_2 (in a sense stated in (3)) and satisfies the conditions $(1)_3$, $(2)_3$. This κ_3 is the only homomorphism $\mathrm{GL}_3(A,q) \to C$ which extends κ_2 and satisfies $(1)_3$.

(5) If $\kappa_2\colon \mathrm{GL}_2(A,q) \to C$ is a Kubota homomorphism attached to Mennicke symbol, then the mapping κ_3, defined in (2), satisfies the condition stated in (4).

We would like to give some commentaries.

First, Bass, Milnor and Serre [5] do not assume A to be a Dedekind ring from the very beginning. They prefer to begin with arbitrary commutative ring A with identity, and to introduce some conditions on A, when necessary. We do not want to go into a discussion of these points. For our aims it is sufficient to notice that, as it is shown just in [5], Dedekind rings satisfy all needed conditions. Also, in many other respects the paper of Bass, Milnor and Serre [5] contains rather more than we covered here.

Our second commentary concerns to the point (5) of the inductive procedure above. We would like to emphasize that the point (5) gives just a sufficient condition for (4). Checking which properties of κ_2 are actually used in the last part of section 10 in [5], we find weaken sufficient condition for (4). That is

(6) Let $\kappa_2\colon \mathrm{GL}_2(A,q) \to C$ be a group homomorphism which satisfies the conditions $(1)_2$, $(2)_2$, and which is attached to some mapping

$$\left[\frac{\cdot}{\cdot}\right]\colon W_q \to C \quad \text{in such a way that}$$

$$\kappa_2(\gamma) = \left[\frac{b}{a}\right] \quad \text{for} \quad \gamma = \begin{pmatrix} a & b \\ c & d \end{pmatrix} \in \mathrm{GL}_2(A,q).$$

If the mapping $W_q \to C$ above satisfies the Mennicke symbol axioms (a), (b), (c) and has the properties (e), (f), (g), (h), then the $\kappa_3\colon \mathrm{GL}_3(A,q) \to C$, defined in (2), satisfies the condition stated in (4), and thus κ_3 is a group homomorphism which extends κ_2 and satisfies $(1)_3$, $(2)_3$.

We can give two examples to show (6) is weaken rather than (5).

First, let A be the ring of integers of the field $\mathbf{Q}(\sqrt{-3})$, i. e., A is our usuall ring $\mathcal{O}$. And let C be the group of the cubic roots of 1, i. e., $C = \{1, \omega, \omega^2\}$. If $q \subset (3)$, we can define $W_q \to C$ setting

$$\left[\frac{b}{a}\right] = \left(\frac{b}{a}\right)_3 \quad \text{for all} \quad (a,b) \in W_q, \quad \text{where} \quad \left(\frac{\cdot}{\cdot}\right)_3$$

is the cubic residue symbol (which we usually write without subscript 3). Notice that under these circumstances $\mathrm{GL}_n(A,q) = \mathrm{SL}_n(A,q) = \Gamma^{(n)}_{\mathrm{princ}}(q)$. The so defined mapping $W_q \to C$ is a Mennicke symbol if and only if $q \subset (9)$. But this mapping satisfies the requirements stated in (6) for any $q \subset (3)$. (To check one should turn to 0.1.3.) As we know already from 0.2.2, $\kappa_2 \colon \mathrm{GL}_2(A,q) \to C$ attached to our $W_q \to C$ is a group homomorphism for any $q \subset (3)$, and it satisfies $(1)_2$, $(2)_2$. Actually, we have $\kappa_2 = \bar{\kappa}$. Thus we have the series of homomorphisms $\kappa_n \colon \mathrm{GL}_n(A,q) \to C$, $n \geq 2$, with $q = (3)$.

We get one more example† assuming A to be the ring of integers of the field $\mathbf{Q}(\sqrt{-1})$, i. e., A is the ring of Gaußian integers. Let $C = \{\pm 1\}$. This case we have (see [41]) biquadratic and quadratic residue simbols, the quadratic one being the square of the biquadratic one. We can define $W_q \to C$ by means of the quadratic residue symbol, just like in the first example. The so defined $W_q \to C$ is a Mennicke symbol if and only if $q \subset (4)$, but it satisfies the requirements stated in (6) for all $q \subset \big(2(1+\sqrt{-1})\big)$. As in the first example, we have homomorphism $\kappa_2 \colon \mathrm{GL}_2(A,q) \to C$ with $q = \big(2(1+\sqrt{-1})\big)$ which satisfies all needed conditions, and thus we have the series of homomorphisms $\kappa_n \colon \mathrm{GL}_n(A,q) \to C$, $n \geq 2$, with $q = \big(2(1+\sqrt{-1})\big)$.

It should be noticed the described opportunity to slightly extend the homomorphisms κ_n occurs without any changes in proofs in [5]. This opportunity, being irrelevant for the aims of the paper [5], is important if we are interested in studing automorphic forms with multiplier system κ_n.

Let $\kappa_n \colon \mathrm{GL}_n(A,q) \to C$, $n \geq 2$, be a series of homomorphisms constructed by means of Bass–Milnor–Serre inductive procedure. If it begins with κ_2 attached to a Mennicke symbol, then, according to [5], we have $\kappa_n(\gamma\delta\gamma^{-1}) = \kappa_n(\delta)$ for all $n \geq 2$, $\delta \in \mathrm{GL}_n(A,q)$ and $\gamma \in \mathrm{GL}_n(A)$. Perhaps is not so if it begins with κ_2 described in (6). However, in the latter case we have: if $\kappa_2(\gamma\delta\gamma^{-1}) = \kappa_2(\delta)$ for all $\delta \in \mathrm{GL}_2(A,q)$, $\gamma \in \mathrm{GL}_2(\mathbf{Z})$, then $\kappa_n(\gamma\delta\gamma^{-1}) = \kappa_n(\delta)$ for all $n \geq 2$, $\delta \in \mathrm{GL}_n(A,q)$ and $\gamma \in \mathrm{GL}_n(\mathbf{Z})$. This can be easily checked by taking δ and γ in suitable choosen sets of generators of $\mathrm{GL}_n(A,q)$ and $\mathrm{GL}_n(\mathbf{Z})$. For those one can turn to [5] or to [4] to find the following: let A and q be as above, $n \geq 3$,

† Compare with [16]. The homomorphism in [16] is the restriction of κ_4 of this example.

then $\mathrm{GL}_n(A,q)$ is generated by elementary matrices and by matrices

$$\begin{pmatrix} a & b & 0 & \cdots & 0 \\ c & d & 0 & \cdots & 0 \\ 0 & 0 & & & \\ \vdots & \vdots & & e_{n-2} & \\ 0 & 0 & & & \end{pmatrix}, \quad \text{where} \quad \begin{pmatrix} a & b \\ c & d \end{pmatrix} \in \mathrm{GL}_2(A,q), \tag{0.2.7}$$

and e_{n-2} is $(n-2)\times(n-2)$ identity matrix. To get generators of $\mathrm{GL}_n(\mathbf{Z})$ one can take $A = q = \mathbf{Z}$.

0.2.4 Bass–Milnor–Serre homomorphisms. In Part 1 and in Part 2 we shall deal with homomorphisms κ_3 and κ_4 defined in the first example in 0.2.3. We find it is reasonable to turn once more to the homomorphisms given by this example and to state, in the theorem below, all their properties we need.

Theorem 0.2.2 *There exists a unique series of homomorphisms*

$$\kappa_n : \Gamma^{(n)}_{\text{princ}}(3) \to \mathbf{C}^*, \qquad n \in \mathbf{Z}, \quad n \geq 2,$$

satisfying the conditions:

(a) $\kappa_2 = \bar{\kappa}$;

(b) $\kappa_n\left(\begin{pmatrix} & & * \\ & \delta & \vdots \\ & & * \\ 0 & \cdots \; 0 & 1 \end{pmatrix}\right) = \kappa_n\left(\begin{pmatrix} 1 & * & \cdots & * \\ 0 & & & \\ \vdots & & \delta & \\ 0 & & & \end{pmatrix}\right) = \kappa_{n-1}(\delta)$ *if* $\delta \in \Gamma^{(n-1)}_{\text{princ}}(3)$, $n \geq 3$;

(c) $\kappa_n(\delta) = 1$ *if* δ *is an elementary matrix*[†] *in* $\Gamma^{(n)}_{\text{princ}}(3)$;

(d) $\kappa_n({}^t\delta)^{-1} = \kappa_n(\delta)$ *for* $\delta \in \Gamma^{(n)}_{\text{princ}}(3)$;

(e) $\kappa_n(\gamma\delta\gamma^{-1}) = \kappa_n(\delta)$ *if* $\delta \in \Gamma^{(n)}_{\text{princ}}(9)$, $\gamma \in \mathrm{SL}(n,\mathcal{O})$;

(f) $\kappa_n(\gamma\delta\gamma^{-1}) = \kappa_n(\delta)$ *if* $\delta \in \Gamma^{(n)}_{\text{princ}}(3)$, $\gamma \in \mathrm{SL}(n,\mathbf{Z})$;

(g) *For each* $q \subset (3)$ *and* $\gamma \in \mathrm{SL}(n, \mathbf{Q}(\sqrt{-3}))$, *there exists an ideal* $q' \subset (3)$ *such that* $\gamma\Gamma^{(n)}_{\text{princ}}(q')\gamma^{-1} \subset \Gamma^{(n)}_{\text{princ}}(q)$ *and* $\kappa_n(\gamma\delta\gamma^{-1}) = \kappa_n(\delta)$ *for all* $\delta \in \Gamma^{(n)}_{\text{princ}}(q')$. ∎

We know already all the assertions of this theorem except (d) and (g). These remaining parts it is sufficient to prove for generators of $\Gamma^{(n)}_{\text{princ}}(3)$ and $\Gamma^{(n)}_{\text{princ}}(q')$, say for those given in 0.2.3. For q' one can take l^2q, where l is the product of the denominaters of the entries of γ. Let us omit the details.

---

† Recall, a matrix $\sigma = (a_{ij})$ is said to be elementary if and only if $a_{ii} = 1$ and there exists at most only one pair i, j with $i \neq j$ such that $a_{ij} \neq 0$; $i, j = 1, \ldots, n$.

0.2.5 Extensions. As it was observed already by Kubota, one can extend homomorphism $\kappa\colon \Gamma^{(2)}_{\text{princ}}(3) \to \mathbf{C}^*$ in Theorem 0.2.1 (that is, $\overline{\kappa}_2$ in Theorem 0.2.2) to be the homomorphism of the group generated by $\Gamma^{(2)}_{\text{princ}}(3)$ and

$$\gamma = \begin{pmatrix} 0 & -1 \\ 1 & 0 \end{pmatrix},$$

just setting $\kappa(\gamma) = 1$. Then Patterson noticed that one can extend Kubota's κ from $\Gamma^{(2)}_{\text{princ}}(3)$ to the group $\widetilde{\Gamma}^{(2)}_{\text{princ}}(3)$ (which is generated by $\Gamma^{(2)}_{\text{princ}}(3)$ and $\mathrm{SL}(2,\mathbf{Z})$) by setting $\kappa(\gamma) = 1$ for all $\gamma \in \mathrm{SL}(2,\mathbf{Z})$. It is easy to show, all the Bass-Milnor-Serre's κ_n in 0.2.4 can be extended in a similar manner.

Theorem 0.2.3 *There exists a unique series of homomorphisms*

$$\kappa_n\colon \widetilde{\Gamma}^{(n)}_{\text{princ}}(3) \to \mathbf{C}^*, \quad n \in \mathbf{Z}, \quad n \geq 2,$$

such that

(a) *The restriction of* κ_n *to* $\Gamma^{(n)}_{\text{princ}}(3)$ *coincides with* κ_n *in* 0.2.4;

(b) $\kappa_n(\gamma) = 1$ *for* $\gamma \in \mathrm{SL}(n,\mathbf{Z})$. □

Proof. Let us remark that, for κ_n in Theorem 0.2.2, one has $\kappa_n(\gamma) = 1$ for $\gamma \in \Gamma^{(n)}_{\text{princ}}(3) \cap \mathrm{SL}(n,\mathbf{Z})$. Clearly, it is sufficitent to check this only for generators of $\Gamma^{(n)}_{\text{princ}}(3) \cap \mathrm{SL}(n,\mathbf{Z})$ (= the principal congruence subgroup mod 3 in $\mathrm{SL}(n,\mathbf{Z})$), say for those given in (0.2.7). In view of Theorem 0.2.2 (b), (c), this reduces to the case $n = 2$, that is, to the known fact stated in 0.1.3 (h).

Now, let $\gamma \in \widetilde{\Gamma}^{(n)}_{\text{princ}}(3)$. We have $\gamma \equiv \delta \pmod q$ for some $\delta \in \mathrm{SL}(n,\mathbf{Z})$, and thus $\gamma = \delta\sigma$ with $\sigma \in \Gamma^{(n)}_{\text{princ}}(3)$. By the remark above, $\kappa_n(\sigma)$ depends on γ only, but not on the choice of δ. Thus we can define a mapping $\kappa_n\colon \widetilde{\Gamma}^{(n)}_{\text{princ}}(3) \to \mathbf{C}^*$ by setting $\kappa_n(\gamma) = \kappa_n(\sigma)$. By means of Theorem 0.2.2 (f) one can easily find that the so defined $\kappa_n\colon \widetilde{\Gamma}^{(n)}_{\text{princ}}(3) \to \mathbf{C}^*$ is a group homomorphism. The properties (a) and (b) are obvious. ■

From now on, by the Kubota homomorphism κ and by the Bass-Milnor-Serre homomorphisms κ_n we shall mean just the extended ones.

0.2.6 Explicit formulae. In some cases it is not a problem to give explicit formulae expressing $\kappa_n(\gamma)$ by means of the residue symbol values at the entries of the matrix γ. Let us give two examples.

Theorem 0.2.4 *If*

$$\gamma = \begin{pmatrix} * & * & * \\ b_1 & b_2 & b_3 \\ c_1 & c_2 & c_3 \end{pmatrix} \in \Gamma^{(3)}_{\text{princ}}(3) \quad \textit{and} \quad \gcd(b_3, c_3) = 1, \tag{0.2.8}$$

then

$$\kappa_3(\gamma) = \left(\frac{-b_3}{c_3}\right)^{-1} \left(\frac{b_1c_3 - b_3c_1}{b_2c_3 - b_3c_2}\right). \quad \square$$

Proof. The conditions (0.2.8) provide the existence of $\begin{pmatrix} a & b \\ c & d \end{pmatrix} \in \Gamma^{(2)}_{\text{princ}}(3)$ such that

$$\begin{pmatrix} a & b \\ c & d \end{pmatrix}\begin{pmatrix} b_3 \\ c_3 \end{pmatrix} = \begin{pmatrix} 0 \\ 1 \end{pmatrix}.$$

Obviously,

$$a = c_3, \quad b = -b_3. \tag{0.2.9}$$

One has $\gamma = \vartheta_1\vartheta_2\vartheta_3\vartheta_4$ with $\vartheta_j \in \Gamma^{(3)}_{\text{princ}}(3)$ of the form

$$\vartheta_1 = \begin{pmatrix} 1 & 0 & 0 \\ 0 & d & -b \\ 0 & -c & a \end{pmatrix}, \quad \vartheta_2 = \begin{pmatrix} 1 & 0 & * \\ 0 & 1 & 0 \\ 0 & 0 & 1 \end{pmatrix}, \quad \vartheta_3 = \begin{pmatrix} * & * & 0 \\ \varepsilon & \zeta & 0 \\ 0 & 0 & 1 \end{pmatrix}, \quad \vartheta_4 = \begin{pmatrix} 1 & 0 & 0 \\ 0 & 1 & 0 \\ * & * & 1 \end{pmatrix},$$

where $\varepsilon = ab_1 + bc_1$, $\zeta = ab_2 + bc_2$. It follows from Theorem 0.2.2 (a), (b) and Theorem 0.2.1 that

$$\kappa_3(\vartheta_1) = \left(\frac{b}{a}\right)^{-1}, \quad \kappa_3(\vartheta_3) = \left(\frac{\varepsilon}{\zeta}\right), \quad \kappa_3(\vartheta_2) = \kappa_3(\vartheta_4) = 1$$

(notice also $bc \equiv -1 \pmod a$, and see 0.1.3). Thus

$$\kappa_3(\gamma) = \left(\frac{b}{a}\right)^{-1}\left(\frac{ab_1 + bc_1}{ab_2 + bc_2}\right),$$

and substituting a, b from (0.2.9) we get the desired result. ∎

Theorem 0.2.5 *If*

$$\gamma = \begin{pmatrix} * & * & * & * \\ * & * & * & * \\ c_1 & c_2 & d_1 & d_2 \\ c_3 & c_4 & d_3 & d_4 \end{pmatrix} \in \Gamma^{(4)}_{\text{princ}}(3) \cap \mathrm{Sp}(4, \mathbf{C}) \quad \textit{and} \quad \gcd(d_2, d_4) = 1, \tag{0.2.10}$$

then

$$\kappa_4(\gamma) = \left(\frac{c_3 d_2 - c_1 d_4}{d_1 d_4 - d_2 d_3}\right)\left(\frac{d_2}{d_4}\right)^{-2}. \quad ∎$$

Proof. (See 2.1.1 for the definiton of the symplectic group.) Let $D = \begin{pmatrix} d_1 & d_2 \\ d_3 & d_4 \end{pmatrix}$. The conditions (0.2.10) provide the existence of the matrices

$$T = \begin{pmatrix} 1 & 0 \\ * & 1 \end{pmatrix} \in \Gamma^{(2)}_{\text{princ}}(3), \quad U = \begin{pmatrix} a & b \\ c & d \end{pmatrix} \in \Gamma^{(2)}_{\text{princ}}(3) \text{ with } UDT = \begin{pmatrix} k & 0 \\ 0 & 1 \end{pmatrix},$$

where $k = \det D = d_1 d_4 - d_2 d_3$, $a = d_4$ and $b = -d_2$. One has $\gamma = \vartheta_1 \gamma' \vartheta_2$ and

$$\kappa_4(\gamma) = \kappa_4(\vartheta_1)\kappa_4(\gamma')\kappa_4(\vartheta_2)$$

with $\gamma', \vartheta_1, \vartheta_2 \in \Gamma^{(4)}_{\text{princ}}(3) \cap \mathrm{Sp}(4, \mathbf{C})$ of the form

$$\vartheta_1 = \begin{pmatrix} {}^tU & \begin{matrix}0&0\\0&0\end{matrix} \\ \begin{matrix}0&0\\0&0\end{matrix} & U^{-1} \end{pmatrix}, \quad \gamma' = \begin{pmatrix} * & * & * & b_2' \\ * & * & * & b_4' \\ c_1' & * & k & 0 \\ c_3' & c_4' & 0 & 1 \end{pmatrix}, \quad \vartheta_2 = \begin{pmatrix} {}^tT & \begin{matrix}0&0\\0&0\end{matrix} \\ \begin{matrix}0&0\\0&0\end{matrix} & T^{-1} \end{pmatrix},$$

where $c_1' = ac_1 + bc_3 = c_1d_4 - c_3d_2$. One can write

$$\vartheta_1 = \begin{pmatrix} {}^tU & \begin{matrix}0&0\\0&0\end{matrix} \\ \begin{matrix}0&0\\0&0\end{matrix} & \begin{matrix}1&0\\0&1\end{matrix} \end{pmatrix}\begin{pmatrix} \begin{matrix}1&0\\0&1\end{matrix} & \begin{matrix}0&0\\0&0\end{matrix} \\ \begin{matrix}0&0\\0&0\end{matrix} & U^{-1} \end{pmatrix}, \quad \vartheta_2 = \begin{pmatrix} {}^tT & \begin{matrix}0&0\\0&0\end{matrix} \\ \begin{matrix}0&0\\0&0\end{matrix} & \begin{matrix}1&0\\0&1\end{matrix} \end{pmatrix}\begin{pmatrix} \begin{matrix}1&0\\0&1\end{matrix} & \begin{matrix}0&0\\0&0\end{matrix} \\ \begin{matrix}0&0\\0&0\end{matrix} & T^{-1} \end{pmatrix},$$

and then to apply Theorem 0.2.2 (a), (b), (d) and Theorem 0.2.1 to get

$$\begin{aligned} \kappa_4(\vartheta_1) &= \kappa_2({}^tU)\kappa_2(U^{-1}) = \kappa_2({}^tU^{-1})^{-2} = \left(\frac{-b}{a}\right)^{-2} = \left(\frac{d_2}{d_4}\right)^{-2}, \\ \kappa_4(\vartheta_2) &= \kappa_2({}^tT)\kappa_2(T^{-1}) = \kappa_2(T)^{-2} = 1. \end{aligned} \tag{0.2.11}$$

So we have $\kappa_4(\gamma) = \left(\frac{d_2}{d_4}\right)^{-2} \kappa_4(\gamma')$, and it remains to evaluate $\kappa_4(\gamma')$. For this we set

$$\vartheta_3 = \begin{pmatrix} 1 & 0 & 0 & b_2' \\ 0 & 1 & b_2' & b_4' \\ 0 & 0 & 1 & 0 \\ 0 & 0 & 0 & 1 \end{pmatrix}, \quad \vartheta_4 = \begin{pmatrix} 1 & 0 & 0 & 0 \\ 0 & 1 & 0 & 0 \\ 0 & 0 & 1 & 0 \\ c_3' & c_4' & 0 & 1 \end{pmatrix},$$

and we notice that $\gamma' = \vartheta_3\gamma''\vartheta_4$, where $\gamma'' \in \Gamma^{(4)}_{\text{princ}}(3)$ is of the form

$$\gamma'' = \begin{pmatrix} \delta & \begin{matrix}0\\0\\0\end{matrix} \\ \begin{matrix}0&0&0\end{matrix} & 1 \end{pmatrix} \quad \text{with} \quad \delta = \begin{pmatrix} * & \xi & * \\ 0 & 1 & 0 \\ c_1' & \eta & k \end{pmatrix} \quad \text{(use (2.1.4))}.$$

In the meantime, ϑ_3 and ϑ_4 are the products of elementary matrices in $\Gamma^{(4)}_{\text{princ}}(3)$, so $\kappa_4(\vartheta_3) = \kappa_4(\vartheta_4) = 1$, and thus $\kappa_4(\gamma') = \kappa_4(\gamma'')$. By Theorem 0.2.2 (b) we have $\kappa_4(\gamma'') = \kappa_3(\delta)$. To evaluate $\kappa_3(\delta)$ we write $\delta = \vartheta_5\vartheta_6$ with

$$\vartheta_5 = \begin{pmatrix} 1 & \xi & 0 \\ 0 & 1 & 0 \\ 0 & \eta & 1 \end{pmatrix}, \quad \vartheta_6 = \begin{pmatrix} * & 0 & * \\ 0 & 1 & 0 \\ c_1' & 0 & k \end{pmatrix}.$$

The matrix ϑ_5 can be written as the product of the elementary matrices in $\Gamma^{(3)}_{\text{princ}}(3)$, so we have $\kappa_3(\vartheta_5) = 1$. The matrix ϑ_6 can be written as

$$\vartheta_6 = \lambda \begin{pmatrix} * & * & 0 \\ -c_1' & k & 0 \\ 0 & 0 & 1 \end{pmatrix} \lambda^{-1} \quad \text{with} \quad \lambda = \begin{pmatrix} 1 & 0 & 0 \\ 0 & 0 & 1 \\ 0 & -1 & 0 \end{pmatrix} \in \mathrm{SL}(3, \mathbf{Z}).$$

So, by Theorem 0.2.2 (a), (b), (f), and by Theorem 0.2.1 we have

$$\kappa_3(\vartheta_6) = \kappa_2\left(\begin{pmatrix} * & * \\ -c_1' & k \end{pmatrix}\right) = \left(\frac{-c_1'}{k}\right) = \left(\frac{c_3 d_2 - c_1 d_4}{d_1 d_4 - d_2 d_3}\right).$$

Thus we have $\kappa_4(\gamma') = \kappa_4(\gamma'') = \kappa_3(\delta) = \kappa_3(\vartheta_5)\kappa_3(\vartheta_6)$

$$= \kappa_3(\vartheta_6) = \left(\frac{c_3 d_2 - c_1 d_4}{d_1 d_4 - d_2 d_3}\right), \quad \text{as claimed.} \quad \blacksquare$$

In [78] we really used the formula of Theorem 0.2.4 for κ_3. In sequel [81], [84] we find it is better not to use formulae of such a type. However, we find the formulae instructive, especially that for the symplectic groups. It is clear, the formulae of the last two theorems will be unchanged if we replace† the cubic Bass–Milnor–Serre homomorphisms and the cubic residue symbol by the quadratic or sixth order ones, which are defined for the field $\mathbf{Q}(\sqrt{-3})$ since it contains the full group of the sixth order roots of 1. It is clear, one has also just the same formulae for the quadratic and biquadratic homomorphisms defined on the congruence subgroups of $\mathrm{SL}(3, \mathbf{Z}[\sqrt{-1}])$ and $\mathrm{Sp}(4, \mathbf{Z}[\sqrt{-1}])$. In the quadratic case one factor in the formula of Theorem 0.2.5 becomes trivial, so this case is distinguished from all others. In this connection we should mention the recent work of Imamoḡlu [40], where explicit formulae for the quadratic Bass–Milnor–Serre homomorphism of the congruence subgroups of $\mathrm{Sp}(2n, \mathbf{Z}[\sqrt{-1}])$ are given. Considering the standard group monomorphism $\mathrm{Sp}(2n, \mathbf{C}) \to \mathrm{Sp}(4n, \mathbf{R})$ Imamoḡlu finds that the classical theta multiplier system of weight 1/2, being restricted from a congruence subgroup of $\mathrm{Sp}(4n, \mathbf{Z})$ to a congruence subgroup of $\mathrm{Sp}(2n, \mathbf{Z}[\sqrt{-1}])$, reduces to the quadratic Bass–Milnor–Serre homomorphism.

0.3 Cubic metaplectic forms and Kubota–Patterson cubic theta function

0.3.1 Basic notions. 0.3.2 Alternative models of $\mathrm{SL}(2, \mathbf{C})/\mathrm{SU}(2)$. 0.3.3 Some facts on discrete subgroups of $\mathrm{SL}(2, \mathbf{C})$. 0.3.4 Fourier expansions. 0.3.5 Eisenstein series $\mathrm{E}_*(w, s)$. 0.3.6 Eisenstein series for $\Gamma^{(2)}_{\mathrm{princ}}(q)$. 0.3.7 One special case. 0.3.8 Cubic theta function. 0.3.9. Involution ι. 0.3.10 Twists. 0.3.11 Shift operators. 0.3.12 Hecke operators T_{p^3}. 0.3.13 Hecke operators U_r. 0.3.14 One more cubic metaplectic form.

0.3.1 Basic notions. The special linear group $\mathrm{SL}(2, \mathbf{C})$ is a simple complex Lie group of type A_1 in Cartan-Killing classification. The group $\mathrm{SU}(2)$ is a maximal compact subgroup of $\mathrm{SL}(2, \mathbf{C})$. Let us look at the symmetric space $\mathrm{SL}(2, \mathbf{C})/\mathrm{SU}(2)$. We have the Iwasawa decomposition $\mathrm{SL}(2, \mathbf{C}) = NAK$ with

$$N = \left\{ \begin{pmatrix} 1 & z \\ 0 & 1 \end{pmatrix} \mid z \in \mathbf{C} \right\}, \quad A = \left\{ \begin{pmatrix} \sqrt{v} & 0 \\ 0 & 1/\sqrt{v} \end{pmatrix} \mid v \in \mathbf{R}_+^* \right\}, \quad K = \mathrm{SU}(2).$$

† We have also to replace $\Gamma^{(3)}_{\mathrm{princ}}(3)$ by $\Gamma^{(3)}_{\mathrm{princ}}(q)$ with appropriately chosen q.

This implies the mapping

$$(z,v) \mapsto \begin{pmatrix} 1 & z \\ 0 & 1 \end{pmatrix} \begin{pmatrix} \sqrt{v} & 0 \\ 0 & 1/\sqrt{v} \end{pmatrix} \mathrm{SU}(2), \qquad z \in \mathbf{C}, \quad v \in \mathbf{R}_+^*,$$

is an isomorphism of SL(2, C)/SU(2) and

$$\mathbf{H} = \mathbf{C} \times \mathbf{R}_+^* \tag{0.3.1}$$

as real analytic manifolds. The action by left translations of the group SL(2, C) on SL(2, C)/SU(2) transfers on **H** via this isomorphism, and we shall write γw for the image of w under γ. For this action one has

$$\gamma w = \left(\frac{(az+b)\overline{(cz+d)} + a\bar{c}v^2}{|cz+d|^2 + |c|^2v^2}, \frac{v}{|cz+d|^2 + |c|^2v^2} \right)$$
$$\text{if } \gamma = \begin{pmatrix} a & b \\ c & d \end{pmatrix} \in \mathrm{SL}(2,\mathbf{C}), \quad w = (z,v) \in \mathbf{H}. \tag{0.3.2}$$

For $w = (z,v) \in \mathbf{H}$, $w' = (z',v') \in \mathbf{H}$, let us set $x = \Re(z)$, $y = \Im(z)$, $x' = \Re(z')$, $y' = \Im(z')$, and let

$$k(w,w') = \frac{(x-x')^2 + (y-y')^2 + (v-v')^2}{2vv'}.$$

The so defined function $k\colon \mathbf{H} \times \mathbf{H} \to \mathbf{R}$ is a point-pair invariant, that is, one has $k(\delta w, \delta w') = k(w,w')$ for all $w, w' \in \mathbf{H}$ and $\delta \in \mathrm{SL}(2,\mathbf{C})$. Following Beltrami [6], we equip the space **H** with the Riemannian metric with the line element $(dx^2 + dy^2 + dv^2)/v^2$, to get the Lobachevsky space, i.e., the 3-dimensional hyperbolic space. This metric can be described directly by $\cosh d(w,w') = k(w,w') + 1$, $w, w' \in \mathbf{H}$, where we write $d(\cdot,\cdot)$ for the hyperbolic distance and cosh for the hyperbolic cosinus. In particular, we have this metric is SL(2, C)-invariant. One knows, due to Poincaré [77], the group of orientation preserving isometrices of **H** is the quotient of the group SL(2, C) by its center. (The center, consisting of identity and minus identity matrices only, acts trivially on **H**.)

The group SL(2, C) acts on complex projective line $\mathbf{P}^1_{\mathbf{C}} = \mathbf{C} \cup \{\infty\}$ by linear fractional transformations —

$$\begin{pmatrix} a & b \\ c & d \end{pmatrix} z = \frac{az+b}{cz+d}, \qquad z \in \mathbf{P}^1_{\mathbf{C}}.$$

One can view $\mathbf{P}^1_{\mathbf{C}}$ as the 'boundary' of **H**. To describe this more precise, let us consider **H** as the half-space bounded by the plane $\{(z,0) \mid z \in \mathbf{C}\}$ in Euclidean space $\mathbf{C} \times \mathbf{R} \simeq \mathbf{R}^3$. The formula (0.3.2) induces (take $v = 0$) the action of SL(2, C) by linear fractional transformations on the extended plane $\{(z,0) \mid z \in \mathbf{C}\} \cup \{\infty\}$, which we identify with $\mathbf{P}^1_{\mathbf{C}}$ by $(z,0) \mapsto z$, $\infty \mapsto \infty$.

In obvious way one can describe $\mathbf{H}$ also as $\mathrm{GL}(2,\mathbf{C})/\mathrm{Z}(2)\mathrm{SU}(2)$, where $\mathrm{Z}(2)$ is the center $\{\,\mathrm{diag}(t,t)\mid t\in\mathbf{C}^*\,\}$ of the general linear group $\mathrm{GL}(2,\mathbf{C})$. This allows to extend the action of $\mathrm{SL}(2,\mathbf{C})$ to the action of $\mathrm{GL}(2,\mathbf{C})$ on $\mathbf{H}$, so that

$$\gamma w=\left(\frac{(az+b)\overline{(cz+d)}+a\bar{c}v^2}{|cz+d|^2+|c|^2v^2},\ \frac{|ad-bc|v}{|cz+d|^2+|c|^2v^2}\right)$$
$$\text{if }\ \gamma=\begin{pmatrix}a&b\\c&d\end{pmatrix}\in\mathrm{GL}(2,\mathbf{C}),\quad w=(z,v)\in\mathbf{H}. \tag{0.3.3}$$

Certainly, the matrices in $\mathrm{Z}(2)$ act trivially on $\mathbf{H}$.

Now let us introduce into consideration a second order differential operator

$$\mathrm{D}_{\text{L-B}}=-v^2\left(\frac{\partial^2}{\partial x^2}+\frac{\partial^2}{\partial y^2}+\frac{\partial^2}{\partial v^2}\right)+v\frac{\partial}{\partial v},\qquad x=\Re(z),\ \ y=\Im(z),$$

known as the Laplace–Beltrami operator. One can characterize $\mathrm{D}_{\text{L-B}}$, up to a constant factor, as the only second order differential operator which is invariant, in the sense that its action on C^∞–functions $\mathbf{H}\to\mathbf{C}$ commute with the action of $\mathrm{SL}(2,\mathbf{C})$ on $\mathbf{H}$. Moreover, the Laplace–Beltrami operator generates the algebra of all invariant differential operators on $\mathbf{H}$.

Let Γ be a discrete subgroup of $\mathrm{SL}(2,\mathbf{C})$. By definition, this means that any $\gamma\in\Gamma$ has a neighbourhood which does not contain any elements of Γ except γ itself. Then let $\epsilon\in\mathbf{P}^1_{\mathbf{C}}$. We write Γ_ϵ for the stabilizer of ϵ in Γ. Thus, $\Gamma_\epsilon=\{\gamma\in\Gamma\mid\gamma\epsilon=\epsilon\}$. Certainly, Γ_ϵ is a subgroup of Γ. One has $\Gamma_\infty=\Gamma\cap B$ and, for any ϵ, $\Gamma_\epsilon=\sigma B\sigma^{-1}\cap\Gamma$, where B is the group of upper triangular matrices in $\mathrm{SL}(2,\mathbf{C})$ and $\sigma\in\mathrm{SL}(2,\mathbf{C})$ is any matrix so that $\sigma\infty=\epsilon$. We say ϵ is a cusp of Γ if

$$\left\{\sigma\begin{pmatrix}1&z\\0&1\end{pmatrix}\sigma^{-1}\ \middle|\ z\in\Lambda_\sigma\right\}\subset\Gamma_\epsilon\qquad\text{for some lattice }\Lambda_\sigma\text{ in }\mathbf{C}.$$

If ϵ is a cusp of Γ, $\delta\in\Gamma$, and $\epsilon'=\delta\epsilon$, then ϵ' is a cusp of Γ also. In such case we say the cusps ϵ and ϵ' are equivalent. Thus we have an action of the group Γ on the set $\mathbf{P}(\Gamma)$ of its cusps, and $\mathbf{P}(\Gamma)$ splits into Γ-orbits consisting of equivalent cusps.

Let $\tilde{\kappa}\colon\Gamma\to\mathbf{C}^*$ be an unitary character[†], Γ being a discrete subgroup of $\mathrm{SL}(2,\mathbf{C})$. We say $f\colon\mathbf{H}\to\mathbf{C}$ is an automorphic form under Γ with multiplier system $\tilde{\kappa}$ if f satisfies the conditions

(a) $f(\gamma w)=\tilde{\kappa}(\gamma)f(w)$ for $\gamma\in\Gamma$, $w\in\mathbf{H}$;

(b) f is an eigenfunction of the Laplace–Beltrami operator $\mathrm{D}_{\text{L-B}}$;

(c) There is $c\in\mathbf{R}$ such that $|f(z,v)|<\{v+(1+|z|^2)v^{-1}\}^c$ for all $(z,v)\in\mathbf{H}$.

(Clearly, for the so defined automorphic forms exist one should have $\tilde{\kappa}(-e_2)=1$ if $-e_2\in\Gamma$, e_2 being 2×2 identity matrix.)

[†] I.e. homomorphism with values in the unit circle.

We say an automorphic form f is square-integrable if

$$\int_{\mathcal{F}} |f(z,v)|^2 \frac{dz\,dv}{v^3} < \infty,$$

where $\mathcal{F}$ is a fundamental domain of Γ. Since the measure $dz\,dv/v^3$ is invariant by $\mathrm{SL}(2,\mathbf{C})$, the integral does not depend on the particular choice of $\mathcal{F}$. By a fundamental domain of Γ is understood any open set $\mathcal{F} \subset \mathbf{H}$ such that $\bigcup_{\gamma\in\Gamma} \gamma\mathcal{F}$ is a disjoint union whose closure coincides with $\mathbf{H}$.

Let $\kappa\colon \widetilde{\Gamma}^{(2)}_{\mathrm{princ}}(3) \to \mathbf{C}^*$ be the (extended) Kubota homomorphism (see 0.2.2 and 0.2.5), and let Γ be a subgroup of finite index in $\widetilde{\Gamma}^{(2)}_{\mathrm{princ}}(3)$. We say $f\colon \mathbf{H} \to \mathbf{C}$ is a cubic metaplectic form under Γ if f is an automorphic form under Γ with multiplier system κ.

Let us write $\mathcal{L}(q,r)$ for the vector space of cubic metaplectic forms f under $\Gamma^{(2)}_{\mathrm{princ}}(q)$ which are the eigenfunctions of the operator $\mathrm{D}_{\text{L-B}}$ with eigenvalue $\lambda = r(2-r)$. The subspace $\mathcal{L}^2(q,r)$ of square-integrable forms in $\mathcal{L}(q,r)$ equipped with the Petersson inner product

$$\langle f_1, f_2\rangle = \int_{\mathcal{F}} f_1(z,v)\overline{f_2(z,v)}\,\frac{dz\,dv}{v^3}. \tag{0.3.4}$$

becomes a finite-dimensional Hilbert space [32].

For $f\colon \mathbf{H} \to \mathbf{C}$ and $\sigma \in \mathrm{SL}(2, \mathbf{Q}(\sqrt{-3}))$ define $f_\sigma\colon \mathbf{H} \to \mathbf{C}$ by $f_\sigma(w) = f(\sigma w)$. Two rather simple facts are useful to know:

(a) If f is a cubic metaplectic form under Γ, $\sigma\Gamma\sigma^{-1} = \Gamma$ and $\kappa(\sigma\gamma\sigma^{-1}) = \kappa(\gamma)$ for all $\gamma \in \Gamma$, then f_σ is a cubic metaplectic form under Γ also.

(b) If $f \in \mathcal{L}(q,r)$, then $f_\sigma \in \mathcal{L}(q',r)$ for some ideal q'. One has also $f_\sigma \in \mathcal{L}^2(q',r)$ if $f \in \mathcal{L}^2(q,r)$.

0.3.2 Alternative models of SL(2, C)/SU(2). Let us write $\mathrm{Hermite}^n_+$ for the space of $n \times n$ positive Hermitian matrices of determinant 1 —

$$\mathrm{Hermite}^n_+ = \{h \in \mathrm{SL}(n,\mathbf{C}) \mid {}^t\overline{h} = h,\ h \text{ is positive}\}, \qquad n \geq 2.$$

Recall that Hermitian matrix h is said to be positive if $uh{}^t\overline{u} > 0$ for all $u \in \mathbf{C}^n$, $u \neq 0$. The mapping

$$\mathrm{SL}(n,\mathbf{C})/\mathrm{SU}(n) \to \mathrm{Hermite}^n_+, \qquad \delta\mathrm{SU}(n) \mapsto \delta{}^t\overline{\delta},$$

identifies $\mathrm{SL}(n,\mathbf{C})/\mathrm{SU}(n)$ with $\mathrm{Hermite}^n_+$ and transfers on $\mathrm{Hermite}^n_+$ the action of the group $\mathrm{SL}(n,\mathbf{C})$ in such a way that the image of the point $h \in \mathrm{Hermite}^n_+$ under the action of $\gamma \in \mathrm{SL}(n,\mathbf{C})$ is $\gamma h{}^t\overline{\gamma}$. This mapping for $n = 2$, being combined with the isomorphism of 0.3.1, gives us the isomorphism

$$\mathbf{H} \to \mathrm{Hermite}^2_+, \qquad (z,v) \mapsto \frac{1}{v}\begin{pmatrix} |z|^2 + v^2 & z \\ \overline{z} & 1 \end{pmatrix}.$$

By using this model it is easy to prove the formula (0.3.2). We should notice also that the function $(z,v) \mapsto v+(1+|z|^2)v^{-1}$, which occurs in the point (c) of automorphic forms definition in 0.3.1, is nothing but the trace of the Hermitian matrix attached above to $(z,v) \in \mathbf{H}$.

To get one more model we consider the division algebra of Hamilton's quaternions $\mathbf{R}[i,j,k]$ generated by i,j,k with

$$i^2 = j^2 = k^2 = -1, \quad ij = k, \quad jk = i, \quad ki = j.$$

One can identify the space $\mathbf{H}$ with $\{x+yi+vj \mid x,y,v \in \mathbf{R},\ v > 0\} \subset \mathbf{R}[i,j,k]$ by means of the mapping $w = (z,v) \mapsto m = x + yi + vj$, $x = \Re(z)$, $y = \Im(z)$. And, if so, the action of $\mathrm{SL}(2,\mathbf{C})$ can be written as

$$\gamma m = (am+b)(cm+d)^{-1} \quad \text{for} \quad \gamma = \begin{pmatrix} a & b \\ c & d \end{pmatrix}.$$

For the quaternion algebra $\mathbf{R}[i,j,k]$ we have a standard representation by the algebra of 2×2 complex matrices, which is defined as

$$x + yi + vj + uk \mapsto \begin{pmatrix} x+iy & -v-ui \\ v-ui & x-iy \end{pmatrix}, \qquad x,y,v,u \in \mathbf{R}.$$

Thus we can identify $\mathrm{SL}(2,\mathrm{C})/\mathrm{SU}(2)$ and $\mathbf{H}$ with the space of matrices

$$w = \begin{pmatrix} z & -v \\ v & \bar{z} \end{pmatrix}, \qquad (z,v) \in \mathbf{H}.$$

In this case, the image of w under the action of $\gamma \in \mathrm{SL}(2,\mathbf{C})$ can be written as

$$\gamma w = (\tilde{a}w + \tilde{b})(\tilde{c}w + \tilde{d})^{-1} \quad \text{if} \quad \gamma = \begin{pmatrix} a & b \\ c & d \end{pmatrix} \quad \text{and} \quad \tilde{t} = \mathrm{diag}(t, \bar{t}) \quad \text{for} \quad t \in \mathbf{C}.$$

More general construction, involving Clifford algebras instead of the quaternions one, proposed by Vahlen [97] to obtain models for the hyperbolic spaces of any dimension. In this connection see also Elstrodt, Grunewald and Mennicke [26].

0.3.3 Some facts on discrete subgroups of SL(2, C). The discrete subgroups of $\mathrm{SL}(2,\mathbf{C})$ were studied extensively in connection with the geometry of 3-dimentional hyperbolic manifolds and the reduction theory of Hermitian forms. One of the first and simplest examples is the group $\mathrm{SL}(2,\mathbf{Z}[\sqrt{-1}])$ considered already by Picard, who described [76] its fundamental domain in $\mathbf{H}$. More generally, let $\mathcal{O}_D$ be the ring of integers of the imaginary quadratic field $\mathbf{Q}(\sqrt{-D})$, $D > 0$, $D \in \mathbf{Z}$. Then $\mathrm{SL}(2,\mathcal{O}_D)$ is a discrete subgroup of $\mathrm{SL}(2,\mathbf{C})$. These groups were studied by Bianchi, Humbert. The reader interested in the discrete subgroups of different type can find examples in [26] and in [64].

Bianchi showed [9], the set of all cusps of $\mathrm{SL}(2,\mathcal{O}_D)$ is $\mathbf{Q}(\sqrt{-D}) \cup \{\infty\}$, and, if h is the class number of $\mathcal{O}_D$, this set splits into h orbits with respect to the action of $\mathrm{SL}(2,\mathcal{O}_D)$. One more remarkable result is Humbert's [39] formula

$$V_D = (2\pi)^{-2}|d|^{3/2}\zeta_{\mathbf{Q}(\sqrt{-D})}(2)$$

for the hyperbolic volume V_D of a fundamental domain $\mathcal{F}_D$ of $\mathrm{SL}(2,\mathcal{O}_D)$. In this formula d denotes the discriminant and $\zeta_{\mathbf{Q}(\sqrt{-D})}$ denotes the Dedekind zeta-function of $\mathbf{Q}(\sqrt{-D})$. The fundamental domain $\mathcal{F}_D$ can be choosen as polyhedron bounded by hemispheres; one can find a lot of examples in [8], [9].

For different proofs of Humbert's formula see Borel [10], Sarnak [86], and Elstrodt, Grunewald, Mennicke [25]. Concerning the generators and relations for $\mathrm{SL}(2,\mathcal{O}_D)$ we can refer again to Bianchi [8], [9] and Humbert [38], and also to more recent researches of Swan [96] and Grunewald, Helling, Mennicke [31].

We turn now to the group $\mathrm{SL}(2,\mathcal{O})$ and its congruence subgroups defined in 0.2.1. These are the only discrete subgroups of $\mathrm{SL}(2,\mathbf{C})$ which we shall deal with. First, we notice the matrices

$$\begin{pmatrix}\omega & 0\\ 0 & \omega^2\end{pmatrix},\quad \begin{pmatrix}0 & -1\\ 1 & 0\end{pmatrix},\quad \begin{pmatrix}1 & 1\\ 0 & 1\end{pmatrix}\qquad (\text{as usual, } \omega=\exp(2\pi i/3))$$

generate $\mathrm{SL}(2,\mathcal{O})$. This is rather simple fact, because of $\mathcal{O}$ is Euclidian ring.

$\Gamma^{(2)}_{\mathrm{princ}}(q)$ is a normal subgroup of finite index in $\mathrm{SL}(2,\mathcal{O})$, $\mathrm{SL}(2,\mathcal{O})/\Gamma^{(2)}_{\mathrm{princ}}(q) \simeq \mathrm{SL}(2,\mathcal{O}/q)$. We have also $\Gamma^{(2)}_{\mathrm{princ}}(q)\subset\Gamma_2(q)\subset\widetilde{\Gamma}_2(q)$ and $\Gamma^{(2)}_{\mathrm{princ}}(q)\subset\widetilde{\Gamma}^{(2)}_{\mathrm{princ}}(q)$ (see 0.1.1 for the definitions).

According to Bianchi, the set of all cusps of $\mathrm{SL}(2,\mathcal{O})$ is $\mathbf{Q}(\sqrt{-3})\cup\{\infty\}$, and all the cusps are equivalent to ∞. If Γ and Γ' are discrete subgroups of $\mathrm{SL}(2,\mathbf{C})$ and $\Gamma\subset\Gamma'$, then $\Gamma_\epsilon\subset\Gamma'_\epsilon$ for any $\epsilon\in\mathbf{P}^1_{\mathbf{C}}$, and so the set of all cusps of Γ is a subset of the set of all cusps of Γ'. In particular, if $\Gamma\subset\mathrm{SL}(2,\mathcal{O})$ and $\mathbf{P}(\Gamma)$ means the set of all its cusps, then $\mathbf{P}(\Gamma)\subset\mathbf{Q}(\sqrt{-3})\cup\{\infty\}$, and any $\epsilon\in\mathbf{P}(\Gamma)$ can be written as $\epsilon=\sigma\infty$ with some $\sigma\in\mathrm{SL}(2,\mathcal{O})$. One can easily check that $\mathbf{P}(\Gamma)=\mathbf{Q}(\sqrt{-3})\cup\{\infty\}$ if Γ is either $\Gamma^{(2)}_{\mathrm{princ}}(q)$ or one of the groups $\widetilde{\Gamma}^{(2)}_{\mathrm{princ}}(q)$, $\Gamma_2(q)$, $\widetilde{\Gamma}_2(q)$.

Let Γ be any subgroup of $\widetilde{\Gamma}^{(2)}_{\mathrm{princ}}(3)$. We have the Kubota homomorphism κ defined on $\widetilde{\Gamma}^{(2)}_{\mathrm{princ}}(3)$ and we say a cusp ϵ of Γ is essential for κ if the restriction of κ to Γ_ϵ is trivial. Given two equivalent cusps of Γ, we have if one of them is essential for κ, then does another one. If Γ is a subgroup of $\Gamma^{(2)}_{\mathrm{princ}}(9)$, then all cusps of Γ are essential for κ (this follows easily from Theorem 0.2.2 (e)).

Let us look closely to the case $q=(3)$, considered in [69]. We have

$$\mathrm{SL}(2,\mathcal{O})/\Gamma^{(2)}_{\mathrm{princ}}(3)\simeq\mathrm{SL}(2,\mathbf{Z}/9\mathbf{Z}),\quad \widetilde{\Gamma}^{(2)}_{\mathrm{princ}}(3)/\Gamma^{(2)}_{\mathrm{princ}}(3)\simeq\mathrm{SL}(2,\mathbf{Z}/3\mathbf{Z}),$$

$$\text{and so}\quad [\mathrm{SL}(2,\mathcal{O}):\Gamma^{(2)}_{\mathrm{princ}}(3)]=648,\quad [\widetilde{\Gamma}^{(2)}_{\mathrm{princ}}(3):\Gamma^{(2)}_{\mathrm{princ}}(3)]=24,$$

$$\text{and}\quad [\mathrm{SL}(2,\mathcal{O}):\widetilde{\Gamma}^{(2)}_{\mathrm{princ}}(3)]=27.$$

The following 27 matrices represent cosets in $\widetilde{\Gamma}^{(2)}_{\mathrm{princ}}(3)\backslash\mathrm{SL}(2,\mathcal{O})$ —

$$\begin{pmatrix}1 & \omega t\\ 0 & 1\end{pmatrix},\quad \begin{pmatrix}\omega & t\\ 0 & \omega^2\end{pmatrix}\quad\text{and}\quad \begin{pmatrix}\omega^2 & t\\ 0 & \omega\end{pmatrix}\qquad\text{with}\quad t=0,\pm1;$$

$$\begin{pmatrix}1 & 0\\ \varepsilon & 1\end{pmatrix}\begin{pmatrix}1 & t\\ 0 & 1\end{pmatrix}\qquad\text{with}\quad t=0,\pm1,\pm\varepsilon,\pm\varepsilon^2,\pm(1-\varepsilon),\quad \varepsilon=\omega^{\pm1};$$

$\mathbf{Q}(\sqrt{-3})\cup\{\infty\}$ splits into $\Gamma^{(2)}_{\mathrm{princ}}(3)$-orbits whose representatives can be chosen as ∞, 0, ± 1, $\pm\omega$, $\pm\omega^2$, $\pm\lambda$, $\pm\lambda^{-1}$, where $\lambda = 1-\omega$. Denote this set of representatives by P. For $\epsilon \in P$ we define σ_ϵ respectively by

$$\begin{pmatrix}1&0\\0&1\end{pmatrix}, \begin{pmatrix}0&-1\\1&0\end{pmatrix}, \begin{pmatrix}1&0\\\pm 1&1\end{pmatrix}, \begin{pmatrix}1&0\\\pm\omega^2&1\end{pmatrix}, \begin{pmatrix}1&0\\\pm\omega&1\end{pmatrix}, \begin{pmatrix}\pm\lambda&-1\\1&0\end{pmatrix}, \begin{pmatrix}1&0\\\pm\lambda&1\end{pmatrix}.$$

With the so defined σ_ϵ we have $\sigma_\epsilon\infty = \epsilon$ and $\Gamma^{(2)}_{\text{princ}}(3)_\epsilon = \sigma_\epsilon\Gamma^{(2)}_{\text{princ}}(3)_\infty\sigma_\epsilon^{-1}$. Only 4 of 12 orbits consist of essential cusps. The representatives of these orbits in P are ∞, 0, ± 1. Also, the set $\mathbf{Q}(\sqrt{-3}) \cup \{\infty\}$ splits into $\widetilde{\Gamma}^{(2)}_{\text{princ}}(3)$-orbits whose representatives can be chosen as ∞, ω, ω^2. Only the orbit represented by ∞ consists of essential cusps. The set

$$\mathcal{F}_* = \left\{(z,v) \in \mathbf{H} \;\middle|\; |z|^2 + v^2 > 1;\ z \in \pm\nabla\right\},$$

where ∇ is the interior of the triangle with vertices 0, $(1-\omega)^{-1}$ and $(1-\omega^2)^{-1}$, is a fundamental domain for $\mathrm{SL}(2,\mathcal{O})$. This fundamental domain is given by Bianchi [8], and it is used by Patterson in [69]. As a fundamental domain $\mathcal{F}$ of any subgroup $\Gamma \subset \mathrm{SL}(2,\mathcal{O})$ one can take $\bigcup_{\gamma\in R}\gamma\mathcal{F}_*$, where R is a set of representatives for $\Gamma\backslash\mathrm{SL}(2,\mathcal{O})$.

0.3.4 Fourier expansions. Let Γ be a subgroup of finite index in $\widetilde{\Gamma}^{(2)}_{\text{princ}}(3)$, and let $\tilde\kappa\colon \Gamma \to \mathbf{C}^*$ be any unitary character of Γ satisfying $\tilde\kappa(-e_2) = 1$ if $-e_2 \in \Gamma$. For example, one can take $\tilde\kappa$ to be Kubota's κ.

As we know, the set $\mathbf{P}(\Gamma)$ of cusps of Γ is a subset of $\mathbf{Q}(\sqrt{-3}) \cup \{\infty\}$ and each point in $\mathbf{P}(\Gamma)$ can be written as $\sigma\infty$ with appropriately choosen $\sigma \in \mathrm{SL}(2,\mathcal{O})$. For any $\sigma \in \mathrm{SL}(2,\mathcal{O})$ we set

$$\begin{aligned}\Lambda_\sigma &= \left\{l \in \mathbf{C} \;\middle|\; \sigma\begin{pmatrix}1&l\\0&1\end{pmatrix}\sigma^{-1} \in \Gamma\right\},\\ \Lambda^*_\sigma &= \left\{k \in \mathbf{C} \;\middle|\; (kl + \overline{kl}) \in \mathbf{Z} \text{ for all } l \in \Lambda_\sigma\right\}.\end{aligned} \tag{0.3.5}$$

Now let us assume that $\epsilon = \sigma\infty$ is a cusp of Γ. Then Λ_σ is a lattice in $\mathbf{C}$ and Λ^*_σ is a lattice in $\mathbf{C}$ dual to Λ_σ. The function $\Lambda_\sigma \to \mathbf{C}^*$ defined by

$$l \mapsto \tilde\kappa\left(\sigma\begin{pmatrix}1&l\\0&1\end{pmatrix}\sigma^{-1}\right)$$

is a homomorphism of the lattice Λ_σ. If this homomorphism is trivial, we shall say ϵ is an essential cusp of Γ for $\tilde\kappa$ (omitting the reference to group and homomorphism in obvious cases). In general, with some $h_\sigma \in \mathbf{C}$ we have

$$\tilde\kappa\left(\sigma\begin{pmatrix}1&l\\0&1\end{pmatrix}\sigma^{-1}\right) = e(h_\sigma l) \qquad \text{for all } \ l \in \Lambda_\sigma. \tag{0.3.6}$$

For essential cusps we can take $h_\sigma = 0$. Let f be an automorphic form under Γ with multiplier system $\tilde\kappa$, and let $w = (z,v) \in \mathbf{H}$. For each $l \in \Lambda_\sigma$, we have

$$\begin{aligned}f(\sigma(z+l,v)) &= f\left(\sigma\begin{pmatrix}1&l\\0&1\end{pmatrix}w\right) = f\left(\sigma\begin{pmatrix}1&l\\0&1\end{pmatrix}\sigma^{-1}\sigma w\right)\\ &= \tilde\kappa\left(\sigma\begin{pmatrix}1&l\\0&1\end{pmatrix}\sigma^{-1}\right)f(\sigma w) = e(h_\sigma l)f(\sigma(z,v)).\end{aligned}$$

This yields immediately a Fourier expansion of the form

$$f(\sigma w) = \sum_{\nu} c_\nu(v)e(\nu z), \quad \text{where}$$
$$c_\nu(v) = \mathrm{vol}(\mathbf{C}/\Lambda_\sigma)^{-1} \int_{\mathbf{C}/\Lambda_\sigma} f(\sigma(z,v))e(\nu z)\,dz, \tag{0.3.7}$$

and the summation is carried out over $\nu \in h_\sigma + \Lambda_\sigma^*$. We have $\mathrm{D}_{\text{L-B}} f = \lambda f$ with some $\lambda \in \mathbf{C}$ and, applying the Laplace–Beltrami operator $\mathrm{D}_{\text{L-B}}$ to both sides of (0.3.7), we find the coefficients c_ν satisfy the second order linear differential equations

$$v^2 c_\nu''(v) - v c_\nu'(v) = \left((4\pi|\nu|v)^2 - \lambda\right)c_\nu(v) \tag{0.3.8}$$

(compare with the Bessel equation in 0.1.6). Let us choose $r \in \mathbf{C}$ in such a way that $\lambda = r(2-r)$. If $\nu = 0$ and $r \neq 1$, the solutions of equation (0.3.8) are linear combinations of $v \mapsto v^r$ and $v \mapsto v^{2-r}$. For $\nu = 0$ and $r = 1$ those are linear combinations of $v \mapsto v\log v$ and $v \mapsto v$. Due to the growth condition† satisfied by f, the coefficients $c_\nu(v)$ have at most polynomial growth as $v \longrightarrow \infty$. Under this restriction, the only solutions of (0.3.8) with $\nu \neq 0$ are $v \mapsto \rho(\nu)vK_{r-1}(4\pi|\nu|v)$, $\rho(\nu) \in \mathbf{C}$, where K_{r-1} is the Bessel–MacDonald function. These observations, taken from [53], [54], and some other facts, noticed in particular in 0.3.3, are summarized in the theorem below.

Theorem 0.3.1 *Let Γ be a subgroup of finite index in $\widetilde{\Gamma}^{(2)}_{\text{princ}}(3)$, and let f be an automorphic form under Γ with miltiplier system $\tilde{\kappa}\colon \Gamma \to \mathbf{C}^*$ so that $\mathrm{D}_{\text{L-B}} f = r(2-r)f$, $r \in \mathbf{C}$. Let $\epsilon = \sigma\infty$ be a cusp of Γ, $\sigma \in \mathrm{SL}(2,\mathcal{O})$ and let Λ_σ, Λ_σ^* and h_σ be as defined in (0.3.5), (0.3.6). Then we have a Fourier expansion*

$$f(\sigma(z,v)) = c_0^\sigma(v) + \sum_{\substack{\nu\neq 0\\ \nu\in h_\sigma+\Lambda_\sigma^*}} \rho(\sigma;\nu)vK_{r-1}(4\pi|\nu|v)e(\nu z), \qquad (z,v)\in \mathbf{H},$$

where $c_0^\sigma(v)$ is either $\rho_+(\sigma)v^r + \rho_-(\sigma)v^{2-r}$ or $\rho_+(\sigma)v\log v + \rho_-(\sigma)v$, according to $r \neq 1$ or $r = 1$; $\rho_+(\sigma)$, $\rho_-(\sigma)$, $\rho(\sigma;\nu) \in \mathbf{C}$.

If ϵ is an essential cusp of Γ, then one can take $h_\sigma = 0$. If ϵ is not an essential cusp of Γ, then $c_0^\sigma(v) = 0$. If Γ is a subgroup of $\Gamma^{(2)}_{\text{princ}}(9)$ and $\tilde{\kappa}$ is Kubota's κ, then all the cusps of Γ are essential ones. If Γ is either $\Gamma_2(q)$ or $\Gamma^{(2)}_{\text{princ}}(q)$, then $\epsilon = \sigma\infty$ is a cusp of Γ for any $\sigma \in \mathrm{SL}(2,\mathcal{O})$, and one has $\Lambda_\sigma = q$, $\Lambda_\sigma^* = q^*$. ∎

We shall refer to $c_0^\sigma(v)$ as the constant term of f, and to $\rho(\sigma;\nu)$ as the νth Fourier coefficients of f about the cusp $\epsilon = \sigma\infty$.

If $c_0^\sigma(v) = 0$ for all cusps σ, then f is said to be a cusp form. The notion of cusp forms is one of fundamental ones. However, the forms that we are mainly

---

† We have in mind the point (c) in the definition of automorphic forms in 0.3.1.

interested in these notes are not cusp forms. Cusps forms decay exponentially about all cusps. This yields, in particular, cusps forms are square-integrable. More generally, we have that an automorphic form f is square-integrable if and only if for all σ (in the notations of Theorem 0.3.1) one has

$$\rho_+(\sigma) = 0 \text{ if } \Re(r) \geq 1 \quad \text{and} \quad \rho_-(\sigma) = 0 \text{ if } \Re(r) \leq 1.$$

Fourier expansions play a central role throughout. Especially often we shall deal with expansions about the cusps ∞ and 0. In these cases to simplify notations we prefer to drop σ and to write ρ_+, ρ_-, $\rho(\nu)$ instead of $\rho_+(\sigma)$, $\rho_-(\sigma)$, $\rho(\sigma;\nu)$. To avoid any possible misunderstandigs and for references convenience it seems reasonable to extract these two cases from the general Theorem 0.3.1 and to state them as corollaries explicitly in a form that will be used. For this let us set

$$\Delta_* = \left\{ \begin{pmatrix} 1 & l \\ 0 & 1 \end{pmatrix} \mid l \in q \right\}. \tag{0.3.9}$$

Notice that Δ_* is the stabilizer of ∞, and ${}^t\Delta_*$ is the stabilizer of 0 as in the group $\Gamma^{(2)}_{\text{princ}}(q)$ as in the group $\Gamma_2(q)$.

Corollary 0.3.2 *Let* Γ *be either* $\Gamma^{(2)}_{\text{princ}}(q)$ *or* $\Gamma_2(q)$, *and let* $\tilde{\kappa}\colon \Gamma \to \mathbf{C}^*$ *be an unitary homomorphism trivial on* Δ_*. *(This case* ∞ *is an essential cusp of* Γ. *In particular, one can take* $\tilde{\kappa}$ *to be the Kubota's* κ.*) Let* f *be an automorphic form under* Γ *with multiplier system* $\tilde{\kappa}$, *and* $\mathrm{D}_{\text{L-B}} f = r(2-r)f$, $r \in \mathbf{C}$. *Then we have a Fourier expansion*

$$f(z,v) = c_0(v) + \sum_{\nu \in q^* \setminus \{0\}} \rho(\nu) v K_{r-1}(4\pi|\nu|v) e(\nu z), \quad (z,v) \in \mathbf{H},$$

where $\rho(\nu) \in \mathbf{C}$, *and, with some* $\rho_+, \rho_- \in \mathbf{C}$, *we have*

$$\textit{either} \quad c_0(v) = \rho_+ v^r + \rho_- v^{2-r} \quad \textit{or} \quad c_0(v) = \rho_+ v \log v + \rho_- v,$$

according to $r \neq 1$ or $r = 1$. ∎

Corollary 0.3.3 *Let* Γ *be either* $\Gamma^{(2)}_{\text{princ}}(q)$ *or* $\Gamma_2(q)$, *and let* $\tilde{\kappa}\colon \Gamma \to \mathbf{C}^*$ *be an unitary homomorphism trivial on* ${}^t\Delta_*$. *(This case* 0 *is an essential cusp of* Γ. *In particular, one can take* $\tilde{\kappa}$ *to be the Kubota's* κ.*) Let* f *be an automorphic form under* Γ *with multiplier system* $\tilde{\kappa}$, *and* $\mathrm{D}_{\text{L-B}} f = r(2-r)f$, $r \in \mathbf{C}$. *Then we have a Fourier expansion*

$$f\left(\begin{pmatrix} 0 & -1 \\ 1 & 0 \end{pmatrix}(z,v)\right) = c_0(v) + \sum_{\nu \in q^* \setminus \{0\}} \rho(\nu) v K_{r-1}(4\pi|\nu|v) e(\nu z), \quad (z,v) \in \mathbf{H},$$

where $\rho(\nu) \in \mathbf{C}$, *and, with some* $\rho_+, \rho_- \in \mathbf{C}$, *we have*

$$\textit{either} \quad c_0(v) = \rho_+ v^r + \rho_- v^{2-r} \quad \textit{or} \quad c_0(v) = \rho_+ v \log v + \rho_- v,$$

according to $r \neq 1$ or $r = 1$. ∎

0.3.5 Eisenstein series $\mathrm{E}_*(w,s)$. For $w=(z,v)\in\mathbf{H}$ let $V(w)=v$ and

$$\mathrm{E}_*(w,s)=\sum_{\gamma}\overline{\kappa}(\gamma)V(\gamma w)^s,\quad s\in\mathbf{C},\tag{0.3.10}$$

where the summation is carried out over a set of matrices γ that contains a unique representative of each coset in $\Delta_*\backslash\Gamma_2(q)$, Δ_* being as in (0.3.9). (It follows from (0.2.1) and (0.3.2) that $\kappa(\gamma)$ and $V(\gamma w)$ do not depend on the choice of representatives.) Let $m=\mathrm{vol}(\mathbf{C}/q)$ and

$$\Psi_*(s,\lambda)=\sum_{c\in q\backslash\{0\}}\frac{S_*(\lambda,c)}{\|c\|^s},\quad \Psi(s,\lambda)=\sum_{\substack{c\equiv 1(3)\\ \gcd(c,q)=1}}\frac{S(\lambda,c)}{\|c\|^s},\tag{0.3.11}$$

where $S_*(\lambda,c)$ and $S(\lambda,c)$ are the Gauß sums defined in 0.1.4, $s\in\mathbf{C}$, and both the series are absolutely convergent if $\Re(s)>3/2$.

Theorem 0.3.4 *The series* (0.3.10) *has the following properties.*

(a) *It converges absolutely and locally uniformly in $\mathbf{H}\times\{s\in\mathbf{C}\mid\Re(s)>2\}$. For each $w\in\mathbf{H}$, the function $\mathrm{E}_*(w,\cdot)$ is regular in $\{s\in\mathbf{C}\mid\Re(s)>2\}$ and has a meromorphic continuation to $\mathbf{C}$. The only singularity of $\mathrm{E}_*(w,\cdot)$ in $\{s\in\mathbf{C}\mid\Re(s)>1\}$ is a simple pole at the point $4/3$.*

(b) *$\mathrm{E}_*(\cdot,s)$ is a cubic metaplectic form under $\Gamma_2(q)$, for any non-singular $s\in\mathbf{C}$, and one has $\mathrm{D}_{\text{L-B}}\mathrm{E}_*(\cdot,s)=s(2-s)\mathrm{E}_*(\cdot,s)$.*

(c) $$\mathrm{E}_*(w,s)=v^s+\phi_*(s)v^{2-s}+\sum_{\lambda\in q^*\backslash\{0\}}\phi_*(s,\lambda)vK_{s-1}(4\pi|\lambda|v)e(\lambda z),$$

where $w=(z,v)\in\mathbf{H}$, $s\in\mathbf{C}$, the coefficients ϕ_* and $\phi_*(\cdot,\lambda)$ are meromorphic functions on $\mathbf{C}$, and

$$\phi_*(s)=\frac{\pi}{m(s-1)}\Psi_*(s,0),\quad \phi_*(s,\lambda)=\frac{(2\pi)^s|\lambda|^{s-1}}{m\Gamma(s)}\Psi_*(s,\lambda)$$

with $\Psi_*(s,\cdot)$ defined for $\Re(s)\geq 3/2$ in (0.3.11).

(d) $$\mathrm{E}_*\left(\begin{pmatrix}0&-1\\1&0\end{pmatrix}w,s\right)=\phi(s)v^{2-s}+\sum_{\lambda\in q^*\backslash\{0\}}\phi(s,\lambda)vK_{s-1}(4\pi|\lambda|v)e(\lambda z),$$

where $w=(z,v)\in\mathbf{H}$, $s\in\mathbf{C}$, the coefficients ϕ and $\phi(\cdot,\lambda)$ are meromorphic functions on $\mathbf{C}$ and

$$\phi(s)=\frac{\pi}{m(s-1)}\Psi(s,0),\quad \phi(s,\lambda)=\frac{(2\pi)^s|\lambda|^{s-1}}{m\Gamma(s)}\Psi(s,\lambda)$$

with $\Psi(s,\cdot)$ defined for $\Re(s)\geq 3/2$ in (0.3.11).

(e) *The series in* (0.3.11) *can be continued to meromorphic functions on $\mathbf{C}$. And the formulae in* (c), (d) *remain valid for all $s\in\mathbf{C}$.* ■

In the sense of Selberg–Langland's theory [87], [88], [59] the series (0.3.10) is the Eisenstein series attached to the group $\Gamma_2(q)$, the multiplier system κ and the cusp ∞. Parts (c) and (d) of the theorem above give us the Fourier expansions for $E_*(\cdot, s)$ about the cusps ∞ and 0.

More generally, given a subgroup Γ of finite index in $\widetilde{\Gamma}^{(2)}_{\text{princ}}(3)$ and its essential cusp ϵ one can attach an Eisenstein series with a similar properties. We shall give basic facts from the Eisenstein series theory for the group $\Gamma^{(2)}_{\text{princ}}(q)$ in the next subsection.

0.3.6 Eisenstein series for $\Gamma^{(2)}_{\text{princ}}(q)$. To simplify notations, let us write Γ for $\Gamma^{(2)}_{\text{princ}}(q)$, $q \subset (3)$. First of all, we choose a family $\epsilon_1, \epsilon_2, \ldots, \epsilon_h$ of cusps of Γ essential for Kubota's κ such that

each essential cusp of Γ is equivalent to one of ϵ_i;

ϵ_i is not equivalent to ϵ_j, if $i \neq j$.

Recall (see 0.3.3) that $\epsilon_i \in \mathbf{Q}(\sqrt{-3}) \cup \{\infty\}$ and that all the points in $\mathbf{Q}(\sqrt{-3}) \cup \{\infty\}$ are equivalent under $\mathrm{SL}(2, \mathcal{O})$. We can and do choose $\sigma_i \in \mathrm{SL}(2, \mathcal{O})$ in such a way that $\sigma_i \infty = \epsilon_i$, $i = 1, \ldots, h$. Let us write Γ_i for the stabilizer of ϵ_i in Γ, and let Δ_* be the subgroup of upper triangular matrices in Γ (as in (0.3.9)). Then Δ_* is the stabilizer of ∞ in Γ, and $\Gamma_i = \sigma_i \Delta_* \sigma_i^{-1}$, for each $i = 1, \ldots, h$.

Let $V(w) = v$ for $w = (z, v) \in \mathbf{H}$. The Eisenstein series attached to ϵ_i is

$$E_{\epsilon_i}(w, s) = \sum_{\gamma} \overline{\kappa}(\gamma) V(\sigma_i^{-1} \gamma w)^s, \qquad s \in \mathbf{C}, \tag{0.3.12}$$

where the summation is carried out over some set of matrices γ that contains a unique representative of each coset in $\Gamma_i \backslash \Gamma$. The series (0.3.12) depends only on the cusp ϵ_i, and not on the choice of $\sigma_i \in \mathrm{SL}(2, \mathcal{O})$. The main properties of these Eisenstein series are summarized in the next theorem. First, we need some notations. We set

$$\begin{aligned} C_{ij} &= \{c \in \mathcal{O} \setminus \{0\} \mid c \text{ is the left low entry of some matrix in } \sigma_i^{-1}\Gamma\sigma_j\}, \\ D_{ij}(c) &= \{d \in \mathcal{O} \mid (c, d) \text{ is the low row of some matrix in } \sigma_i^{-1}\Gamma\sigma_j\}, \\ R_{ij}(c) &= D_{ij}(c) \cap \mathfrak{c}(c), \quad \mathfrak{c}(c) \text{ being a complete residue system mod } cq. \end{aligned}$$

For each $c \in C_{ij}$ and $d \in R_{ij}(c)$, choose somehow a matrix $\gamma_{c,d} \in \sigma_i^{-1}\Gamma\sigma_j$ whose low row is (c, d). Then, for $i, j = 1, \ldots, h$ and $\lambda \in q^*$ we set

$$\begin{aligned} \Psi_{ij}(s, \lambda) &= \sum_{c \in C_{ij}} \frac{S_{ij}(\lambda, c)}{\|c\|^s}, \quad \text{where} \\ S_{ij}(\lambda, c) &= \sum_{d \in R_{ij}(c)} \overline{\kappa}(\sigma_i \gamma_{c,d} \sigma_j^{-1}) e(\lambda d / c), \quad s \in \mathbf{C}, \ \Re(s) > 2. \end{aligned} \tag{0.3.13}$$

The so defined $S_{ij}(\lambda, c)$ and $\Psi_{ij}(s, \lambda)$ do not depend on the choice of $\gamma_{c,d}$ and $\mathfrak{c}(c)$. To do make this more clear, notice that one can define an equivalence relation on $\sigma_i^{-1}\Gamma\sigma_j$ assuming $\gamma' \in \sigma_i^{-1}\Gamma\sigma_j$ is equivalent to $\gamma'' \in \sigma_i^{-1}\Gamma\sigma_j$ if and only if $\gamma' \in \Delta_*\gamma''\Delta_*$. And then we see that each $\gamma' \in \sigma_i^{-1}\Gamma\sigma_j$ is equivalent either to one of $\gamma_{c,d}$ with $c \in \mathrm{C}_{ij}$, $d \in \mathrm{R}_{ij}(c)$ or to the identity matrix (the latter case can has place only if $i = j$).

Theorem 0.3.5 *The Eisenstein series* (0.3.12) *attached to the group* $\Gamma^{(2)}_{\text{princ}}(q)$ *and its cusps* $\epsilon_1, \epsilon_2, \ldots, \epsilon_h$ *have the following properties.*

(a) *They converge absolutely and locally uniformly in* $\mathbf{H} \times \{ s \in \mathbf{C} \mid \Re(s) > 2 \}$. *Given* $w \in \mathbf{H}$, *the functions* $\mathrm{E}_{\epsilon_i}(w, \cdot)$ *are regular in* $\{ s \in \mathbf{C} \mid \Re(s) > 2 \}$ *and have meromorphic continuations to* $\mathbf{C}$. *The only singularity of* $\mathrm{E}_{\epsilon_i}(w, \cdot)$ *in* $\{ s \in \mathbf{C} \mid \Re(s) > 1 \}$ *is a simple pole at the point* 4/3.

(b) $\mathrm{E}_{\epsilon_i}(\cdot, s)$ *are cubic metaplectic forms under* $\Gamma^{(2)}_{\text{princ}}(q)$, *for non-singular* $s \in \mathbf{C}$, *and one has* $\mathrm{D}_{\text{L-B}}\mathrm{E}_{\epsilon_i}(\cdot, s) = s(2 - s)\mathrm{E}_{\epsilon_i}(\cdot, s)$.

(c) *Let* $\delta_{ij} = 1$ *for* $i = j$, $\delta_{ij} = 0$ *for* $i \neq j$. *The Fourier expansion of the Eisenstein series attached to the cusp* ϵ_i *about the cusp* ϵ_j *is*

$$\mathrm{E}_{\epsilon_i}(\sigma_j w, s) = \delta_{ij} v^s + \phi_{ij}(s) v^{2-s} + \sum_{\lambda \in q^* \setminus \{0\}} \phi_{ij}(s, \lambda) v K_{s-1}(4\pi|\lambda|v) e(\lambda z),$$

where $w = (z, v) \in \mathbf{H}$, $s \in \mathbf{C}$, *the coefficients* ϕ_{ij} *and* $\phi_{ij}(\cdot, \lambda)$ *are meromorphic functions on* $\mathbf{C}$ *and*

$$\phi_{ij}(s) = \frac{\pi}{m(s-1)} \Psi_{ij}(s, 0), \quad \phi_{ij}(s, \lambda) = \frac{(2\pi)^s |\lambda|^{s-1}}{m\Gamma(s)} \Psi_{ij}(s, \lambda)$$

with $\Psi_{ij}(s, \cdot)$ *defined for* $\Re(s) > 2$ *in* (0.3.13), $m = \mathrm{vol}(\mathbf{C}/q)$.

(d) *Let*

$$\mathcal{E}(w, s) = \begin{pmatrix} \mathrm{E}_{\epsilon_1}(w, s) \\ \mathrm{E}_{\epsilon_2}(w, s) \\ \vdots \\ \mathrm{E}_{\epsilon_h}(w, s) \end{pmatrix} \quad \text{and} \quad \Phi(s) = \begin{pmatrix} \phi_{11}(s) & \phi_{12}(s) & \cdots & \phi_{1h}(s) \\ \phi_{21}(s) & \phi_{22}(s) & \cdots & \phi_{2h}(s) \\ \vdots & \vdots & & \vdots \\ \phi_{h1}(s) & \phi_{h2}(s) & \cdots & \phi_{hh}(s) \end{pmatrix}.$$

Then we have

$\Phi(1 + it)$ *is an unitary matrix if* $t \in \mathbf{R}$,

$\Phi(s)\Phi(2 - s)$ *is the identity matrix for any* $s \in \mathbf{C}$,

$\mathcal{E}(w, s) = \Phi(s)\mathcal{E}(w, 2 - s)$ *(the functional equation).* ■

The matrix $\Phi(s)$ is known as a scattering matrix. The main tool to get meromorphic continuation and the functional equation for the Eisenstein series is the Maaß–Selberg formula. For more detailed explanation we refer to Kubota [53], [54], Patterson [69], [70] (in the context under consideration) and to the fundamental works of Selberg [87], [88] and Maaß [61], [62].

0.3.7 One special case. For the group $\Gamma^{(2)}_{\text{princ}}(3)$, we can give explicit formulae for the scattering matrix and the functional equation of Theorem 0.3.5. It was mentioned already in 0.3.3 that there are 4 $\Gamma^{(2)}_{\text{princ}}(3)$-orbits consisting of essential cusps, and that representatives of these orbits can be taken as ∞, 0, ± 1. Thus we have 4 Eisenstein series attached to the group $\Gamma^{(2)}_{\text{princ}}(3)$. It is instructive to look at them more closely.

Theorem 0.3.6 *Let* $\mathrm{E}_\epsilon(w,s)$ *be the Eisenstein series attached to the group* $\Gamma^{(2)}_{\text{princ}}(3)$ *and to its essential cusp* $\epsilon = \infty, 0, \pm 1$, *and let*

$$\mathcal{E}(w,s) = \begin{pmatrix} \mathrm{E}_\infty(w,s) \\ \mathrm{E}_0(w,s) \\ \mathrm{E}_1(w,s) \\ \mathrm{E}_{-1}(w,s) \end{pmatrix}, \quad \Phi(s) = \frac{2\pi\zeta_*(3s-3)}{3^{5/2}(s-1)\zeta_*(3s-2)} \begin{pmatrix} n_s & 1 & 1 & 1 \\ 1 & n_s & 1 & 1 \\ 1 & 1 & n_s & 1 \\ 1 & 1 & 1 & n_s \end{pmatrix},$$

$$\widetilde{\mathcal{E}}(w,s) = \mathrm{E}_\infty(w,s) + \mathrm{E}_0(w,s) + \mathrm{E}_1(w,s) + \mathrm{E}_{-1}(w,s), \quad n_s = 2(3^{3s-3}-1)^{-1},$$

where $w \in \mathbf{H}$ and $s \in \mathbf{C}$. *Then we have*

$$\mathcal{E}(w,s) = \Phi(s)\mathcal{E}(w,2-s),$$

$$\widetilde{\mathcal{E}}(w,s) = \frac{2\pi\zeta_*(3s-3)}{3^{5/2}(s-1)\zeta_*(3s-2)}(n_s+3)\widetilde{\mathcal{E}}(w,2-s),$$

and also $\mathrm{E}_\infty(w,s) = \mathrm{E}_\epsilon(\sigma_\epsilon w, s)$ *with* σ_ϵ *as in* 0.3.3. ∎

Here the matrix $\Phi(s)$ is the scattering matrix of Theorem 0.3.5. The functional equation for $\widetilde{\mathcal{E}}(w,\cdot)$ follows immediately from that for $\mathcal{E}(w,\cdot)$. See [69].

All given above Fourier expansions of the Eisenstein series are expansions about essential cusps. The next proposition gives us examples of Fourier expansions about cusps which are not essential.

Proposition 0.3.7 *Let* $\mathrm{E}_\infty(w,s)$ *be the Eisenstein series attached to the group* $\Gamma^{(2)}_{\text{princ}}(3)$ *and to its cusp* ∞. *Let either* $\varepsilon = \omega$, $\xi = -1/9$ *or* $\varepsilon = \omega^2$, $\xi = 1/9$. *Then we have Fourier expansions*

$$\mathrm{E}_\infty\left(\begin{pmatrix} 1 & 0 \\ \varepsilon & 1 \end{pmatrix} w, s\right) = \frac{2^{s+1}\pi^s}{3^{5/2}\Gamma(s)} \sum_\lambda \Psi(s, 27\varepsilon\lambda) e(\bar{\varepsilon}\lambda) |\lambda|^{s-1} v K_{s-1}(4\pi|\lambda|v) e(\lambda z),$$

where $w = (z,v) \in \mathbf{H}$, $s \in \mathbf{C}$, $\Psi(\cdot, 27\varepsilon\lambda)$ *is as in* (0.3.11) (*with* $q = (3)$), *and the summation is carried out over* $\lambda \in \xi + (\sqrt{-3})^{-3}\mathcal{O}$. ∎

0.3.8 Cubic theta function. Following Kubota [54], we introduce now the cubic theta function $\Theta_{\text{K-P}}\colon \mathbf{H} \to \mathbf{C}$ as a residue of the Eisenstein series in 0.3.5 at the exceptional (maximal) pole. More precisely, assuming $q = (3)$, we set

$$\Theta_{\text{K-P}} = 2^{-3} 3^{9/2} \pi^{-2} \zeta_*(2) \operatorname*{Res}_{s=4/3} \mathrm{E}_*(\cdot, s). \tag{0.3.14}$$

This function should be considered as a cubic analogue of the classical quadratic theta function. It turns out to be difficult problem to find its Fourier coefficients.

The form of final results were suggested and some techniques needed have been introduced by Kubota [54], [56], [57]. Nevertheless, the problem remained open and it was solved by Patterson [69]. The technique which he has used is the 'converse theorem' of the Hecke theory in a version close to that of Maaß [61].

To state the properties of the cubic theta function we need in Patterson's τ-function, which we define for $\nu \in \mathbf{Q}(\sqrt{-3})$ as follows:

$$\tau(\nu) = \overline{S(1,c)}\left|\frac{d}{c}\right|3^{(n-1)/2}\begin{cases} \left(\frac{3}{c}\right) & \text{if } \nu = \pm(\sqrt{-3})^{3n-1}cd^3, \\ \left(\frac{3\omega}{c}\right)\xi^{-1} & \text{if } \nu = \pm\omega(\sqrt{-3})^{3n-1}cd^3, \\ \left(\frac{3\omega^2}{c}\right)\xi & \text{if } \nu = \pm\omega^2(\sqrt{-3})^{3n-1}cd^3, \\ 1 & \text{if } \nu = \pm(\sqrt{-3})^{3n-3}cd^3, \\ 0 & \text{for all others } \nu, \end{cases} \tag{0.3.15}$$

where $c, d \in \mathcal{O}$, $c \equiv d \equiv 1 \pmod 3$, c is square-free, $n \in \mathbf{Z}$, $\omega = \exp(2\pi i/3)$ and $\xi = \exp(2\pi i/9)$.

Part (c) in the next theorem gives us the Fourier expansion of the cubic theta function $\Theta_{\text{K-P}}$ about the cusp ∞, and this is nothing but the main result in [69]. Part (d) gives us Fourier expansions of $\Theta_{\text{K-P}}$ about the cusps ω and ω^2. These expansions, in essence, also found in [69], though were not written explicitly. (To get them one should combine Proposition 0.3.7 with Theorem 0.4.3 (b) below.)

Theorem 0.3.8 *The cubic theta function* $\Theta_{\text{K-P}}\colon \mathbf{H} \to \mathbf{C}$, *defined by* (0.3.14), *has the following properties.*

(a) $\Theta_{\text{K-P}}$ *is a cubic metaplectic form under* $\tilde{\Gamma}_2(3)$;

(b) $D_{\text{L-B}}\Theta_{\text{K-P}} = \frac{8}{9}\Theta_{\text{K-P}}$;

(c) $\Theta_{\text{K-P}}(z,v) = 2^{-1}3^{-1/2}v^{2/3} + \sum_{\nu} \tau(\nu) v K_{1/3}(4\pi|\nu|v)e(\nu z)$,

where the summation is carried out over $\nu \in (\sqrt{-3})^{-3}\mathcal{O} \setminus \{0\}$;

(d) $\Theta_{\text{K-P}}\left(\begin{pmatrix} 1 & 0 \\ \varepsilon & 1 \end{pmatrix}(z,v)\right) = \sum_{\nu} \tau(\varepsilon\nu)e(\bar{\varepsilon}\nu) v K_{1/3}(4\pi|\nu|v)e(\nu z)$,

where the summation is carried out over $\nu \in \xi + (\sqrt{-3})^{-3}\mathcal{O}$ *and either* $\varepsilon = \omega$, $\xi = -1/9$ *or* $\varepsilon = \omega^2$, $\xi = 1/9$. ∎

We refer to $\Theta_{\text{K-P}}$ as the Kubota–Patterson cubic theta function. It should be emphasized, the point (a) of the theorem above yields $\Theta_{\text{K-P}}(\gamma w) = \Theta_{\text{K-P}}(w)$ for any $\gamma \in \mathrm{SL}(2,\mathbf{Z})$, though the Eisenstein series $\mathrm{E}_*(\cdot, s)$ involved in the definition (0.3.14) is not invariant under $\mathrm{SL}(2,\mathbf{Z})$. We know $\Gamma_2(3) = \Gamma^{(2)}_{\text{princ}}(3)$, and the Eisenstein series $\mathrm{E}_*(\cdot, s)$ in (0.3.14) is nothing but the Eisenstein series $\mathrm{E}_\infty(\cdot, s)$

in 0.3.7. In view of Theorem 0.3.6 and Theorem 0.3.8 (a), one can replace $E_*(\cdot, s)$ in the definition (0.3.14) by $E_\epsilon(\cdot, s)$, $\epsilon = \infty, 0, \pm 1$, and this gives rise just to the same function.

Notice that the τ-function, just like the constant in (0.3.14), differs by the factor 27 from those in [69]; this makes it more convenient to state its multiplicative properties.

Proposition 0.3.9 *One has*

$$\tau(ab) = \left(\frac{a}{b}\right)\tau(a)\tau(b),$$

where $a \in (\sqrt{-3})^{-3}\mathcal{O}$, $b \in \mathcal{O}$, $b \equiv 1 \pmod 3$, *and* $\gcd(a,b) = 1$. □

Proof. This follows easily from the definition (0.3.15) and from the multiplicative properties of the Gauß sums stated in 0.1.4 (g) (take $\lambda = c_1' = c_2' = 1$). ■

Corollary 0.3.10 *Let* $c = \xi(\sqrt{-3})^t c'$ *with* $\xi \in \mathcal{O}$, $t \in \mathbf{Z}$, $c' \equiv 1 \pmod 3$, *and let* $\omega_p = \operatorname{ord}_p c$ *for each prime* $p \equiv 1 \pmod 3$. *Then we have*

$$\tau(c) = k(c) \prod_{\substack{p \equiv 1(3) \\ p \text{ is prime}}} \tau(p^{\omega_p}), \qquad \textit{where}$$

$$k(c) = \tau(cc'^{-1})\,\Omega(c')\left(\frac{cc'^{-1}}{c'}\right) = \tau(\xi(\sqrt{-3})^t)\,\Omega(c') \prod_{\substack{p \equiv 1(3) \\ p \text{ is prime}}} \left(\frac{\xi(\sqrt{-3})^t}{p}\right)^{\omega_p}$$

with

$$\Omega(c') = \prod_{\substack{p \neq p' \\ p,p' \equiv 1(3) \text{ are prime}}} \left(\frac{p'}{p}\right)^{-\omega_p \omega_{p'}}. \qquad □$$

Proof. We take $a = \xi(\sqrt{-3})^t$ and $b = c'$ in Proposition 0.3.9 to get

$$\tau(c) = \tau(\xi(\sqrt{-3})^t)\tau(c')\left(\frac{\xi(\sqrt{-3})^t}{c'}\right) = \tau(\xi(\sqrt{-3})^t)\tau(c') \prod_{\substack{p \equiv 1(3) \\ p \text{ is prime}}} \left(\frac{\xi(\sqrt{-3})^t}{p}\right)^{\omega_p}.$$

Then, again by means of Proposition 0.3.9, we express $\tau(c')$ as $\Omega(c')$ times the product of $\tau(p^{\omega_p})$ taken over all prime $p \equiv 1 \pmod 3$. ■

We would like now to give some remarks. First, let us write $\tilde\tau(\nu)$ for $\tau(\nu)|\nu|^{-1/3}$ and $\tilde\Theta_{\text{K-P}}$ for $2^{1/3}3^{-1/6}\pi^{-2/3}\Theta_{\text{K-P}}$. Replacing $K_{1/3}$ by the Airy function in accordance with (0.1.26) we find Theorem 0.3.8 (c) is equivalent to

$$\tilde\Theta_{\text{K-P}}(z,v) = (6\pi)^{-2/3}v^{2/3} + v^{2/3}\sum_\nu \tilde\tau(\nu)Ai((6\pi|\nu|v)^{2/3})e(\nu z), \tag{0.3.16}$$

$(z, v) \in \mathbf{H}$, where the summation is carried out over $\nu \in (\sqrt{-3})^{-3}\mathcal{O} \setminus \{0\}$. In the meantime, in the notations of (0.3.15), $\tilde\tau(\nu)$ does not depend of d. At least for this reason it seems preferable to write the Fourier expansion of the Kubota–Patterson cubic theta function as it is given in (0.3.16) instead of the more traditional form given in Theorem 0.3.8 (c). In what follows we shall find more evidences to involve the Airy function instead of the Bessel–MacDonald functions $K_{1/3}$ and $K_{2/3}$ dealing with cubic theta functions.† Our second remark concernes the Laurent expansion of the Eisenstein series $\mathrm{E}_*(w,\cdot)$ about the point 4/3. We have

$$\mathrm{E}_*(w,s) = \Theta_{-1}(w)(s-4/3)^{-1} + \Theta_0(w) + \Theta_1(w)(s-4/3) + \ldots,$$

where the coeffitients Θ_n are functions $\mathbf{H} \to \mathbf{C}$, and the Θ_{-1} is just the $\Theta_{\text{K-P}}$, up to the constant involved in the definition (0.3.14). Expressing the Θ_n by the Cauchy formula one can find all of them satisfy the conditions (a) and (c) of cubic metaplectic forms (under $\Gamma_2(3)$) definition given in 0.3.1. In the meantime, the Θ_{-1} only is egenfunction of the Laplace–Beltrami operator $\mathrm{D}_{\text{L-B}}$ while other Θ_n are egenfunctions of some operators $P_n(\mathrm{D}_{\text{L-B}})$, P_n being polynomial, but not of the Laplace–Beltrami operator itself. To clarify the last point, let us substitute the Laurent expansion above to the identity

$$\mathrm{D}_{\text{L-B}}\mathrm{E}_*(\cdot,s) = s(2-s)\mathrm{E}_*(\cdot,s).$$

Comparing coefficients we get

$$\mathrm{D}_{\text{L-B}}\Theta_{-1} = \frac{8}{9}\Theta_{-1}, \quad \mathrm{D}_{\text{L-B}}\Theta_0 = \frac{8}{9}\Theta_0 - \frac{2}{3}\Theta_{-1},$$
$$\mathrm{D}_{\text{L-B}}\Theta_n = \frac{8}{9}\Theta_n - \frac{2}{3}\Theta_{n-1} - \Theta_{n-2} \quad \text{if } n \geq 1,$$

and then

$$\left(\mathrm{D}_{\text{L-B}} - \frac{8}{9}\right)^{n+2}\Theta_n = 0 \quad \text{for all } \ n \geq -1.$$

So, if we change the requirement (b) in the definition in 0.3.1 by requirement

(b′) $P(\mathrm{D}_{\text{L-B}})f = 0$ for some polynomial P,

then we shall have all the Θ_n are cubic metaplectic forms on $\mathbf{H}$ under $\Gamma_2(3)$ in this extended sense. As far as we know, the Θ_n with $n \neq -1$ never have been studied. Actually, (b′) is in agreement with standard requirements accepted in the general automorphic forms theory, while (b) is too strong. However, all automorphic forms on $\mathbf{H}$ we deal with in these notes satisfy (b).

0.3.9 Involution ι. Let us define an automorphism, say ι, of $\mathrm{SL}(2,\mathbf{C})$ by $\gamma \mapsto {}^{\iota}\gamma = \sigma^t\gamma^{-1}\sigma^{-1}$, where

$$\sigma = \begin{pmatrix} 0 & 1 \\ 1 & 0 \end{pmatrix}, \quad \text{and so} \quad {}^{\iota}\gamma = \begin{pmatrix} a & -b \\ -c & d \end{pmatrix} \quad \text{for any} \quad \gamma = \begin{pmatrix} a & b \\ c & d \end{pmatrix} \in \mathrm{SL}(2,\mathbf{C}).$$

† In a similar way, the Bessel–MacDonald functions $K_{1/n}$, which occur in Fourier expansions of degree $n \geq 4$ theta functions, can be replaced by generalized Airy functions (in the sense of Hardy, see [101]). Patterson [73] has considered generalized p-adic Airy integrals in framework of his effort to determine Fourier coefficents of degree $n \geq 4$ theta functions.

The so defined ι is an involution of SL(2, **C**), and the groups N, A and $K =$ SU(2) in the Iwasawa decomposition (see 0.3.1) are invariant under ι. Passing to the quotient we get an involution of $\mathbf{H} \simeq \mathrm{SL}(2,\mathbf{C})/\mathrm{SU}(2)$, which we again will denote by ι. Certainly, the image of the point $w = (z, v) \in \mathbf{H}$ under ι is ${}^{\iota}w = (-z, v)$. One has also ${}^{\iota}\gamma{}^{\iota}w = {}^{\iota}(\gamma w)$ for all $w \in \mathbf{H}$, $\gamma \in \mathrm{SL}(2,\mathbf{C})$. We will write ${}^{\iota}f$ for the composition $f\colon \mathbf{H} \to \mathbf{C}$ with ι, so that ${}^{\iota}f(z, v) = f(-z, v)$ for all $(z, v) \in \mathbf{H}$.

Proposition 0.3.11 *The Eisenstein series* $\mathrm{E}_*(\cdot, s)$ *and the Kubota-Patterson cubic theta function* $\Theta_{\text{K-P}}$ *are invariant under* ι. □

Proof. Immediately from the definitions, see (0.3.10), we have

$$ {}^{\iota}\mathrm{E}_*(w,s) = \sum_{\gamma\in\Omega} \overline{\kappa}(\gamma)V(\gamma{}^{\iota}w)^s, \qquad w \in \mathbf{H}, \quad s \in \mathbf{C}, \quad \Re(s) > 2, $$

where Ω is any subset of $\Gamma_2(3)$ which contains a unique representative of each coset in $\Delta_*\backslash\Gamma_2(3)$. Since the set ${}^{\iota}\Omega = \{{}^{\iota}\gamma \mid \gamma \in \Omega\}$ also contains a unique representative of each coset in $\Delta_*\backslash\Gamma_2(3)$, we have

$$ {}^{\iota}\mathrm{E}_*(w,s) = \sum_{\gamma\in\Omega} \overline{\kappa}({}^{\iota}\gamma)V({}^{\iota}\gamma{}^{\iota}w)^s, \qquad w \in \mathbf{H}, \quad s \in \mathbf{C}, \quad \Re(s) > 2. $$

Next notice that $\kappa({}^{\iota}\gamma) = \kappa(\gamma)$ and $V({}^{\iota}\gamma{}^{\iota}w) = V({}^{\iota}(\gamma w)) = V(\gamma w)$. These notices yield ${}^{\iota}\mathrm{E}_*(\cdot, s) = \mathrm{E}_*(\cdot, s)$, if $s \in \mathbf{C}$, $\Re(s) > 2$. The restriction $\Re(s) > 2$ can be dropped by analytic continuation. The invariance of the cubic theta function follows from that of the Eisenstein series, and also can be seen from the Fourier expansion in Theorem 0.3.8 (c), since $\tau(\nu) = \tau(-\nu)$ for all ν. ■

0.3.10 Twists. Let $f\colon \mathbf{H} \to \mathbf{C}$ be a cubic metaplectic form under the group $\Gamma^{(2)}_{\text{princ}}(q)$ with a Fourier expansion

$$ f(z,v) = c_0(v) + \sum_{\nu\in q^*\backslash\{0\}} \rho(\nu)vK_{r-1}(4\pi|\nu|v)e(\nu z), \qquad (z,v) \in \mathbf{H}, $$

For any periodic function $\varepsilon\colon q^* \to \mathbf{C}$ with the lattice of periods nq^*, $n \in \mathcal{O}\backslash\{0\}$, we put

$$ f*\varepsilon(z,v) = c_0(v)\varepsilon(0) + \sum_{\nu\in q^*\backslash\{0\}} \rho(\nu)\varepsilon(\nu)vK_{r-1}(4\pi|\nu|v)e(\nu z), \qquad (z,v) \in \mathbf{H}. $$

The so defined function $f*\varepsilon\colon \mathbf{H} \to \mathbf{C}$ is known[†] as the twist of f by ε.

Theorem 0.3.12 *The twisting by* ε *is a linear operator* $\mathcal{L}(q,r) \to \mathcal{L}(n^2q, r)$. *Moreover, if* $f \in \mathcal{L}^2(q,r)$, *then* $f*\varepsilon \in \mathcal{L}^2(n^2q, r)$. □

---

† See, for example, Serre and Stark [89].

Proof. Let $\widehat{\varepsilon}\colon q \to \mathbf{C}$ be the Fourier transform of ε, defined by

$$\widehat{\varepsilon}(\mu) = \frac{1}{\|n\|} \sum_{\nu \in q^*/nq^*} \varepsilon(\nu) e(-\mu\nu/n), \qquad \mu \in q.$$

We then have

$$\varepsilon(\nu) = \sum_{\mu \in q/nq} \widehat{\varepsilon}(\mu) e(\mu\nu/n), \qquad \nu \in q^*,$$

hence

$$f * \varepsilon(z, v) = \sum_{\mu \in q/nq} \widehat{\varepsilon}(\mu) f(z + \mu/n, v).$$

Next let us look at the functions $\tilde{f}\colon \mathbf{H} \to \mathbf{C}$ defined by $\tilde{f}(z, v) = f(z + \mu/n, v)$, where $\mu \in q$. The theorem will be proved if we show that $\tilde{f} \in \mathcal{L}(n^2 q, r)$ for each $f \in \mathcal{L}(q, r)$, and $\tilde{f} \in \mathcal{L}^2(n^2 q, r)$ for each $f \in \mathcal{L}^2(q, r)$. To examine $\tilde{f}$ let us set

$$\gamma = \begin{pmatrix} a & b \\ c & d \end{pmatrix}, \quad \sigma = \begin{pmatrix} 1 & \mu/n \\ 0 & 1 \end{pmatrix},$$

$$\delta = \sigma\gamma\sigma^{-1} = \begin{pmatrix} a + c\mu/n & b - (a-d)\mu/n - c\mu^2/n^2 \\ c & d - c\mu/n \end{pmatrix}.$$

By (0.3.2) we have $(z + \mu/n, v) = \sigma(z, v)$. This yields (see remark at the end of 0.3.1) $\tilde{f} \in \mathcal{L}(q', r)$ with some ideal q', and also $\tilde{f} \in \mathcal{L}^2(q', r)$ if $f \in \mathcal{L}^2(q, r)$. To show that one can take $q' = n^2 q$, first notice that

$$\tilde{f}(\gamma w) = f(\sigma\gamma w) = f(\delta\sigma w). \tag{0.3.17}$$

If $\gamma \in \Gamma^{(2)}_{\text{princ}}(n^2 q)$, then $\delta \in \Gamma^{(2)}_{\text{princ}}(q)$, $c\mu/n \equiv 0 \pmod 9$, $c\mu/n \equiv 0 \pmod p$ for any prime $p \mid c$, $p \equiv 1 \pmod 3$, and thus

$$\kappa(\delta) = \left(\frac{c}{d - c\mu/n}\right) = \left(\frac{c}{d}\right) = \kappa(\gamma),$$

see 0.1.3 and 0.2.2. Consequently,

$$f(\delta\sigma w) = \kappa(\delta) f(\sigma w) = \kappa(\gamma) f(\sigma w) = \kappa(\gamma) \tilde{f}(w).$$

Combining with (0.3.17), we get $\tilde{f}(\gamma w) = \kappa(\gamma) \tilde{f}(w)$, as required. ■

0.3.11 Shift operators. For $l \in \mathbf{C}^*$ one associates so called shift operator S_l defined by

$$\mathrm{S}_l f(z, v) = f(lz, |l| v), \qquad (z, v) \in \mathbf{H}, \tag{0.3.18}$$

for any function $f\colon \mathbf{H} \to \mathbf{C}$. By (0.3.3), for $m_l = \mathrm{diag}(l, 1)$ one has $m_l(z, v) = (lz, |l| v)$, and thus

$$\mathrm{S}_l f(w) = f(m_l w), \qquad w \in \mathbf{H}. \tag{0.3.19}$$

This yields immediately

$$\mathrm{S}_l f(\gamma(z,v)) = f\big((m_l\gamma m_l^{-1})(lz,|l|v)\big) \quad \text{for } (z,v)\in \mathbf{H},\ \ \gamma\in \mathrm{SL}(2,\mathbf{C});$$
$$\mathrm{S}_l \mathrm{D}_{\text{L-B}} f = \mathrm{D}_{\text{L-B}} \mathrm{S}_l f \quad \text{if } \mathrm{D}_{\text{L-B}} f \text{ is well defined.} \qquad (0.3.20)$$

Theorem 0.3.13 *If $l \in \mathcal{O}$ and $f \in \mathcal{L}(q,r)$, $q \subset (9)$, then $\mathrm{S}_l f \in \mathcal{L}(lq,r)$. Moreover, if $f \in \mathcal{L}^2(q,r)$, then $\mathrm{S}_l f \in \mathcal{L}^2(lq,r)$.* □

Proof. If

$$\gamma = \begin{pmatrix} a & b \\ c & d \end{pmatrix} \in \Gamma^{(2)}_{\text{princ}}(lq), \quad \text{then} \quad m_l\gamma m_l^{-1} = \begin{pmatrix} a & bl \\ c/l & d \end{pmatrix}, \in \Gamma^{(2)}_{\text{princ}}(q)$$

and, since $d \equiv 1 \pmod{9l}$, one has $\left(\frac{l}{d}\right) = 1$, see 0.1.3. So,

$$\kappa(m_l\gamma m_l^{-1}) = \left(\frac{c/l}{d}\right)^{-1} = \left(\frac{l}{d}\right)\left(\frac{c}{d}\right)^{-1} = \left(\frac{c}{d}\right)^{-1} = \kappa(\gamma),$$

see 0.2.2. It is clear just from the definition (0.3.18) that if f then $\mathrm{S}_l f$ satisfies the growth condition (c) in the definition given in 0.3.1. Now, let $f \in \mathcal{L}(q,r)$. The preceding remarks and (0.3.20) show that $\mathrm{S}_l f \in \mathcal{L}(q,r)$, and it remains only to check that $\mathrm{S}_l f$ is square-integrable form for square-integrable form f. For this we have to examine $\mathrm{S}_l f(\sigma(z,v))$ as $v \to \infty$, where $\sigma \in \mathrm{SL}(2,\mathbf{C})$ and $\epsilon = \sigma\infty$ is a cusp of $\Gamma^{(2)}_{\text{princ}}(q)$, i.e., $\epsilon \in \mathbf{Q}(\sqrt{-3}) \cup \{\infty\}$. For such σ choose $\sigma' \in \mathrm{SL}(2,\mathbf{C})$ in such a way that $\sigma'^{-1} m_l \sigma$ is an upper triangular matrix in $\mathrm{SL}(2,\mathbf{C})$, say

$$m_l\sigma = \sigma' \begin{pmatrix} a & b \\ 0 & l/a \end{pmatrix}, \qquad a,b \in \mathbf{C}.$$

Then $\sigma'\infty = m_l\epsilon \in \mathbf{Q}(\sqrt{-3}) \cup \{\infty\}$, i.e., $\sigma'\infty$ is a cusp of $\Gamma^{(2)}_{\text{princ}}(q)$. This yields that for the square-integrable form f we have that $|f(\sigma'(z,v))|$ is majorized by v^c as $v \to \infty$, with some $c \in \mathbf{R}$, $c < 1$. By (0.3.19) and (0.3.3) one has

$$\mathrm{S}_l f(\sigma(z,v)) = f(m_l\sigma(z,v)) = f\big(\sigma'((az+b)a/l, |a^2/l|v)\big).$$

It is clear now that $|\mathrm{S}_l f(\sigma(z,v))|$ is majorized by v^c as $v \to \infty$ with the same $c < 1$. This just means that $\mathrm{S}_l f$ is a square-integrable form if f is a square-integrable one. ■

0.3.12 Hecke operators T_{p^3}. The Hecke operators in context of metaplectic forms on $\mathbf{H}$ were studied and used by Kubota in [54]. There are different ways to introduce them, we adopt the one described by Shimura [91]. This allows us to attach a linear operator, say $\mathrm{T}_\sigma\colon \mathcal{L}(q,r) \to \mathcal{L}(q,r)$, to any 2×2 matrices σ with entries in $\mathcal{O}$, $\det\sigma \neq 0$, whenever $\kappa(\sigma^{-1}h\sigma) = \kappa(h)$ for all $h \in \Gamma^{(2)}_{\text{princ}}(q) \cap \sigma\Gamma^{(2)}_{\text{princ}}(q)\sigma^{-1}$. By definition, if $f \in \mathcal{L}(q,r)$, then

$$\mathrm{T}_\sigma f(w) = \sum_{j=1,\dots,n} \overline{\kappa}(\delta_j) f(\sigma\delta_j w), \qquad w \in \mathbf{H},$$

where the matrices $\delta_1, \ldots, \delta_n \in \Gamma^{(2)}_{\text{princ}}(q)$ should be choosen in such a way that the classes $\Gamma^{(2)}_{\text{princ}}(q)\sigma\delta_j$ and $\Gamma^{(2)}_{\text{princ}}(q)\sigma\delta_k$ are different† if $j \neq k$, and

$$\Gamma^{(2)}_{\text{princ}}(q)\sigma\Gamma^{(2)}_{\text{princ}}(q) = \bigcup_{j=1,\ldots,n} \Gamma^{(2)}_{\text{princ}}(q)\sigma\delta_j .$$

Equivalently, one can say the family $\{\delta_j\}_{j=1,\ldots,n}$ should contain a unique representative of each coset in $(\Gamma^{(2)}_{\text{princ}}(q) \cap \sigma^{-1}\Gamma^{(2)}_{\text{princ}}(q)\sigma)\backslash\Gamma^{(2)}_{\text{princ}}(q)$. To check this definition is correct, i.e., $\mathrm{T}_\sigma f$ depends only on σ and f, but not on the choice of $\{\delta_j\}_{j=1,\ldots,n}$, let us to be given one more family, say $\{\delta'_j\}_{j=1,\ldots,n}$, with the similar properties. Clearly, for any j we have $\Gamma^{(2)}_{\text{princ}}(q)\sigma\delta_j = \Gamma^{(2)}_{\text{princ}}(q)\sigma\delta'_k$ with some k, and, changing the order if necessary, we can and do assume $k = j$. Then let $h_j = (\sigma\delta_j)(\sigma\delta'_j)^{-1}$, $j = 1, \ldots, n$. We have $h_j \in \Gamma^{(2)}_{\text{princ}}(q) \cap \sigma\Gamma^{(2)}_{\text{princ}}(q)\sigma^{-1}$, $\kappa(\sigma^{-1}h_j\sigma) = \kappa(h_j)$, and then

$$\sum_{j=1,\ldots,n} \overline{\kappa}(\delta'_j)f(\sigma\delta'_j w) = \sum_{j=1,\ldots,n} \overline{\kappa}(\sigma^{-1}h_j^{-1}\sigma\delta_j)f(h_j^{-1}\sigma\delta_j w)$$

$$= \sum_{j=1,\ldots,n} \kappa(h_j^{-1})\overline{\kappa}(\sigma^{-1}h_j^{-1}\sigma)\overline{\kappa}(\delta_j)f(\sigma\delta_j w) = \mathrm{T}_\sigma f(w), \quad w \in \mathbf{H},$$

as claimed. One more point to check is that $\mathrm{T}_\sigma f \in \mathcal{L}(q, r)$ for any $f \in \mathcal{L}(q, r)$. It is clear just from definition of T_σ and from the invariance of Laplace–Beltrami operator $\mathrm{D}_{\text{L-B}}$ that $\mathrm{T}_\sigma f$ is eigenfunction of $\mathrm{D}_{\text{L-B}}$. Also, $\mathrm{T}_\sigma f$ satisfies the necessary growth conditions — this can be proved by arguments quite similar to that in the preceding subsection. Then we have only to prove $\mathrm{T}_\sigma f(\gamma w) = \kappa(\gamma)\mathrm{T}_\sigma f(w)$ for $\gamma \in \Gamma^{(2)}_{\text{princ}}(q)$, $w \in \mathbf{H}$. For this, let $\{\delta_j\}_{j=1,\ldots,n}$ be as in the definition of T_σ, and let $\delta'_j = \delta_j\gamma$, $\gamma \in \Gamma^{(2)}_{\text{princ}}(q)$, $j = 1, \ldots, n$. It is easy to see that the so defined family $\{\delta'_j\}_{j=1,\ldots,n}$ satisfies all necessary conditions to be used in the definition of T_σ instead of the original family $\{\delta_j\}_{j=1,\ldots,n}$. So, we have

$$\mathrm{T}_\sigma f(w) = \sum_{j=1,\ldots,n} \overline{\kappa}(\delta'_j)f(\sigma\delta'_j w) = \sum_{j=1,\ldots,n} \overline{\kappa}(\delta_j\gamma)f(\sigma\delta_j\gamma w), \qquad w \in \mathbf{H}.$$

On the other hand, just by definition we have

$$\mathrm{T}_\sigma f(\gamma w) = \sum_{j=1,\ldots,n} \overline{\kappa}(\delta_j)f(\sigma\delta_j\gamma w), \qquad w \in \mathbf{H}.$$

Comparing, we get $\mathrm{T}_\sigma f(\gamma w) = \kappa(\gamma)\mathrm{T}_\sigma f(w)$, as required.

It is clear, one has $\mathrm{T}_{\sigma'} = \mathrm{T}_\sigma$ if the matrices σ' and σ are such that $\Gamma^{(2)}_{\text{princ}}(q)\sigma'\Gamma^{(2)}_{\text{princ}}(q) = \Gamma^{(2)}_{\text{princ}}(q)\sigma\Gamma^{(2)}_{\text{princ}}(q)$. Then standard arguments show we get all the Hecke operators considering only diagonal matrices $\sigma = \mathrm{diag}(1, n)$,

† This requirement yields $\Gamma^{(2)}_{\text{princ}}(q)\sigma\delta_j \cap \Gamma^{(2)}_{\text{princ}}(q)\sigma\delta_k$ is empty for $j \neq k$.

$n \in \mathcal{O}\setminus\{0\}$. Let $\sigma_p = \mathrm{diag}(1, p^3)$, with prime $p \equiv 1 \pmod 3$, $p \nmid q$. It occurs further the Hecke operators T_σ are either trivial ones or polynomials of the operators T_{σ_p}. Let us not go into more details and restrict our attention by the operators attached to the matrices σ_p. We shall write T_{p^3} for $c_p \mathrm{T}_{\sigma_p}$, where we involve the factor $c_p = |p|^{-3}$ to get more symmetric the forthcoming formulae. It is not difficult to find the needed family of matrices $\{\delta_j\}_{j=1,\dots,n}$ to get the description of T_{p^3} as

$$\mathrm{T}_{p^3} f(w) = \frac{1}{|p|^3}\Big\{ f\Big(\begin{pmatrix} p^3 & 0 \\ 0 & 1 \end{pmatrix} w\Big) + \sum_{\substack{m\,(p^3 q)\\ m\in q}} f\Big(\begin{pmatrix} 1 & m \\ 0 & p^3 \end{pmatrix} w\Big)$$

$$+ \sum_{\substack{m\,(p^2 q)\\ \gcd(m,p)=1,\ m\in q}} f\Big(\begin{pmatrix} p & m \\ 0 & p^2 \end{pmatrix} w\Big)\Big(\frac{m}{p}\Big) + \sum_{\substack{m\,(pq)\\ \gcd(m,p)=1,\ m\in q}} f\Big(\begin{pmatrix} p^2 & m \\ 0 & p \end{pmatrix} w\Big)\Big(\frac{m}{p}\Big)^{-1}\Big\},$$

or equivalently, see (0.3.3), as

$$\mathrm{T}_{p^3} f(w) = \frac{1}{|p|^3}\Big\{ f(p^3 z, |p|^3 v) + \sum_{\substack{m\,(p^3 q)\\ m\in q}} f((z+m)/p^3, v/|p|^3)$$

$$+ \sum_{\substack{m\,(p^2 q)\\ \gcd(m,p)=1,\ m\in q}} f(z/p + m/p^2, v/|p|)\Big(\frac{m}{p}\Big) + \sum_{\substack{m\,(pq)\\ \gcd(m,p)=1,\ m\in q}} f(pz + m/p, |p|v)\Big(\frac{m}{p}\Big)^{-1}\Big\};$$

in both formulae $w = (z, v) \in \mathbf{H}$. Some properties of the operators T_{p^3} are summarized in the next two theorems.

Theorem 0.3.14 *The Hecke operators $\mathrm{T}_{p^3}\colon \mathcal{L}(q,r) \to \mathcal{L}(q,r)$, $p \nmid q$, commute one with another and map square-integrable forms to square-integrable ones. The Eisenstein series and Kubota–Patterson cubic theta function $\Theta_{\text{K-P}}$ are eigenfunctions of the Hecke operators. More precisely, one has*

$$\mathrm{T}_{p^3}\mathrm{E}_\epsilon(\cdot, s) = \big(|p|^{3s-3} + |p|^{3-3s}\big)\mathrm{E}_\epsilon(\cdot, s), \qquad s \in \mathbf{C},$$

for any cusp ϵ of $\Gamma^{(2)}_{\text{princ}}(q)$, and any prime $p \nmid q$;

$$\mathrm{T}_{p^3}\Theta_{\text{K-P}} = \big(|p| + |p|^{-1}\big)\Theta_{\text{K-P}} \quad \textit{for all prime} \quad p \equiv 1 \pmod 3). \qquad \blacksquare$$

Theorem 0.3.15 *Let $f \in \mathcal{L}(q,r)$ has a Fourier expansion*

$$f(z,v) = c_0(v) + \sum_{\nu \in q^*\setminus\{0\}} \rho(\nu) v K_{r-1}(4\pi|\nu|v) e(\nu z), \qquad (z,v) \in \mathbf{H}.$$

Then

$$\mathrm{T}_{p^3} f(z,v) = \tilde{c}_0(v) + \sum_{\mu \in q^* \setminus \{0\}} \tilde{\rho}(\mu) v K_{r-1}(4\pi|\mu|v) e(\mu z), \quad (z,v) \in \mathbf{H},$$

where $\tilde{c}_0(v) = \left(|p|^{3r-3} + |p|^{3-3r}\right) c_0(v)$, *and*

$$\tilde{\rho}(\mu) = \rho(\mu p^3) + \begin{cases} \rho(\mu p)|p|^{-2} S(\mu, p) & \textit{if } \operatorname{ord}_p \mu = 0, \\ \rho(\mu/p)|p|^{-2} S(\mu/p, p) & \textit{if } \operatorname{ord}_p \mu = 1, \\ 0 & \textit{if } \operatorname{ord}_p \mu = 2, \\ \rho(\mu/p^3) & \textit{if } \operatorname{ord}_p \mu \geq 3. \end{cases} \quad \blacksquare$$

0.3.13 Hecke operators U_n. Following Atkin and Lehner [1], one can define on $\mathcal{L}(q,r)$ so called 'exceptional' Hecke operators U_n for any $n \mid q$, $n \in \mathcal{O}$. By definition, for any $f \in \mathcal{L}(q,r)$ one has

$$\mathrm{U}_n f(w) = \frac{1}{|n|} \sum_{\substack{m\,(nq) \\ m \in q}} f\left(\begin{pmatrix} 1 & m \\ 0 & n \end{pmatrix} w\right) = \frac{1}{|n|} \sum_{\substack{m\,(nq) \\ m \in q}} f((z+m)/n, v/|n|),$$

$w = (z,v) \in \mathbf{H}$. The next theorem is obvious from the definition. Notice that the case $n \in \mathcal{O}^*$ is not excluded and that, for this case, U_n agrees with the Shift operator $\mathrm{S}_{\bar{n}}$.

Theorem 0.3.16 *Under the condition $n \mid q$, $q \subset (9)$, defined above U_n are linear operators $\mathcal{L}(q,r) \to \mathcal{L}(q,r)$. Each two of them commute one with another. Let $f \in \mathcal{L}(q,r)$ has a Fourier expansion*

$$f(z,v) = c_0(v) + \sum_{\nu \in q^* \setminus \{0\}} \rho(\nu) v K_{r-1}(4\pi|\nu|v) e(\nu z),$$

where $c_0(v)$ is either $\rho_+ v^r + \rho_- v^{2-r}$ or $\rho_+ v \log v + \rho_- v$, according to $r \neq 1$ or $r = 1$. Then

$$\mathrm{U}_n f(z,v) = \tilde{c}_0(v) + \sum_{\mu \in q^* \setminus \{0\}} \tilde{\rho}(\mu) v K_{r-1}(4\pi|\mu|v) e(\mu z), \quad (z,v) \in \mathbf{H},$$

where $\tilde{\rho}(\mu) = \rho(\mu n)$ and either $\tilde{c}_0(v) = |n|^{1-r} \rho_+ v^r + |n|^{r-1} \rho_- v^{2-r}$ or $\tilde{c}_0(v) = \rho_+ v \log v + (\rho_- - \rho_+ \log|n|) v$ according to $r \neq 1$ and $r = 1$. ■

0.3.14 One more cubic metaplectic form. Let us turn to 0.3.7 and look at the scattering matrices Φ. We have

$$(s-1)\zeta_*(3s-2)\big|_{s=1} = 2\pi/3^{7/2},$$

$$\zeta_*(3s-3)(3^{3s-3}-1)^{-1}\big|_{s=1} = 3^{-3s+3}\zeta_{\mathbf{Q}(\sqrt{-3})}(3s-3)\big|_{s=1} = -1/6$$

(see 0.1.5), and then we find easily $\Phi(1)$ is the minus identity matrix. Then the functional equation stated in Theorem 0.3.6 yields $\mathcal{E}(\cdot,1) = 0$, and thus the cubic metaplectic Eisenstein series attached to the essential cusps of the group $\Gamma^{(2)}_{\text{princ}}(3)$ vanish at the point 1. Let us take that of these series attached to the cusp 0 and introduce into consideration the function

$$\Theta_!(w) = 3^{-3/2}(\log 3)^{-1}\frac{\partial}{\partial s}\mathrm{E}_0(w,s)\big|_{s=1}, \qquad w=(z,v)\in \mathbf{H}. \tag{0.3.21}$$

The so defined $\Theta_!\colon \mathbf{H}\to\mathbf{C}$ has pleasant properties stated in the next two theorems. (The factor $3^{-3/2}(\log 3)^{-1}$ we involve in the definition to simplify the formulae in 2.5.6)

Let Ψ, Ψ_* be the functions defined in (0.3.11), and let Ψ', Ψ'_* be their derivatives over first argument. We assume $q=(3)$ as in (0.3.11) as in what follows. So, $q^*=(\sqrt{-3})^{-3}\mathcal{O}$ and $m=\mathrm{vol}(\mathbf{C}/q)=2^{-1}3^{5/2}$.

Theorem 0.3.17 *$\Theta_!$ is a cubic metaplectic form under $\Gamma^{(2)}_{\text{princ}}(3)$. One has*

(a) $\mathrm{D}_{\text{L-B}}\Theta_! = \Theta_!$;

(b) $\mathrm{T}_{p^3}\Theta_! = 2\Theta_!$ *for each prime* $p\equiv 1 \pmod 3$;

(c) $\Theta_!(w) = \eta(0)v + \sum_{\nu\in q^*\setminus\{0\}} \eta(\nu)vK_0(4\pi|\nu|v)e(\nu z)$, *where*

$$\eta(0) = \frac{2\pi}{3^4\log 3}\frac{\partial}{\partial s}\left\{\frac{1}{s-1}\Psi(s,0)\right\}\Big|_{s=1} \quad \textit{and}$$
$$\eta(\nu) = \frac{4\pi}{3^4\log 3}\Psi'(1,\nu) \quad \textit{for } \nu\neq 0;$$

(d) $\Theta_!\left(\begin{pmatrix}0&-1\\1&0\end{pmatrix}w\right) = \widehat{\eta}_+(0)v\log v + \widehat{\eta}_-(0)v + \sum_{\nu\in q^*\setminus\{0\}}\widehat{\eta}(\nu)vK_0(4\pi|\nu|v)e(\nu z)$

with $\widehat{\eta}_+(0) = 2/(3^{3/2}\log 3)$,

$$\widehat{\eta}_-(0) = \frac{2\pi}{3^4\log 3}\frac{\partial}{\partial s}\left\{\frac{1}{s-1}\Psi_*(s,0)\right\}\Big|_{s=1} \quad \textit{and}$$
$$\widehat{\eta}(\nu) = \frac{4\pi}{3^4\log 3}\Psi'_*(1,\nu) \quad \textit{for } \nu\neq 0;$$

(e) *For any $\varepsilon\in\mathbf{R}^*_+$, the Fourier coefficients $\eta(\nu)$ and $\widehat{\eta}(\nu)$ are majorized by $\|\nu\|^{1+\varepsilon}$ as $\|\nu\|\to\infty$.* □

Proof. For any $\gamma\in\Gamma^{(2)}_{\text{princ}}$ and $w\in\mathbf{H}$ we have $\Theta_!(\gamma w)=\kappa(\gamma)\Theta_!(w)$ just by the definition of $\Theta_!$ and by the respective property of the Eisenstein series. It is also clear that $\Theta_!$ satisfies the needed growth conditions, because of the Eisenstein series satisfies. For (a), we take $\mathrm{D}_{\text{L-B}}\mathrm{E}_0(\cdot,s) = s(2-s)\mathrm{E}_0(\cdot,s)$ from Theorem

0.3.5 (b), and we find

$$\begin{aligned}
D_{\text{L-B}} \frac{\partial}{\partial s} E_0(\cdot, s)\Big|_{s=1} &= \frac{\partial}{\partial s}\{D_{\text{L-B}} E_0(\cdot, s)\}\Big|_{s=1} \\
&= \frac{\partial}{\partial s}\{s(2-s)E_0(\cdot, s)\}\Big|_{s=1} \\
&= \frac{\partial}{\partial s}\{s(2-s)\}\Big|_{s=1} E_0(\cdot, 1) + \frac{\partial}{\partial s} E_0(\cdot, s)\Big|_{s=1} \\
&= \frac{\partial}{\partial s} E_0(\cdot, s)\Big|_{s=1}.
\end{aligned}$$

Part (b) follows in just a similar way from Theorem 0.3.14. For (c) and (d), first we should notice that the Eisenstein series $E_\infty(w, s)$ attached to the group $\Gamma^{(2)}_{\text{princ}}(3)$ coincides with the Eisenstein series $E_*(w, s)$ defined in 0.3.5 for the group $\Gamma_2(3)$. Then, in view of Theorem 0.3.6, we have

$$\begin{aligned}
E_0(w, s) &= E_\infty\left(\begin{pmatrix} 0 & -1 \\ 1 & 0 \end{pmatrix} w, s\right) = E_*\left(\begin{pmatrix} 0 & -1 \\ 1 & 0 \end{pmatrix} w, s\right), \\
E_0\left(\begin{pmatrix} 0 & -1 \\ 1 & 0 \end{pmatrix} w, s\right) &= E_\infty(w, s) = E_*(w, s).
\end{aligned} \tag{0.3.22}$$

Thus, it occurs Theorem 0.3.4 gives us the Fourier expansions of $E_0(\cdot, s)$ about the cusps ∞ and 0. Since $E_0(\cdot, 1) = 0$, we have $\Psi(1, \lambda) = \Psi_*(1, \lambda) = 0$ for all $\lambda \in q^*$. Also, $\Psi(s, 0) = \zeta_*(3s-3)/\zeta_*(3s-2)$, and hence $\Psi(\cdot, 0)$ has a second order zero at 1 (see 0.1.5 and Theorem 0.4.1 (b) with $n = 0$). Next, by Theorem 0.4.3 (d) and Theorem 0.4.1 (c), for each $\varepsilon \in \mathbf{R}_+^*$ the Fourier coefficients $\Psi(s, \nu)$ and $\Psi_*(s, \nu)$, with s in a sufficiently small neighbourhood of 1, are majorized by $\|\nu\|^{1+\varepsilon}$ as $\|\nu\| \to \infty$. By the Cauchy formula† this yields part (e) and also allows us to differentiate term-by-term the expansions given in Theorem 0.3.4. Differentiating these expansions and taking into account (0.3.22) we receive (c) and (d). ■

Now let us consider the associated L-functions

$$\widehat{L}(s) = \sum_{\nu \in q^* \setminus \{0\}} \frac{\widehat{\eta}(\nu)}{\|\nu\|^s}, \qquad L(s) = \sum_{\nu \in q^* \setminus \{0\}} \frac{\eta(\nu)}{\|\nu\|^s}, \tag{0.3.23}$$

and the functions

$$\widehat{G}(s) = 2^{-1-2s}\pi^{1-2s}\Gamma(s)^2 \widehat{L}(s), \quad G(s) = 2^{-1-2s}\pi^{1-2s}\Gamma(s)^2 L(s),$$

where $s \in \mathbf{C}$ and the series are absolutely convergent whenever $\Re(s) > 2$.

---

† We have in mind that one: $F'(z) = \dfrac{1}{2\pi i}\displaystyle\int_{\nabla} \dfrac{F(s)\,ds}{(s-z)^2}$ for each function F defined and regular in some neighborhood of $z \in \mathbf{C}$, ∇ being sufficiently small circle with center in z.

Theorem 0.3.18 *The functions $\widehat{L}$ and L have meromorphic continuations to all of* $\mathbf{C}$. *One has the functional equation*

$$(2\pi)^{-2s}\Gamma(s)^2\widehat{L}(s) = (2\pi)^{-2(1-s)}\Gamma(1-s)^2 L(1-s).$$

The function $\widehat{G}$ has no singularities except a simple pole at 1 *and a second order pole at* 0. *The function G has no singularities except a simple pole at* 0 *and a second order pole at* 1. □

Proof. Set for convenience

$$\widehat{\Theta}_!(w) = \Theta_!\left(\begin{pmatrix} 0 & -1 \\ 1 & 0 \end{pmatrix} w\right), \qquad w \in \mathbf{H}.$$

Immediately from the definitions and (0.3.2) follows

$$\widehat{\Theta}_!(0,v) = \Theta_!(0, 1/v), \qquad v \in \mathbf{R}_+^*. \tag{0.3.24}$$

Integrating the Fourier expansions term-by-term we find

$$\widehat{G}(s) = \int_0^\infty \{\widehat{\Theta}_!(0,v) - \widehat{\eta}_+(0) v \log v - \widehat{\eta}_-(0) v\} v^{2s-2}\, dv,$$

$$G(s) = \int_0^\infty \{\Theta_!(0,v) - \eta(0) v\} v^{2s-2}\, dv$$

if $\Re(s)$ is sufficiently large. Splitting $(0,\infty)$ as $(0,1] \cup (1,\infty)$, we get

$$\widehat{G}(s) = \widehat{A}(s) + \widehat{B}(s), \quad G(s) = A(s) + B(s),$$

where

$$\widehat{A}(s) = \int_0^1 \{\widehat{\Theta}_!(0,v) - \widehat{\eta}_+(0) v \log v - \widehat{\eta}_-(0) v\} v^{2s-2}\, dv,$$

$$\widehat{B}(s) = \int_1^\infty \{\widehat{\Theta}_!(0,v) - \widehat{\eta}_+(0) v \log v - \widehat{\eta}_-(0) v\} v^{2s-2}\, dv,$$

$$A(s) = \int_0^1 \{\Theta_!(0,v) - \eta(0) v\} v^{2s-2}\, dv,$$

$$B(s) = \int_1^\infty \{\Theta_!(0,v) - \eta(0) v\} v^{2s-2}\, dv.$$

Here the integrands decay exponentially as $v \to \infty$, and we see that B and $\hat{B}$ are entire functions on $\mathbf{C}$. Then, changing variable $v \mapsto 1/v$ and taking into account (0.3.24), we transform $\widehat{A}(s)$ as

$$\begin{aligned}\widehat{A}(s) &= \int_1^\infty \{\widehat{\Theta}_!(0, v^{-1}) - \widehat{\eta}_+(0)v^{-1}\log(v^{-1}) - \widehat{\eta}_-(0)v^{-1}\}v^{-2s}\,dv \\ &= \int_1^\infty \{\Theta_!(0, v) - \widehat{\eta}_+(0)v^{-1}\log(v^{-1}) - \widehat{\eta}_-(0)v^{-1}\}v^{-2s}\,dv \\ &= \int_1^\infty \{\Theta_!(0, v) - \eta(0)v\}v^{-2s}\,dv \\ &\qquad + \int_1^\infty \{\eta(0)v - \widehat{\eta}_+(0)v^{-1}\log(v^{-1}) - \widehat{\eta}_-(0)v^{-1}\}v^{-2s}\,dv \\ &= \frac{\widehat{\eta}_+(0)}{4s^2} - \frac{\widehat{\eta}_-(0)}{2s} - \frac{\eta(0)}{2-2s} + B(1-s).\end{aligned}$$

Combining the last expression for $\widehat{A}(s)$ with the original one for $\widehat{B}(s)$ we get

$$\widehat{G}(s) = \frac{\widehat{\eta}_+(0)}{4s^2} - \frac{\widehat{\eta}_-(0)}{2s} - \frac{\eta(0)}{2-2s} + \widehat{B}(s) + B(1-s). \tag{0.3.25}$$

Since $\widehat{B}$ and B are entire functions, we get $\widehat{G}$ is a meromorphic function on $\mathbf{C}$ with only a simple pole at the point $s = 1$ and a second order pole at the point $s = 0$. Quite similar computation with $G(s)$ instead of $\widehat{G}(s)$ gives us

$$G(s) = -\frac{\eta(0)}{2s} - \frac{\widehat{\eta}_-(0)}{2-2s} + \frac{\widehat{\eta}_+(0)}{4(1-s)^2} + B(s) + \widehat{B}(1-s). \tag{0.3.26}$$

We see that G is a meromorphic function on $\mathbf{C}$ with only a simple pole at the point $s = 0$ and a second order pole at the point $s = 1$. Moreover, the right-hand side of (0.3.26) with $1-s$ instead of s coincides with the right-hand side of (0.3.25). Thus, $\widehat{G}(s) = G(1-s)$, that agrees with the stated functional equation. ∎

Let us notice that the functional equation is essentially the same as that for the square of Dedekind zeta function (see 0.1.5) and for the squares of L-functions attached to classical regular modular forms of the weight 1 (see [19]). However we are far from any hypothesis.

The significance of the function $\Theta_!$ will be clear in Section 2.5, where we shall evaluate the Fourier coefficients of the cubic symplectic theta function $\Theta_\natural$. We shall find that some of these coefficients are nothing but $\eta(\nu)$. For this reason, it would be important to find explicit formulae for $\eta(\nu)$, something like Patterson's formulae (0.3.15) for $\tau(\nu)$.

0.4 On Dirichlet series associated with cubic Gauß sums

0.4.1 On the series Ψ. 0.4.2 On the series Q. 0.4.3 On the series P.

0.4.1 On the series Ψ. The first consequence of the cubic metaplectic forms theory is that the Dirichlet series (0.3.11) have meromorphic continuation to all of $\mathbf{C}$. More general and precise results were obtained by Kubota [54], [55] and Patterson [69], [70] by using matrix metaplectic Eisenstein series on $\mathrm{SL}(2,\mathbf{C})$, the functional equations for the Eisenstein series and some other things. We would like to state here some of these results. With this aim let us introduce some notations. First, for $l \in \mathbf{Z}$ and $c \in \mathcal{O}$, $c \neq 0$, we set $\phi_l(c) = (\bar{c}/|c|)^l$, and then we set

$$\zeta_*(s,n) = \sum_{c\equiv 1(3)} \frac{1}{\|c\|^s}\phi_{3n}(c), \tag{0.4.1}$$

where $n \in \mathbf{Z}$, $s \in \mathbf{C}$ and, for convergence, it is assumed $\Re(s) > 1$. So, $\zeta_*(\cdot,n)$ is, up to a simple factor, the well-known Hecke series with Größencharakter ϕ_{3n}. One has $\zeta_*(\cdot,0) = \zeta_*$, see 0.1.5. For $n \neq 0$ the function $\zeta_*(\cdot,n)$ can be continued as a regular function to $\mathbf{C}$. Next let

$$\Psi_*(s,\nu,n) = \sum_{\substack{c\equiv 0(3)\\ c\neq 0}} \frac{S_*(\nu,c)}{\|c\|^s}\phi_n(c), \quad \Psi(s,\nu,n) = \sum_{c\equiv 1(3)} \frac{S(\nu,c)}{\|c\|^s}\phi_n(c),$$

$$\Lambda_*(s,\nu,n) = (2\pi)^{-2s}\Gamma\Big(s+\frac{|n|}{2}-\frac{1}{3}\Big)\Gamma\Big(s+\frac{|n|}{2}-\frac{2}{3}\Big)\zeta_*(3s-2,n)\Psi_*(s,\nu,n),$$

$$\Lambda(s,\nu,n) = (2\pi)^{-2s}\Gamma\Big(s+\frac{|n|}{2}-\frac{1}{3}\Big)\Gamma\Big(s+\frac{|n|}{2}-\frac{2}{3}\Big)\zeta_*(3s-2,n)\Psi(s,\nu,n),$$

where $n \in \mathbf{Z}$, $s \in \mathbf{C}$, $\Re(s) > 3/2$, and $S_*(\nu,c)$, $S(\nu,c)$ are the Gauß sums defined in (0.1.3) and (0.1.5). Throughout this subsection we assume

$$q = (3), \quad \nu \in q^* = (\sqrt{-3})^{-3}\mathcal{O}, \quad \nu \neq 0, \quad s \in \mathbf{C}.$$

Under these assumptions the series defined in (0.3.11) are nothing but $\Psi_*(s,\nu,0)$ and $\Psi(s,\nu,0)$. Now let us state some properties of Ψ_* and Ψ which can be proved by an elementary computation under the assumption $\Re(s) > 3/2$ and still valid for any $s \in \mathbf{C}$ by analytic continuation.

Theorem 0.4.1 *In the notations above one has*

(a) $\displaystyle \Psi_*(s,0,n) = \frac{1+(-1)^n}{3^{3s-3}i^n - 1}\,\frac{\zeta_*(3s-3,n)}{\zeta_*(3s-2,n)}$,

(b) $\displaystyle \Psi(s,0,n) = (-1)^n\frac{\zeta_*(3s-3,n)}{\zeta_*(3s-2,n)}$,

(c) $\displaystyle \Psi_*(s,\nu,n) = \sum_{\substack{\varepsilon\in\mathcal{O}^*\\ l\in\mathbf{Z},\ l\geq 2}} 3^{-ls}(\varepsilon i^l)^{-n} S_*\big(\nu,\varepsilon(\sqrt{-3})^l\big)\,\Psi\big(s,\varepsilon(\sqrt{-3})^l\nu,n\big)$.

■

The series in part (c) is actually finite because of the Gauß sum involved is equal to 0 for all sufficiently large l, see 0.1.3. The next theorem describes the analytic properties of the series Ψ_* and Ψ.

Theorem 0.4.2 *In the notations above one has*

(a) *The Dirichlet series $\Psi_*(\cdot,\nu,n)$ and $\Psi(\cdot,\nu,n)$ can be continued as a meromorphic function to* $\mathbf{C}$;

(b) *The functions $\Lambda(\cdot,\nu,n)$ and $\Lambda_*(\cdot,\nu,n)$ are entire if $n \neq 0$. The only possible singularities of $\Lambda(\cdot,\nu,0)$ and $\Lambda_*(\cdot,\nu,0)$ are simple poles at $4/3$ and at $2/3$;*

(c) *$\Lambda(s,\nu,n)$ and $\Lambda_*(s,\nu,n)$ are bounded when $|\Im(s)|$ is large in vertical strips of finite width and satisfy the functional equation*

$$\Lambda(s,\nu,n) = \Omega(s,n)\|\nu\|^{1-s}\phi_n(\nu)\{\Lambda_*(2-s,\nu,-n) + \nabla(s,\nu,n)\Lambda(2-s,\nu,-n)\},$$

where $\nabla(s,\nu,n) = e(\nu) + (-1)^n e(-\nu) + \dfrac{1+(-1)^n}{3^{3s-3}i^n - 1}$, *and*

$$\Omega(s,n) = \begin{cases} (\sqrt{-3})^{-1}3^{12(1-s)} & \text{if } n \text{ is odd,} \\ 3^{8-9s}i^{|n|}\dfrac{1-3^{3-3s}i^n}{1-3^{3s-4}i^n} & \text{otherwise.} \end{cases}$$ ■

For reference convenience let us state explicitly some particular properties of the series $\Psi(\cdot,\nu) = \Psi(\cdot,\nu,0)$. Recall once more that $\nu \neq 0$.

Theorem 0.4.3 *In the notations above one has*

(a) *The only possible singularity of the function $s \mapsto \zeta_*(3s-2)\Psi(s,\nu)$ is a simple pole at the point $4/3$;*

(b) $2^{4/3}3^{3/2}\pi^{-5/3}\Gamma(2/3)\zeta_{\mathbf{Q}(\sqrt{-3})}(2) \operatorname*{Res}_{s=4/3} \Psi(s,\nu) = \dfrac{\tau(\nu)}{|\nu|^{1/3}}$;

(c) $\Psi(1,\nu) = 0$;

(d) *For any $\varepsilon \in \mathbf{R}_+^*$, there exists $c \in \mathbf{R}_+^*$ such that $|\Psi(s,\nu)| \leq c\|\nu\|^{1+\varepsilon}$ for all ν whenever $|s-1| < 1/10$.* ■

Part (b) is essentially equivalent to Theorem 0.3.4. Part (d) follows from the functional equation of Theorem 0.4.2 (c) by means of the Phragmén–Lindelöf principle. Part (c) follows from the functional equation for the Eisenstein series, as we saw already in 0.3.14.

The functions $\Psi(\cdot,\nu,n)$ were used by Heath-Brown and Patterson [37] to study the distribution of cubic Gauß sums at prime arguments. An impressive combination of the consequences of the Kubota theory with the methods of estimating the trigonometrical sums over primes, developed in the analytic number theory, gives rise in [37] to the solution of the old problem stated already by Kummer. For the history see [34], [41].

0.4.2 On the series Q. Patterson [70] has studied the series

$$Q(s,\nu,n)=\sum_{c\equiv 1(3)}\frac{S(\nu,c)^2}{\|c\|^{1/2+s}}\phi_n(c),\quad n\in\mathbf{Z},\ s\in\mathbf{C},\ \Re(s)>3/2. \qquad (0.4.2)$$

The method employed is that of integral of Rankin–Selberg type and is analogous to that used by Shimura in [93]. For further needs we would like to describe this in more details.

Let us define the function $\Theta_\star\colon \mathbf{H}\to\mathbf{C}$ by

$$\Theta_\star(w)=\frac{1}{\sqrt{3}}\left\{\Theta_{\text{K-P}}(w)+\omega\Theta_{\text{K-P}}\left(\begin{pmatrix}1&\omega\\0&1\end{pmatrix}w\right)+\omega^2\Theta_{\text{K-P}}\left(\begin{pmatrix}1&2\omega\\0&1\end{pmatrix}w\right)\right\} \qquad (0.4.3)$$

and the function $\tau_\star$ by

$$\tau_\star(\nu)=\overline{S(1,c)}\left|\frac{d}{c}\right|\begin{cases}1 & \text{if } \nu=(\sqrt{-3})^{-3}cd^3,\\ 0 & \text{for all others } \nu\in\mathbf{Q}(\sqrt{-3}),\end{cases} \qquad (0.4.4)$$

where $c,d\in\mathcal{O}$, $c,d\equiv 1 \pmod 3$, c is square-free. Then one can easily deduce from Theorem 0.3.8 the following.

Proposition 0.4.4 *The function $\Theta_\star$ is a square-integrable cubic metaplectic form under $\Gamma^{(2)}_{\text{princ}}(9)$. Its Fourier expansion about the cusp ∞ is*

$$\Theta_\star(w)=\sum_{\nu\in(\sqrt{-3})^{-3}\mathcal{O}\setminus\{0\}}\tau_\star(\nu)vK_{1/3}(4\pi|\nu|v)e(\nu z),\quad w=(z,v)\in\mathbf{H}. \qquad \blacksquare$$

(However, $\Theta_\star$ is not a cusp form. By Flicker, there are no at all cuspidal cubic theta functions on $\mathbf{H}$.)

We have $\Theta_\star\in\mathcal{L}^2(q,4/3)$ with any $q\subset(9)$. Next let $q\subset(9)$ and $t\in\mathbf{C}$. For $f\in\mathcal{L}^2(q,t)$ one has a Fourier expansion

$$f(w)=\rho_+v^t+\rho_-v^{2-t}+\sum_{\mu\in q^*\setminus\{0\}}\rho(\mu)vK_{t-1}(4\pi|\mu|v)e(\mu z),\quad w=(z,v)\in\mathbf{H},$$

where $\rho_+,\rho_-,\rho(\mu)\in\mathbf{C}$. Since f is assumed square-integrable, we have $\rho_+=0$ if $\Re(t)\ge 1$, and $\rho_-=0$ if $\Re(t)\le 1$. With these notations, we set

$$L_f(s,n)=\sum_{\mu\in(\sqrt{-3})^{-3}\mathcal{O}\setminus\{0\}}\phi_n(\mu)\frac{\tau(\mu)\rho(-\mu)}{\|\mu\|^{s-1/2}},$$

$$\nabla(q;s)=\sum_{c\in q\setminus\{0\}}\frac{1}{\|c\|^s}\sum_{\substack{d(cq)\\ d\equiv 1(q),\ \gcd(c,d)=1}}\left(\frac{c}{d}\right), \qquad (0.4.5)$$

and then

$$\Lambda_f(s,n) = (2\pi)^{-2s}\Gamma(2s+|n|-1;t)\,L_f(s,n), \tag{0.4.6}$$

where $n \in \mathbf{Z}$, $s \in \mathbf{C}$, and we write $\Gamma(s;t)$ for

$$\Gamma\Big(\frac{s+t}{2}-\frac{1}{3}\Big)\Gamma\Big(\frac{s+t}{2}-\frac{2}{3}\Big)\Gamma\Big(\frac{s-t}{2}+\frac{1}{3}\Big)\Gamma\Big(\frac{s-t}{2}+\frac{2}{3}\Big)\Gamma(s)^{-1}. \tag{0.4.7}$$

Observe that $\nabla(q;s)$ in (0.4.5) is nothing but $\Psi_*(s,0)$ in (0.3.11).

Now we can state one of the basic results of [70].

Theorem 0.4.5 *The series defining $L_f(s,n)$ converges if $\Re(s) > 3/2$ and has a continuation to the entire complex plane as a meromorphic function. If $n \neq 0$, $\Lambda_f(\cdot,n)$ has no singularities in $\{s \in \mathbf{C} \mid \Re(s) \geq 1\}$, but $\Lambda_f(\cdot,0)$ may have poles at the points s_0 (lying on the real axis), where $2s_0-1$ is a pole of $\nabla(q;\cdot)$, and at $t_0 = \max\{t/2+1/6, 11/6-t/2\}$. The poles are simple except a possible pole of order 2 at t_0 if t_0 coincides with one of the s_0. There are no other singularities of $\Lambda_f(\cdot,0)$ in $\{s \in \mathbf{C} \mid \Re(s) \geq 1\}$.* ∎

For $q = (9)$ one has

$$\nabla(q;s) = 6\frac{3^{4-4s}}{1-3^{1-s}}\frac{\zeta_*(3s-3)}{\zeta_*(3s-2)}.$$

Consequently, the only singularity of $\nabla(q;\cdot)$ in $\{s \in \mathbf{C} \mid \Re(s) \geq 1\}$ is a simple pole at the point 4/3 and the only possible singularities of $\Lambda_f(\cdot,0)$ are simple poles at the points 7/6 and t_0 if $t_0 \neq 7/6$, and a pole of order ≤ 2 at the point 7/6 if $t_0 = 7/6$. One can take $f = \Theta_\star$ and for this choice one has $q = (9)$, $t_0 = 7/6$ and

$$L_f(s,n) = i^{-n}3^{3s-2}\zeta_*(3s-5/2,n)Q(s,1,n).$$

This notice allows Patterson to deduce from Theorem 0.4.5 the following result for $\nu = 1$.

Theorem 0.4.6 *The series $Q(\cdot,\nu,n)$ can be continued to a meromorphic function on $\mathbf{C}$ and it is regular in the half-plane $\{s \in \mathbf{C} \mid \Re(s) \geq 7/6\}$ except for a possible simple pole at 7/6 if $n = 0$.* ∎

It is noticed also in [70] that the method can be extended to study the series $Q(s,\nu,n)$ with $\nu \neq 1$. Another approach to the same question is through applying the theory of Kubota to sextic reciprocity and using the Davenport–Hasse relation between the Gauß sum of the order six and the square of the cubic Gauß sum. This approach is also mentioned in [70] and explained in details by Patterson in a private letter to the author. For this we believe the preceding theorem as known for all ν.

Now let us observe that, for any $q \subset (9)$, one can express $\nabla(q;s)$ as

$$\frac{6\|q\|^{1-s}}{1-3^{1-s}}\frac{\zeta_*(3s-3)}{\zeta_*(3s-2)}\prod_{\substack{p\equiv 1(3)\\ p|q,\ p \text{ is prime}}}\left(1+\frac{1}{\|p\|^{s-1}}+\frac{1}{\|p\|^{2s-2}}\right)\left(1-\frac{1}{\|p\|^{3s-2}}\right)^{-1}, \quad (0.4.8)$$

and so, like as in the particular case $q = (9)$, $\nabla(q;\cdot)$ has no singularities in $\{s \in \mathbf{C} \mid \Re(s) \geq 1\}$ except a simple pole at $4/3$. Thus we have the following supplement to Theorem 0.4.5.

Proposition 0.4.7 $\Lambda_f(\cdot,0)$ *has no singularities in* $\{s \in \mathbf{C} \mid \Re(s) \geq 1\}$ *other than simple poles at the points* $7/6$ *and* t_0 *if* $t_0 \neq 7/6$, *and a pole of order* ≤ 2 *at the point* $7/6$ *if* $t_0 = 7/6$. □

Proof. We have only to get the expression (0.4.8) for $\nabla(q;s)$. For this let us represent q as $(9l)$ with $l \in \mathcal{O}$. Turning to (0.4.5), notice that c can be expressed uniquely as $c = c'c''$ with $c', c'' \in \mathcal{O}$ under the conditions

$$(9l) \mid c', \quad c' \mid (9l)^\infty, \quad \gcd(c',c'') = 1, \quad c'' \equiv 1 \pmod 3.$$

Then we express d as $9lc'x + c''y$ with $x, y \in \mathcal{O}$ and we observe that the summation over d under the conditions of (0.4.5) is equivalent to the summation over $x \pmod{c''}$, $y \pmod{9lc'}$ under the conditions

$$c''y \equiv 1 \pmod{9l}, \qquad \gcd(x, c'') = 1.$$

Notice also that the terms in (0.4.5) do not depend on y. More precisely, using the properties of the residue symbol listed in 0.1.2 we find

$$\left(\frac{c}{d}\right) = \left(\frac{c'}{d}\right)\left(\frac{c''}{d}\right) = \left(\frac{c''}{d}\right) = \left(\frac{d}{c''}\right) = \left(\frac{9lc'x + c''y}{c''}\right)$$
$$= \left(\frac{9lc'x}{c''}\right) = \left(\frac{9lc'}{c''}\right)\left(\frac{x}{c''}\right).$$

After this, noticing that there are exactly $\|c'\|$ different $y \pmod{9lc'}$ which satisfies $c''y \equiv 1 \pmod{9l}$, we obtain

$$\nabla(q;s) = \sum_{c'}\frac{1}{\|c'\|^{s-1}}\sum_{c''}\frac{1}{\|c''\|^s}\left(\frac{9lc'}{c''}\right)\sum_{\substack{x(c'')\\ \gcd(x,c'')=1}}\left(\frac{x}{c''}\right),$$

where c' and c'' are as explained above. Now, the interior sum (over x) is equal to 0 except the case $c'' = m^3$ with some $m \in \mathcal{O}$, $m \equiv 1 \pmod 3$, when we have

$$\sum_{\substack{x(c'')\\ \gcd(x,c'')=1}}\left(\frac{x}{c''}\right) = \tilde{\varphi}(m^3) = \|m\|^2\tilde{\varphi}(m) \quad \text{and} \quad \left(\frac{9lc'}{c''}\right) = 1.$$

($\tilde{\varphi}$ is the Euler totient function, see 0.1.4.) Thus we get $\nabla(q;s) = AB$ with

$$A = \sum_{\substack{c' \in \mathcal{O}\setminus\{0\} \\ (9l)|c'|(9l)^\infty}} \frac{1}{\|c'\|^{s-1}} = \|q\|^{1-s} \sum_{\substack{\tilde{c} \in \mathcal{O}\setminus\{0\} \\ \tilde{c}|(9l)^\infty}} \frac{1}{\|\tilde{c}\|^{s-1}}$$

$$= 6 \frac{\|q\|^{1-s}}{1 - 3^{1-s}} \prod_{\substack{p \equiv 1(3) \\ p|l,\ p \text{ is prime}}} \left(1 - \frac{1}{\|p\|^{s-1}}\right)^{-1},$$

$$B = \sum_{\substack{m \equiv 1(3) \\ \gcd(m,l)=1}} \frac{\tilde{\varphi}(m)}{\|m\|^{3s-2}} = \prod_{\substack{p \equiv 1(3) \\ p \nmid l,\ p \text{ is prime}}} \sum_{k \geq 0} \frac{\tilde{\varphi}(p^k)}{\|p\|^{(3s-2)k}}$$

$$= \prod_{\substack{p \equiv 1(3) \\ p \nmid l,\ p \text{ is prime}}} \left(1 - \frac{1}{\|p\|^{3s-3}}\right)^{-1} \left(1 - \frac{1}{\|p\|^{3s-2}}\right)$$

$$= \frac{\zeta_*(3s-3)}{\zeta_*(3s-2)} \prod_{\substack{p \equiv 1(3) \\ p|l,\ p \text{ is prime}}} \left(1 - \frac{1}{\|p\|^{3s-3}}\right)\left(1 - \frac{1}{\|p\|^{3s-2}}\right)^{-1},$$

as required. ∎

0.4.3 On the series P. For $r, l \in \mathcal{O} \setminus \{0\}$, $r \equiv 1 \pmod 3$ and $n \in \mathbf{Z}$, $s \in \mathbf{C}$, we set

$$P_n(s; l, r) = \sum_{\substack{m \equiv 1(3) \\ \gcd(m, rl)=1}} \frac{\overline{S(1,m)^2}}{\|m\|^s} \left(\frac{l}{m}\right) \phi_n(m). \tag{0.4.9}$$

These series converge absolutely and locally uniformly in $\{ s \in \mathbf{C} \mid \Re(s) > 2 \}$. It seems, to study the series $P_n(s; l, r)$ one can express them in terms of the series $Q(s, \nu, n)$ and then use Theorem 0.4.6. However, we find it more convenient to give an independent exposition based on Theorem 0.4.5.

First let us reduce the problem to the case $r = 1$. One has

$$P_n(s; l, r) = P_n(s; lr^3, 1) \tag{0.4.10}$$

and, if $\tilde{\mu}$ denotes the Möbius function defined in 0.1.4 and $\gcd(r, l) = 1$,[†]

$$P_n(s; l, r) = \sum_{\substack{d \equiv 1(3) \\ d|r}} \tilde{\mu}(d) \frac{\overline{S(1,d)^2}}{\|d\|^s} \left(\frac{l}{d}\right) \phi_n(d) P_n(s; ld^2, 1). \tag{0.4.11}$$

† We can assume $\gcd(r, l) = 1$ without lost of generality because the right-hand side of (0.4.9) is unchanged by the substitution rh instead of r, if $h \mid l^\infty$.

In both the formulae $s \in \mathbf{C}$, $\Re(s) > 2$. The formula (0.4.10) is obvious, and it is sufficient for our present purposes. Nevertheless, let us prove (0.4.11). First, we notice that

$$S(1,dk) = \begin{cases} S(1,d)S(1,k)\left(\dfrac{k}{d}\right)^2 & \text{if } \gcd(k,d)=1, \\ 0 & \text{otherwise,} \end{cases}$$

for $k, d \in \mathcal{O}$, $k \equiv d \equiv 1 \pmod 3$, and that

$$\sum_{\substack{t\equiv 1(3)\\ \gcd(t,r)=1}} h(t) = \sum_{\substack{d\equiv 1(3)\\ d\mid r}} \tilde{\mu}(d) \sum_{\substack{c\equiv 1(3)\\ d\mid c}} h(c)$$

for any complex-valued function h defined on $\{\, t \in \mathcal{O} \mid t \equiv 1 \pmod 3 \,\}$, provided that the series converge absolutely. After this one has

$$\begin{aligned} P_n(s;l,r) &= \sum_{\substack{d\equiv 1(3)\\ d\mid r}} \tilde{\mu}(d) \sum_{\substack{c\equiv 1(3)\\ \gcd(c,l)=1,\ d\mid c}} \frac{\overline{S(1,c)}^2}{\|c\|^s}\left(\frac{l}{c}\right)\phi_n(c) \\ &= \sum_{\substack{d\equiv 1(3)\\ d\mid r}} \tilde{\mu}(d) \sum_{\substack{k\equiv 1(3)\\ \gcd(k,l)=1}} \frac{\overline{S(1,dk)}^2}{\|dk\|^s}\left(\frac{l}{dk}\right)\phi_n(dk) \\ &= \sum_{\substack{d\equiv 1(3)\\ d\mid r}} \tilde{\mu}(d) \frac{1}{\|d\|^s}\left(\frac{l}{d}\right)\phi_n(d) \sum_{\substack{k\equiv 1(3)\\ \gcd(k,l)=1}} \frac{\overline{S(1,dk)}^2}{\|k\|^s}\left(\frac{l}{k}\right)\phi_n(k) \\ &= \sum_{\substack{d\equiv 1(3)\\ d\mid r}} \tilde{\mu}(d) \frac{\overline{S(1,d)}^2}{\|d\|^s}\left(\frac{l}{d}\right)\phi_n(d) \sum_{\substack{k\equiv 1(3)\\ \gcd(k,ld)=1}} \frac{\overline{S(1,k)}^2}{\|k\|^s}\left(\frac{ld^2}{k}\right)\phi_n(k) \\ &= \sum_{\substack{d\equiv 1(3)\\ d\mid r}} \tilde{\mu}(d) \frac{\overline{S(1,d)}^2}{\|d\|^s}\left(\frac{l}{d}\right)\phi_n(d) P_n(s;ld^2,1), \quad \text{as required.} \end{aligned}$$

To examine $P_n(s;l,1)$ one can apply Theorem 0.4.5 to the twist $\Theta_\star * \varepsilon$ with $\varepsilon\colon q^* \to \mathbf{C}$ given by

$$\varepsilon(k) = \begin{cases} \left(\dfrac{l}{h}\right) & \text{if } k = (\sqrt{-3})^{-3}h,\ h \equiv 1(3),\ \gcd(h,l)=1, \\ 0 & \text{for all other } k \in q^*. \end{cases} \tag{0.4.12}$$

Here $l \in \mathcal{O}$, $q = (9)$ and, so, $q^* = (\sqrt{-3})^{-5}\mathcal{O}$. One can easily show, see 0.1.2, that the so defined ε is a periodic function and that $3^3 l q^*$ is its lattice of periods.

It follows from Theorem 0.3.12 and Proposition 0.4.4 that $\Theta_\star * \varepsilon \in \mathcal{L}^2(q, 4/3)$, if $q = (3^8 l^2)$.

Lemma 0.4.8 *For $f = \Theta_\star * \varepsilon$ with ε as in (0.4.12), and $n \in \mathbf{Z}$ one has*

$$L_f(s,n) = 3^{3s-2} i^n P_n(s+1/2; l, 1)\,\zeta_*(3s-5/2, n) \prod_{\substack{p\equiv 1(3)\\ p|l,\ p \text{ is prime}}} \left(1 - \frac{\phi_{3n}(p)}{\|p\|^{3s-5/2}}\right)^{-1}. \quad \square$$

Proof. It follows immediately from (0.4.5), Proposition 0.4.4 and the definition of the twists, given in 0.3.10, that

$$L_f(s,n) = \sum_{\nu \in (\sqrt{-3})^{-3}\mathcal{O}\setminus\{0\}} \frac{\tau(\nu)\tau_\star(-\nu)\varepsilon(-\nu)}{\|\nu\|^{s-1/2}} \phi_n(\nu).$$

Notice that the νth term in the series in the right-hand side is non-zero only if $\nu = -(\sqrt{-3})^{-3} cd^3$ with some $c \equiv d \equiv 1 \pmod 3$, $\gcd(cd, l) = 1$, $c, d \in \mathcal{O}$, and that for such ν one has

$$\tau_\star(-\nu) = 3^{1/2}\tau(\nu) = \overline{S(1,c)}\left|\frac{d}{c}\right|, \quad \varepsilon(-\nu) = \left(\frac{l}{c}\right), \quad \phi_n(\nu) = i^n \phi_n(c)\phi_{3n}(d),$$

see (0.3.15), (0.4.4), (0.4.12). With these expressions we get

$$L_f(s,n) = 3^{3s-2} i^n P_n(s+1/2; l, 1) \sum_{\substack{d\equiv 1(3)\\ \gcd(d,l)=1}} \frac{1}{\|d\|^{3s-5/2}} \phi_{3n}(d),$$

and it remains only to notice that the sum over d is

$$\zeta_*(3s-5/2, n) \prod_{\substack{p\equiv 1(3)\\ p|l,\ p \text{ is prime}}} \left(1 - \frac{\phi_{3n}(p)}{\|p\|^{3s-5/2}}\right)^{-1}. \quad \blacksquare$$

Theorem 0.4.9 *For all $n \in \mathbf{Z}$, $r, l \in \mathcal{O}\setminus\{0\}$, $r \equiv 1 \pmod 3$ the series $P_n(\cdot; l, r)$ can be continued to a meromorphic function on $\mathbf{C}$. The function $s \mapsto \zeta_*(3s-4, n) P_n(s; l, r)$ has no singularities in $\{ s \in \mathbf{C} \mid \Re(s) \geq 3/2 \}$, except possible pole of order ≤ 2 at the point $5/3$ if $n = 0$.* $\square$

Proof. It is sufficient for us to consider the case $r = 1$ due to either (0.4.10) or (0.4.11). Let $f = \Theta_\star * \varepsilon$, ε being defined by (0.4.12). The associated function $L_f(\cdot, n)$ is described by Lemma 0.4.8. To apply Theorem 0.4.5, notice that the necessary conditions are satisfied by f and $q = (3^8 l^2)$, see Theorem 0.3.12. It follows immediately from Theorem 0.4.5 and Lemma 0.4.8 that $P_n(\cdot; l, r)$ has a meromorphic continuation to $\mathbf{C}$. To describe the singularities let us look at the function

$$\Lambda_f(s,n) = (2\pi)^{-2s}\Gamma(2s+|n|-1; 4/3)\,L_f(s,n). \qquad (0.4.13)$$

By (0.4.7) and by the triplication and the duplication formulae, we have

$$\begin{aligned}\Gamma(2s'-1;4/3) &= \Gamma(s'-5/6)\Gamma(s'-1/6)\Gamma(s'-1/2)^2\Gamma(2s'-1)^{-1} \\ &= 2\pi 3^{3-3s'}\Gamma(3s'-5/2)\Gamma(s'-1/2)\Gamma(2s'-1)^{-1} \\ &= 2^{3-2s'}3^{3-3s'}\pi^{3/2}\Gamma(3s'-5/2)\Gamma(s')^{-1},\end{aligned}$$

where $s' = s+|n|/2$. We see that the gamma-factor in (0.4.13) has no poles and zeros in the half-plane $\{ s \in \mathbf{C} \mid \Re(s) > 5/6-|n|/2 \}$. Thus, $L_f(\cdot,n)$ has in this half-plane exactly the same poles as $\Lambda_f(\cdot,n)$. By Theorem 0.4.5 and Proposition 0.4.7, $\Lambda_f(\cdot,n)$ is regular in $\{ s \in \mathbf{C} \mid \Re(s) > 1 \}$ except for a possible pole of order ≤ 2 at the point 7/6 if $n=0$. Consequently, we have the same for $L_f(\cdot,n)$ and, look at Lemma 0.4.8, for the function $s \mapsto \zeta_*(3s-5/2,n)P_n(s+1/2;l,1)$. Changing $s+1/2$ by s we conclude the proof. ∎

Part 1

1.1 Group SL(3, C)

1.1.1 Group SL(3, C). The special linear group $\mathrm{SL}(3,\mathbf{C})$ is the group of all 3×3 matrices of determinant 1 with entries in $\mathbf{C}$. It can be viewed as a complex Lie group, and as that it is the simple group of rank 2, dimension 8, and of type A_2 in Cartan–Killing classification. Also one can view $\mathrm{SL}(3,\mathbf{C})$ as a linear algebraic or as a real Lie group. The general automorphic functions theory (as it exposed in [32], for example) deals with real semi-simple connected Lie groups with finite centre. $\mathrm{SL}(3,\mathbf{C})$ is one of such groups. Some more data on $\mathrm{SL}(3,\mathbf{C})$ are collected below.

1.1.2 Iwasawa decomposition. One has $\mathrm{SL}(3,\mathbf{C}) = NAK$, where

$$N \text{ is the group of } n(l) = \begin{pmatrix} 1 & l_3 & l_2 \\ 0 & 1 & l_1 \\ 0 & 0 & 1 \end{pmatrix} \text{ with } l = (l_1, l_2, l_3) \in \mathbf{C}^3, \tag{1.1.1}$$

$$A \text{ is the group of } \mathrm{diag}(t_1, t_2, t_1^{-1}t_2^{-1}) \text{ with } t_1, t_2 \in \mathbf{R}_+^*, \tag{1.1.2}$$

$$K = \mathrm{SU}(3) \text{ is a maximal compact subgroup of } \mathrm{SL}(3,\mathbf{C}).$$

More precisely, each matrix in $\mathrm{SL}(3,\mathbf{C})$ can be factored uniquely as the product of the matrix in N, of the matrix in A, and of the matrix in K. One can prove this fact, known from the linear algebra, by means of Gram–Schmidt ortonormalization process.

One has also that the product mapping $N \times A \times K \to NAK$ is a real analytic manifolds isomorphism (= an analytic diffeomorphism). With this supplement, the decomposition $\mathrm{SL}(3,\mathbf{C}) = NAK$ is a special case of the Iwasawa decomposition.

1.1.3 Weyl group. Let T be the subgroup of all diagonal matrices in $\mathrm{SL}(3,\mathbf{C})$, i.e. of all matrices of the form (1.1.2) with $t_1, t_2 \in \mathbf{C}^*$. The Weyl group, denoted W, should be defined as the factor group $Norm(T)/Centr(T)$, where $Norm(T)$ and $Centr(T)$ denotes, respectivly, the normalizer and the centralizer of T in $\mathrm{SL}(3,\mathbf{C})$. In the meantime, it is clear that $Centr(T)$ is the group T itself. Also, for any $\gamma \in \mathrm{SL}(3,\mathbf{C})$ we have that $\gamma \in Norm(T)$ if and only if each row and each column of γ has exactly one non-zero entry. This yields further that W is a group of order 6 isomorphic to the symmetric group S_3 (= the group of the permutations of three points). It is also clear that representatives in $Norm(T)$ of the cosets in $Norm(T)/Centr(T)$ one can choose to be matrices whose non-zero entries are ± 1. This means, in particular, representatives can be choosen in $Norm(T) \cap K$.

The Weyl group acts on T by conjugation. More precisely, given $\delta \in T$ and $\varepsilon \in W$, the image of δ under ε is defined to be $\delta^\varepsilon = \varepsilon'^{-1}\delta\varepsilon'$, where ε' means any representative of ε in $Norm(T)$. The group $A \subset T$ is invariant under W, and, by restriction, we get an action of W on A. This action induces the action of the Weyl group on the group of characters of A. More precisely, given a character (= a homomorphism) $\Phi\colon A \to \mathbf{C}^*$ and $\varepsilon \in W$, the image ${}^\varepsilon\Phi\colon A \to \mathbf{C}^*$ of Φ under ε is defined to be ${}^\varepsilon\Phi(a) = \Phi(\varepsilon'^{-1}a\varepsilon')$ for any $a \in A$.

1.1.4 Bruhat decomposition. Let B be the subgroup of all upper triangular matrices in $\mathrm{SL}(3,\mathbf{C})$. The group B is a minimal parabolic (or Borel) subgroup of $\mathrm{SL}(3,\mathbf{C})$, and N is its unipotent radical. It is not difficult to find that $B = TN = NT$ (— semidirect product of groups). One has the decomposition of $\mathrm{SL}(3,\mathbf{C})$ into a disjoint union, which was known already to Gauß, and which is a special case of the Bruhat decomposition —

$$\mathrm{SL}(3,\mathbf{C}) = \bigcup_{r \in S} BrB, \tag{1.1.3}$$

S being a representative system in K of the Weyl group. One can take

$$S = \{e_3,\ p,\ \tilde{p},\ p\tilde{p},\ \tilde{p}p,\ \sigma\}, \tag{1.1.4}$$

where e_3 is 3×3 identity matrix,

$$p = \begin{pmatrix} 0 & -1 & 0 \\ 1 & 0 & 0 \\ 0 & 0 & 1 \end{pmatrix}, \quad \tilde{p} = \begin{pmatrix} 1 & 0 & 0 \\ 0 & 0 & 1 \\ 0 & -1 & 0 \end{pmatrix}, \quad \sigma = p\tilde{p}p = -\begin{pmatrix} 0 & 0 & 1 \\ 0 & 1 & 0 \\ 1 & 0 & 0 \end{pmatrix}. \tag{1.1.5}$$

1.1.5 Involution ι. $\mathrm{SL}(3,\mathbf{C})$ possess the Cartan involution $\gamma \mapsto {}^t\gamma^{-1}$. Conjugating by σ in (1.1.5) (notice that $\sigma^2 = e_3$) we get one more involution which we shall denote by ι. So, ι maps γ to ${}^\iota\gamma = \sigma\, {}^t\gamma^{-1}\sigma^{-1}$. The groups N, A and K in the Iwasawa decomposition are invariant under ι, i.e.

$${}^\iota N = N, \qquad {}^\iota A = A, \qquad {}^\iota K = K. \tag{1.1.6}$$

Here and in sequel we write ${}^\iota R$ for $\{\,{}^\iota\gamma \mid \gamma \in R\}$, R being a subset of $\mathrm{SL}(3,\mathbf{C})$.

1.1.6 On the group N. We have that N is a complex Heisenberg group of dimension 3. For any $l = (l_1, l_2, l_3) \in \mathbf{C}^3$, $l' = (l'_1, l'_2, l'_3) \in \mathbf{C}^3$ one has

$$n(l)n(l') = n(t_1, t_2, t_3), \quad n(l)^{-1} = n(r_1, r_2, r_3)$$

with

$$\begin{aligned} t_1 &= l_1 + l'_1, & t_2 &= l_2 + l'_2 + l'_1 l_3, & t_3 &= l_3 + l'_3, \\ r_1 &= -l_1, & r_2 &= -l_2 + l_1 l_3, & r_3 &= -l_3. \end{aligned}$$

Given any $\mu, \nu \in \mathbf{C}$, the mapping

$$n(l_1, l_2, l_3) \mapsto e(\mu l_1 + \nu l_3), \tag{1.1.7}$$

is an unitary character of N, i.e. a homomorphism $N \to \mathbf{C}^*$ with the values in the unit circle. These are the only unitary characters of N, as one can easily check by the formulae above. The character (1.1.7) is said to be degenerate if $\mu\nu = 0$, and non-degenerate otherwise.

1.1.7 Embeddings $\mathrm{SL}(2, \mathbf{C}) \to \mathrm{SL}(3, \mathbf{C})$. For $\gamma = \begin{pmatrix} a & b \\ c & d \end{pmatrix} \in \mathrm{SL}(2, \mathbf{C})$ we set

$$h_1(\gamma) = \begin{pmatrix} 1 & 0 & 0 \\ 0 & a & b \\ 0 & c & d \end{pmatrix}, \quad h_2(\gamma) = \begin{pmatrix} a & 0 & b \\ 0 & 1 & 0 \\ c & 0 & d \end{pmatrix}, \quad h_3(\gamma) = \begin{pmatrix} a & b & 0 \\ c & d & 0 \\ 0 & 0 & 1 \end{pmatrix}.$$

Proposition 1.1.1 *Each of h_j is a group monomorphism* $\mathrm{SL}(2, \mathbf{C}) \to \mathrm{SL}(3, \mathbf{C})$. *One has also ${}^{\iota}h_1(\gamma) = h_3({}^{\iota}\gamma)$, ${}^{\iota}h_2(\gamma) = h_2({}^{\iota}\gamma)$, ${}^{\iota}h_3(\gamma) = h_1({}^{\iota}\gamma)$ for any $\gamma \in$* $\mathrm{SL}(2, \mathbf{C})$ *and with ι defined in* 1.1.5, 0.3.9. □

Proof. Obvious. ■

One can extend each of h_j to be the homomorphism $\mathrm{GL}(2, \mathbf{C}) \to \mathrm{SL}(3, \mathbf{C})$ assuming $\gamma \in \mathrm{GL}(2, \mathbf{C})$ and replacing 1 by $(\det \gamma)^{-1}$ in the definitions above.

In the framework of Lie groups and algebras theory, or Chevalley groups theory, the h_j are associated with so called $sl(2)$-triples in the Lie algebra of $\mathrm{SL}(3, \mathbf{C})$, see for example [13]. These h_j have essential role in our approach to the problem of computation Eisenstein series Fourier coefficients, this will be seen in Section 1.2 and Section 1.4.

1.1.8 Maximal parabolic subgroups. Let us write

P for the group of all matrices $\begin{pmatrix} * & * & * \\ * & * & * \\ 0 & 0 & * \end{pmatrix} \in \mathrm{SL}(3, \mathbf{C})$,

$\widetilde{P}$ for the group of all matrices $\begin{pmatrix} * & * & * \\ 0 & * & * \\ 0 & * & * \end{pmatrix} \in \mathrm{SL}(3, \mathbf{C})$,

N_P for the group $\{ n(l_1, l_2, 0) \mid l_1, l_2 \in \mathbf{C} \} \subset N \subset P$,

$N_{\widetilde{P}}$ for the group $\{ n(0, l_2, l_3) \mid l_2, l_3 \in \mathbf{C} \} \subset N \subset \widetilde{P}$.

The so defined P and $\widetilde{P}$ are maximal parabolic subgroups (associated with two different positive simple roots) of SL(3, C). The groups N_P and $N_{\widetilde{P}}$ are their unipotent radicals. In the notations of 1.1.4 we have: P is generated by B and p; $\widetilde{P}$ is generated by B and $\tilde{p}$. Notice also that the involution ι, defined in 1.1.5, transfers each of these groups into another one. More precisely, one has ${}^{\iota}B = B$, ${}^{\iota}p = \tilde{p}$, ${}^{\iota}\tilde{p} = p$. In particular we have that $\widetilde{P}$ is isomorphic to P.

Proposition 1.1.2 *Each $\gamma \in P$ and $\tilde{\gamma} \in \widetilde{P}$ can be factored in* SL(3, C) *uniquely as*

$$\gamma = \begin{pmatrix} 1 & 0 & * \\ 0 & 1 & * \\ 0 & 0 & 1 \end{pmatrix}\begin{pmatrix} * & * & 0 \\ * & * & 0 \\ 0 & 0 & * \end{pmatrix}, \quad \tilde{\gamma} = \begin{pmatrix} 1 & * & * \\ 0 & 1 & 0 \\ 0 & 0 & 1 \end{pmatrix}\begin{pmatrix} * & 0 & 0 \\ 0 & * & * \\ 0 & * & * \end{pmatrix}. \quad \square \qquad (1.1.8)$$

Proof. Obvious. ∎

1.1.9 An auxiliary decomposition. Coming back to the Bruhat decomposition, we should say it is not difficult to give an explicit description of each coset BrB, $r \in S$. We will need in one particular fact given in the proposition below.

Proposition 1.1.3 *Let*

$$\gamma = \begin{pmatrix} a_1 & a_2 & a_3 \\ b_1 & b_2 & b_3 \\ c_1 & c_2 & c_3 \end{pmatrix} \in \mathrm{SL}(3, \mathbf{C}). \qquad (1.1.9)$$

If

$$c_3 \neq 0, \quad b_2c_3 - b_3c_2 \neq 0 \qquad (1.1.10)$$

and σ is as in (1.1.5), *then one has the decomposition*

$$\gamma\sigma = n(l_1, l_2, l_3)\,\mathrm{diag}(t_1, t_2, t_1^{-1}t_2^{-1})\,\sigma\, n(l'_1, l'_2, l'_3), \qquad (1.1.11)$$

where

$$l_3 = \frac{a_2c_3 - a_3c_2}{b_2c_3 - b_3c_2}, \quad l_2 = \frac{a_3}{c_3}, \quad l_1 = \frac{b_3}{c_3},$$
$$t_1 = \frac{1}{b_2c_3 - b_3c_2}, \quad t_2 = \frac{b_2c_3 - b_3c_2}{c_3}, \qquad (1.1.12)$$
$$l'_3 = \frac{c_2}{c_3}, \quad l'_2 = \frac{c_1}{c_3}, \quad l'_1 = \frac{b_1c_3 - b_3c_1}{b_2c_3 - b_3c_2}. \quad \square$$

Proof. Multiplying the matrices in (1.1.11), we find (1.1.11) is equivalent to

$$\begin{pmatrix} a_3 & a_2 & a_1 \\ b_3 & b_2 & b_1 \\ c_3 & c_2 & c_1 \end{pmatrix} = t_1^{-1}t_2^{-1}\begin{pmatrix} l_2 & t_1t_2^2l_3 + l_2l'_3 & * \\ l_1 & t_1t_2^2 + l_1l'_3 & t_1t_2^2l'_1 + l_1l'_2 \\ 1 & l'_3 & l'_2 \end{pmatrix}.$$

Comparing the entries we get the equations which lead to (1.1.12). ∎

1.2 Discrete subgroups

1.2.1 Preliminaries. Let $q \subset (3)$ be an ideal of the integers ring $\mathcal{O}$ of the field $\mathbf{Q}(\sqrt{-3})$ and let $\Gamma_n(q)$ be the congruence subgroup of $\mathrm{SL}(n,\mathcal{O})$, as it was defined in 0.2.1. We shall prove in this section several propositions that describe the group $\Gamma_3(q)$, some related groups and cosets, and the Bass–Milnor–Serre homomorphism $\psi = \bar{\kappa}_3$ (κ_3 being as in 0.2.4, 0.2.5). Our approach is based on consideration of the three monomorphisms $\Gamma_2(q) \to \Gamma_3(q)$ obtained by restriction of h_1, h_2, h_3 in 1.1.7 from $\mathrm{SL}(2,\mathbf{C})$ to $\Gamma_2(q)$.

To simplify notations, let us write from now on $\Gamma(q)$ instead of $\Gamma_3(q)$.

1.2.2 Groups Δ, $\Delta_\times$, $\widetilde{\Delta}_\times$. Let B be the minimal parabolic subgroup of $\mathrm{SL}(3,\mathbf{C})$ defined in 1.1.4. Let P and $\widetilde{P}$ be the maximal parabolic subgroups of $\mathrm{SL}(3,\mathbf{C})$ defined in 1.1.8. We set

$$\Delta = \Gamma(q) \cap B, \quad \Delta_\times = \Gamma(q) \cap P, \quad \widetilde{\Delta}_\times = \Gamma(q) \cap \widetilde{P}.$$

One may easily verify that $\Delta = \Gamma(q) \cap N$. Also, we have that $\Delta_\times$ is the subgroup of all matrices in $\Gamma(q)$ with the bottom row of the form $(0,0,1)$, while $\widetilde{\Delta}_\times$ is the subgroup of all matrices in $\Gamma(q)$ with the first column ${}^t(1,0,0)$. Obviously, any matrix in either $\Delta_\times$ or $\widetilde{\Delta}_\times$ can be factored as in Proposition 1.1.2 with the factors in $\Gamma(q)$.

Proposition 1.2.1 *One has*

(a) $\Delta = \{\, n(l_1, l_2, l_3) \mid l_1, l_2, l_3 \in q \,\}$;

(b) *Each* $\gamma \in \Delta_\times$ *can be factored uniquely as* $\gamma'\gamma''$ *with*

$$\gamma' = n(*,*,0) \in \Delta, \ \gamma'' = h_3(\lambda), \ \lambda \in \Gamma_2(q);$$

(c) $\Delta_\times = \Delta h_3(\Gamma_2(q))$. □

Proof. Assertions (a) and (b) are obvious. Assertion (c) follows from (b) and $\Delta \subset \Delta_\times$, $h_3(\Gamma_2(q)) \subset \Delta_\times$. ■

Notice that $\Gamma_2(q)$ and $\Gamma_3(q)$ are invariant under the involutions ι defined in 0.3.9 and in 1.1.5. Applying the involution ι to both sides of (c) in the preceding proposition we get $\widetilde{\Delta}_\times = \Delta h_1(\Gamma_2(q))$, it since $\widetilde{\Delta}_\times = {}^\iota\Delta_\times$, $\Delta = {}^\iota\Delta$ and ${}^\iota h_3(\Gamma_2(q)) = h_1(\Gamma_2(q))$ (see Proposition 1.1.1). One can do the same in many other cases in what follows, and this allows us to deal with the group $\Delta_\times$ only, and not with the group $\widetilde{\Delta}_\times$, having in mind to apply ι to obtain an information concerning $\widetilde{\Delta}_\times$ if necessary.

1.2.3 Bottom row. The following proposition describes the bottom rows of matrices in $\Gamma(q)$.

Proposition 1.2.2 (a) *If the vector*

$$(c_1, c_2, c_3) \quad \textit{with} \quad c_j \in \mathcal{O} \tag{1.2.1}$$

is the bottom row of some matrix $\gamma \in \Gamma(q)$, *then*

$$c_1, c_2 \in q, \qquad c_3 \equiv 1 \pmod 3), \tag{1.2.2}$$

$$\gcd(c_1, c_2, c_3) = 1. \tag{1.2.3}$$

(b) *If the vector* (1.2.1) *satisfies the conditions* (1.2.2), (1.2.3), *then there exist matrices*

$$\varepsilon_j = h_j(\lambda_j) \in \Gamma(q), \quad \lambda_j \in \Gamma_2(q), \quad j = 1, 2, \tag{1.2.4}$$

such that the vector (1.2.1) *is the bottom row of the matrix* $\gamma = \varepsilon_2 \varepsilon_1$. □

Proof. (a) For (1.2.2) we refer just to the definition of $\Gamma(q)$. For (1.2.3) we should observe only that each common divisor of the entries of the vector (1.2.1) divides $\det \gamma = 1$.

(b) Let $t = \gcd(c_2, c_3)$. Then we have

$$\gcd(c_1, t) = \gcd(c_2 t^{-1}, c_3 t^{-1}) = 1,$$
$$t \equiv 1 \pmod 3), \quad c_3 t^{-1} \equiv 1 \pmod 3), \quad c_1 \in q, \quad c_2 t^{-1} \in q.$$

Consequently, there exist matrices

$$\lambda_2 = \begin{pmatrix} * & * \\ c_1 & t \end{pmatrix} \in \Gamma_2(q), \quad \lambda_1 = \begin{pmatrix} * & * \\ c_2 t^{-1} & c_3 t^{-1} \end{pmatrix} \in \Gamma_2(q).$$

The bottom row of the matrix $h_2(\lambda_2) h_1(\lambda_1)$ is (1.2.1), as required. ■

1.2.4 On the group $\Gamma(q)$.

Proposition 1.2.3 *For each matrix* $\gamma \in \Gamma(q)$ *there exist matrices* $\delta \in \Delta$ *and* $\lambda_j \in \Gamma_2(q)$ *such that*

$$\gamma = \delta \varepsilon_3 \varepsilon_2 \varepsilon_1 \quad \textit{with} \quad \varepsilon_j = h_j(\lambda_j), \quad j = 1, 2, 3. \quad \square \tag{1.2.5}$$

Proof. It follows from Proposition 1.2.2 that there exist matrices $\lambda_j \in \Gamma_2(q)$ such that the matrix $\gamma' = \varepsilon_2 \varepsilon_1$ with $\varepsilon_j = h_j(\lambda_j)$ has the same bottom row as γ does. Obviously, $\gamma\gamma'^{-1} \in \Delta_\times$. Then, by Proposition 1.2.1 (b), there exist matrices $\delta \in \Delta$ and $\lambda_3 \in \Gamma_2(q)$ such that $\gamma\gamma'^{-1} = \delta\varepsilon_3$, $\varepsilon_3 = h_3(\lambda_3)$. Thus we get $\gamma = \delta\varepsilon_3\gamma' = \delta\varepsilon_3\varepsilon_2\varepsilon_1$, as required. ■

It is seen from the proof of Proposition 1.2.2 one has the following complement to Proposition 1.2.3. (e_3 is 3×3 identity matrix.)

Proposition 1.2.4 *If a vector like $(*,0,*)$ is the bottom row of $\gamma \in \Gamma(q)$, then γ can be factored as in (1.2.5) with $\varepsilon_1 = e_3$.* ∎

1.2.5 A special remark.

Proposition 1.2.5 *Let*

$$\gamma = \begin{pmatrix} a_1 & a_2 & a_3 \\ b_1 & b_2 & b_3 \\ c_1 & c_2 & c_3 \end{pmatrix} \in \mathrm{SL}(3,\mathbf{C}). \tag{1.2.6}$$

If there exist matrices

$$\varepsilon_j = h_j(\lambda_j), \quad \lambda_j = \begin{pmatrix} k_j & l_j \\ m_j & n_j \end{pmatrix} \in \mathrm{SL}(3,\mathbf{C}), \quad j = 1,2, \tag{1.2.7}$$

such that $\gamma = \varepsilon_2\varepsilon_1$, then

(a) $a_1 = k_2, \quad a_2 = l_2 m_1, \quad a_3 = l_2 n_1;$
(b) $b_1 = 0, \quad b_2 = k_1, \quad b_3 = l_1;$
(c) $c_1 = m_2, \quad c_2 = n_2 m_1, \quad c_3 = n_2 n_1;$
(d) $b_1 c_3 - b_3 c_1 = -m_2 l_1, \quad b_2 c_3 - b_3 c_2 = n_2.$ □

Proof. We only have to multiply the matrices ε_j in (1.2.7) and compare the product with the matrix (1.2.6). ∎

1.2.6 Cosets $\Delta_\times\backslash\Gamma(q)$. Let Λ be the set of all vectors

$$c = (c_1, c_2, c_3) \tag{1.2.8}$$

that satisfy the conditions (1.2.2), (1.2.3). According to Proposition 1.2.2, Λ is the set of the bottom rows of all matrices in $\Gamma(q)$. For each $c \in \Lambda$ we choose and denote by γ_c some matrix in $\Gamma(q)$ having the vector c as the bottom row.

Proposition 1.2.6 *The set $\{\,\gamma_c \mid c \in \Lambda\,\}$ contains a unique representative of each coset in $\Delta_\times\backslash\Gamma(q)$.* □

Proof. It is enough to notice that $\gamma' \in \Gamma(q)$ have the same bottom rows as $\gamma \in \Gamma(q)$ if and only if $\gamma' = \delta\gamma$ with some $\delta \in \Delta_\times$. ∎

Given $\gamma \in \Gamma(q)$, we denote by Δ^γ the subgroup of Δ consisting of all matrix $\delta \in \Delta$ such that the bottom row of the matrix $\gamma^t\delta$ is equal to the bottom row of γ. By means of simple computation we find, for γ with the bottom row as in (1.2.8),

$$\Delta^\gamma = \{\, n(0, -c_2 u/l_\gamma, c_3 u/l_\gamma) \mid u \in q \,\}, \tag{1.2.9}$$

where $l_\gamma = \gcd(c_2, c_3)$. With the same notations, let $\vartheta = c_3/l_\gamma$ and

$$\mathcal{F}_\gamma = \mathbf{C}/\vartheta q \times \mathbf{C}^2, \tag{1.2.10}$$

$\mathbf{C}/\vartheta q$ being a fundamental domain of the lattice ϑq. Also, for $\gamma \in \Gamma(q)$ we denote by Λ_γ the set containing a unique representative in ${}^t\Delta$ of each coset in ${}^t\Delta^\gamma \backslash {}^t\Delta$, and we write σ for the matrix defined in (1.1.5).

Proposition 1.2.7 *Let a set $\Omega \subset \Gamma(q)$ contains a unique representative of each double coset in $\Delta_\times \backslash \Gamma(q)/{}^t\Delta$, and let $\mathcal{P} = \{ (\gamma, h) \mid \gamma \in \Omega,\ h \in \Lambda_\gamma \}$. One has*

(a) *The mapping $\mathcal{P} \to \Delta_\times \backslash \Gamma(q)$ defined by $(\gamma, h) \mapsto \Delta_\times \gamma h$ is bijective;*

(b) *$\sigma^{-1}\Lambda_\gamma\sigma$ contains a unique representative of each coset in $\sigma^{-1}\,{}^t\Delta^\gamma\sigma\backslash\Delta$;*

(c) *$n(\mathcal{F}_\gamma) = \{ n(l_1, l_2, l_3) \mid l_1 \in \mathbf{C}/\vartheta q,\ l_2 \in \mathbf{C},\ l_3 \in \mathbf{C} \}$ contains a unique representative of each coset in $\sigma^{-1}\,{}^t\Delta^\gamma\sigma\backslash N$.* □

Proof. (a) For short, let us prove only injectivity. Let $\gamma, \gamma' \in \Omega$, $h \in \Lambda_\gamma$, $h' \in \Lambda_{\gamma'}$. If the images of the pairs (γ, h), (γ', h') are the same, then we have

$$\delta\gamma h = \gamma' h' \tag{1.2.11}$$

with some $\delta \in \Delta_\times$. This can be rewritten as $\gamma' = \delta\gamma h h'^{-1}$, where $\delta \in \Delta_\times$, $hh'^{-1} \in {}^t\Delta$, and so γ and γ' lie in the same double coset in $\Delta_\times\backslash\Gamma(q)/{}^t\Delta$. Hence, $\gamma' = \gamma$. Let c be the bottom row of the matrix γ. From (1.2.11) we get $ch' = ch$. This yields $hh'^{-1} \in {}^t\Delta^\gamma$, and then, since $h, h' \in \Lambda_\gamma$, we obtain $h = h'$.

(b) It is evident from the definition of Λ_γ that the set $\sigma^{-1}\Lambda_\gamma\sigma$ contains a unique representative of each coset in $\sigma^{-1}\,{}^t\Delta^\gamma\sigma\backslash\sigma^{-1}\,{}^t\Delta\sigma$. It remains to notice (for this see (1.1.6)) that $\sigma^{-1}\,{}^t\Delta\sigma = \Delta$.

(c) Follows from (1.2.9) and the formulae in 1.1.6. ∎

1.2.7 Homomorphism ψ: $\widetilde{\Gamma}^{(3)}_{\mathrm{princ}}(3) \to \mathbf{C}^*$. Let κ be the Kubota homomorphism, and let κ_n, $n \geq 2$, be the Bass–Milnor–Serre homomorphisms (see 0.2.2, 0.2.4, 0.2.5). We put $\psi = \bar{\kappa}_3$. We shall prove two propositions to describe the restrictions of the homomorphism ψ to some subgroups of $\Gamma(q) \subset \Gamma^{(3)}_{\mathrm{princ}}(3)$.

Proposition 1.2.8 *One has*

(a) *$\psi(\gamma) = 1$ for $\gamma \in \Delta$ and for $\gamma \in {}^t\Delta$;*

(b) *If $\gamma = \gamma'\gamma'' \in \Delta_\times$ with $\gamma' \in \Delta$ and $\gamma'' = h_3(\lambda) \in \Delta_\times$, $\lambda \in \Gamma_2(q)$, then $\psi(\gamma) = \kappa(\lambda)$;*

(c) *$\psi({}^t\gamma) = \psi(\gamma)$ for all $\gamma \in \Gamma(q)$.* □

Proof. (a) For $\gamma \in \Delta$ we get $\psi(\gamma) = 1$ applying Theorem 0.2.2 (b), (c). Then one can apply Theorem 0.2.2 (d) to get $\psi(\gamma) = 1$ for $\gamma \in {}^t\Delta$.

(b) By (a) we have $\psi(\gamma) = \psi(\gamma'')$. Then, by Theorem 0.2.2 (a), (b)

$$\psi(\gamma'') = \bar{\kappa}_3\left(\begin{pmatrix} \lambda & & 0 \\ & & 0 \\ 0 & 0 & 1 \end{pmatrix}\right) = \bar{\kappa}_2(\lambda) = \kappa(\lambda).$$

(c) This follows from the definition in 1.1.5 and Theorem 0.2.2 (d), (f). ∎

Proposition 1.2.9 *Let $\varepsilon_j = h_j(\lambda)$ with $\lambda \in \Gamma_2(q)$, $j = 1, 2, 3$. Then*

$$\psi(\varepsilon_j) = \kappa(\lambda). \quad \square$$

Proof. The case $j = 3$ already considered in Proposition 1.2.8 (b). For $j = 1$ Theorem 0.2.2 (a), (b) yields

$$\psi(\varepsilon_3) = \bar{\kappa}_3\left(\begin{pmatrix} 1 & 0 & 0 \\ 0 & & \\ 0 & & \lambda \end{pmatrix}\right) = \bar{\kappa}_2(\lambda) = \kappa(\lambda).$$

For $j = 2$ we find, by means of Theorem 0.2.2 (f), that

$$\psi(\varepsilon_2) = \bar{\kappa}_3\left(\begin{pmatrix} a & 0 & b \\ 0 & 1 & 0 \\ c & 0 & d \end{pmatrix}\right) = \bar{\kappa}_3\left(\gamma \begin{pmatrix} a & 0 & b \\ 0 & 1 & 0 \\ c & 0 & d \end{pmatrix} \gamma^{-1}\right) = \bar{\kappa}_3\left(\begin{pmatrix} a & b & 0 \\ c & d & 0 \\ 0 & 0 & 1 \end{pmatrix}\right),$$

if

$$\lambda = \begin{pmatrix} a & b \\ c & d \end{pmatrix}, \quad \gamma = \begin{pmatrix} 1 & 0 & 0 \\ 0 & 0 & 1 \\ 0 & -1 & 0 \end{pmatrix}.$$

Applying Theorem 0.2.2 (a), (b) we can continie the computation to get

$$\psi(\varepsilon_2) = \bar{\kappa}_2\left(\begin{pmatrix} a & b \\ c & d \end{pmatrix}\right) = \bar{\kappa}_2(\lambda) = \kappa(\lambda), \quad \text{as required.} \quad \blacksquare$$

1.2.8 Representative system for $\Delta_x \backslash \Gamma(q) / {}^t\Delta$. Let n denote a pair (n_1, n_2) with $n_j \equiv 1 \pmod 3$, $\gcd(n_j, q) = 1$. For each such n we choose subsets $\mathfrak{m}_1(n)$ and $\mathfrak{l}_1(n)$ of q such that each of them contains a unique representative of each class $\bmod n_1$. For each n and $m_1, l_1 \in q$ we choose subsets $\mathfrak{m}_2(n; m_1, l_1)$ and $\mathfrak{l}_2(n; m_1, l_1)$ of q such that each of them contains a unique representative of each class $\bmod n_2$ coprime with n_2.

The next proposition gives us very convenient representative system for the double cosets in $\Delta_x \backslash \Gamma(q) / {}^t\Delta$, which we shall use in 1.4.4 to evaluate the Fourier coefficients of the Eisenstein series (and which can be used to evaluate the Fourier coefficients of some other automorphic functions not considered in these notes).

Theorem 1.2.10 *Each double coset in $\Delta_x \backslash \Gamma(q) / {}^t\Delta$ contains a unique matrix γ which can be factored as $\gamma = \varepsilon_2 \varepsilon_1$, where*

$$\varepsilon_j = h_j(\lambda_j), \quad \lambda_j = \begin{pmatrix} k_j & l_j \\ m_j & n_j \end{pmatrix} \in \Gamma_2(q),$$

and

$$m_1 \in \mathfrak{m}_1(n), \quad l_1 \in \mathfrak{l}_1(n), \quad m_2 \in \mathfrak{m}_2(n; m_1, l_1),$$
$$l_2 \in \mathfrak{l}_2(n; m_1, l_1), \quad n = (n_1, n_2). \quad \square$$

We shall prove this statement in 1.2.10.

1.2.9 Two auxiliary propositions.

Proposition 1.2.11 *The matrices*

$$\gamma = \begin{pmatrix} * & * & * \\ * & * & * \\ c_1 & c_2 & c_3 \end{pmatrix} \in \Gamma(q), \quad \gamma' = \begin{pmatrix} * & * & * \\ * & * & * \\ c'_1 & c'_2 & c'_3 \end{pmatrix} \in \Gamma(q)$$

lie in one double coset in $\Delta_\times\backslash\Gamma(q)/{}^t\Delta$ *if and only if*

$$\begin{gathered} c_3 = c'_3, \quad \gcd(c_2, c_3) = \gcd(c'_2, c'_3), \quad c_2 \equiv c'_2 \pmod{c_3 q}, \\ c_1 \equiv c'_1 \pmod{\gcd(c_2, c_3) q}. \quad \square \end{gathered} \tag{1.2.12}$$

Proof. Let $c(\gamma)$ denotes the bottom row of the matrix γ and $c(\gamma')$ denotes the bottom row of the matrix γ'. The matrices γ and γ' lie in one double coset in $\Delta_\times\backslash\Gamma(q)/{}^t\Delta$ if and only if

$$c(\gamma') = c(\gamma)\delta \quad \text{with some} \quad \delta = {}^t n(l_1, l_2, l_3), \quad l_1, l_2, l_3 \in q. \tag{1.2.13}$$

Obviously, (1.2.13) is equivalent to

$$c'_1 = c_1 + c_2 l_3 + c_3 l_2, \quad c'_2 = c_2 + c_3 l_1, \quad c'_3 = c_3, \quad \text{with some} \quad l_1, l_2, l_3 \in q.$$

This is equivalent to (1.2.12). ∎

Proposition 1.2.12 *If the matrices* γ *and* γ' *lie in one double coset in* $\Delta_\times\backslash\Gamma(q)/{}^t\Delta$ *and can be factored as* $\gamma = \varepsilon_2\varepsilon_1$, $\gamma' = \varepsilon'_2\varepsilon'_1$ *with* $\varepsilon_j = h_j(\lambda_j)$, $\varepsilon'_j = h_j(\lambda'_j)$,

$$\lambda_j = \begin{pmatrix} k_j & l_j \\ m_j & n_j \end{pmatrix} \in \Gamma_2(q), \quad \lambda'_j = \begin{pmatrix} k'_j & l'_j \\ m'_j & n'_j \end{pmatrix} \in \Gamma_2(q),$$

then $n_1 = n'_1$, $n_2 = n'_2$, $m_1 \equiv m'_1 \pmod{n_1 q}$, $m_2 \equiv m'_2 \pmod{n_2 q}$. □

Proof. From Proposition 1.2.11 and Proposition 1.2.5 (c) follows

$$\begin{gathered} n_2 n_1 = n'_2 n'_1, \quad \gcd(n_2 m_1, n_2 n_1) = \gcd(n'_2 m'_1, n'_2 n'_1), \\ n_2 m_1 \equiv n'_2 m'_1 \pmod{n_1 n_2 q}, \quad m_2 \equiv m'_2 \pmod{\gcd(n_2 m_1, n_2 n_1) q}. \end{gathered}$$

Since $\gcd(m_1, n_1) = \gcd(m'_1, n'_1) = 1$, this implies the desired relations. ∎

1.2.10 Proof of Theorem 1.2.10. Let Z means a double coset in $\Delta_\times\backslash\Gamma(q)/{}^t\Delta$. It follows from Proposition 1.2.1 (c) and Proposition 1.2.3 that Z contains the matrix γ which can be factored as $\gamma = \nu_2\nu_1$ with

$$\nu_j = h_j(\mu_j), \quad \mu_j = \begin{pmatrix} * & * \\ * & n_j \end{pmatrix} \in \Gamma_2(q).$$

Let $n = (n_1, n_2)$. Put $\varepsilon_1 = \alpha\nu_1\beta$ with

$$\alpha = h_1\left(\begin{pmatrix} 1 & \xi' \\ 0 & 1 \end{pmatrix}\right), \quad \beta = h_1\left(\begin{pmatrix} 1 & 0 \\ \xi & 1 \end{pmatrix}\right), \quad \xi, \xi' \in q.$$

With suitably chosen ξ, ξ', the matrix ε_1 satisfies the conditions of Theorem 1.2.10, so

$$\varepsilon_1 = h_1(\lambda_1), \quad \lambda_1 = \begin{pmatrix} k_1 & l_1 \\ m_1 & n_1 \end{pmatrix} \in \Gamma_2(q), \quad m_1 \in \mathfrak{m}_1(n), \quad l_1 \in \mathfrak{l}_1(n).$$

Since $\beta \in {}^t\Delta$, the matrix $\nu_2\alpha^{-1}\varepsilon_1$ lies in the same double coset Z as γ does. The bottom row of the matrix $\nu_2\alpha^{-1}$ looks like $(*, 0, *)$. So, by Proposition 1.2.3 and Proposition 1.2.1 (c), we have $\nu_2\alpha^{-1} = \delta\nu_2'$ with some $\delta \in \Delta_\times$ and

$$\nu_2' = h_2(\mu_2'), \quad \mu_2' = \begin{pmatrix} * & * \\ * & n_2' \end{pmatrix} \in \Gamma_2(q). \tag{1.2.14}$$

We see the double coset Z contains the matrix $\nu_2'\varepsilon_1$ where ε_1 satisfies the conditions of Theorem 1.2.10 and ν_2' is as in (1.2.14). It follows from Proposition 1.2.12 that $n_2' = n_2$. Now, let $\varepsilon_2 = \alpha\nu_2'\beta$ with

$$\alpha = h_2\left(\begin{pmatrix} 1 & \vartheta' \\ 0 & 1 \end{pmatrix}\right), \quad \beta = h_2\left(\begin{pmatrix} 1 & 0 \\ \vartheta & 1 \end{pmatrix}\right), \quad \vartheta, \vartheta' \in q. \tag{1.2.15}$$

With suitably chosen ϑ, ϑ', the matrix ε_2 satisfies the conditions of Theorem 1.2.10, so

$$\varepsilon_2 = h_2(\lambda_2), \quad \lambda_2 = \begin{pmatrix} k_2 & l_2 \\ m_2 & n_2 \end{pmatrix} \in \Gamma_2(q),$$
$$m_2 \in \mathfrak{m}_2(n; m_1, l_1), \quad l_2 \in \mathfrak{l}_2(n; m_1, l_1).$$

Since $\alpha \in \Delta_\times$, we see the double coset Z contains the matrix $\varepsilon_2\beta^{-1}\varepsilon_1$ with ε_1, ε_2 as in Theorem 1.2.10 and with β as in (1.2.15). Then we easily find that $\beta^{-1}\varepsilon_1 \in \varepsilon_1{}^t\Delta$. Consequently, the double coset Z contains $\varepsilon_2\varepsilon_1$, as required.

To prove uniqueness we can apply Proposition 1.2.12. Let Z be a double coset in $\Delta_\times\backslash\Gamma(q)/{}^t\Delta$ and let matrices $\gamma, \gamma' \in Z$ are factored as indicated in Theorem 1.2.10, so $\gamma = \varepsilon_2\varepsilon_1$, $\gamma' = \varepsilon_2'\varepsilon_1'$ with $\varepsilon_j = h_j(\lambda_j)$, $\varepsilon_j' = h_j(\lambda_j')$,

$$\lambda_j = \begin{pmatrix} k_j & l_j \\ m_j & n_j \end{pmatrix} \in \Gamma_2(q), \quad \lambda_j' = \begin{pmatrix} k_j' & l_j' \\ m_j' & n_j' \end{pmatrix} \in \Gamma_2(q),$$

$$m_1 \in \mathfrak{m}_1(n), \quad l_1 \in \mathfrak{l}_1(n), \quad m_2 \in \mathfrak{m}_2(n; m_1, l_1), \quad l_2 \in \mathfrak{l}_2(n; m_1, l_1),$$
$$m_1' \in \mathfrak{m}_1(n'), \quad l_1' \in \mathfrak{l}_1(n'), \quad m_2' \in \mathfrak{m}_2(n'; m_1', l_1'), \quad l_2' \in \mathfrak{l}_2(n'; m_1', l_1'),$$
$$n = (n_1, n_2), \quad n' = (n_1', n_2').$$

By Proposition 1.2.12, $n_1' = n_1$, $n_2' = n_2$. So, $n' = n$ and $m_1, m_1' \in \mathfrak{m}_1(n)$,

$l_1, l'_1 \in \mathfrak{l}_1(n)$. Applying Proposition 1.2.12 once more we find $m'_1 = m_1$ and $l'_1 = l_1$ (notice that $m_1 l_1 \equiv m'_1 l'_1 \equiv -1 \pmod{n_1}$). Now we have $m_2, m'_2 \in \mathfrak{m}_2(n; m_1, l_1)$, $l_2, l'_2 \in \mathfrak{l}_2(n; m_1, l_1)$ and, applying Proposition 1.2.12 once more, we get $m'_2 = m_2$ and $l'_2 = l_2$. So, $\varepsilon'_1 = \varepsilon_1$, $\varepsilon'_2 = \varepsilon_2$, as required.

1.3 Cubic metaplectic forms on $\mathbf{X} \simeq \mathrm{SL}(3, \mathbf{C})/\mathrm{SU}(3)$

1.3.1 Space $\mathbf{X}$. 1.3.2 Involution ι. 1.3.3 Cubic metaplectic forms.
1.3.4 The action of upper triangular matrices. 1.3.5 The action of p and $\tilde{p}$.
1.3.6 The mappings Q and $\widetilde{Q}$. 1.3.7 Eisenstein series.
1.3.8 Fourier expansion according to Piatetski-Shapiro and Shalika.
1.3.9 Invariant differential operators. 1.3.10 Whittaker functions.

1.3.1 Space X. We set $\mathbf{X} = \mathbf{C}^3 \times \mathbf{R}_+^{*2}$, considering the right-hand side as a product of real analytic manifolds. For

$$w = (z_1, z_2, z_3, u, v) \in \mathbf{X} \tag{1.3.1}$$

we set (see 1.1.2)

$$n(w) = n(z_1, z_2, z_3), \quad a(w) = \mathrm{diag}(u^{1/3}v^{2/3}, u^{1/3}v^{-1/3}, u^{-2/3}v^{-1/3}),$$

$$U(w) = u, \quad V(w) = v, \quad Z_j(w) = z_j, \quad \widetilde{Q}(w) = (z_1, u), \quad Q(w) = (z_3, v).$$

We consider $Q(w)$ and $\widetilde{Q}(w)$ as points in $\mathbf{H}$. According to Iwasawa we have the real analytic manifolds isomorphism

$$\mathbf{X} \to \mathrm{SL}(3, \mathbf{C})/\mathrm{SU}(3), \qquad w \mapsto n(w)a(w)\mathrm{SU}(3),$$

which transfers on $\mathbf{X}$ the action by left translations of the group $\mathrm{SL}(3, \mathbf{C})$ on $\mathrm{SL}(3, \mathbf{C})/\mathrm{SU}(3)$. We shall write γw for the image of the point $w \in \mathbf{X}$ under the action of $\gamma \in \mathrm{SL}(3, \mathbf{C})$. This action will be described in 1.3.4, 1.3.5 and 1.3.6.

The space $\mathbf{X}$, being equiped with a Riemannian $\mathrm{SL}(3, \mathbf{C})$-invariant metric (which is determined uniquely up to a constant factor), is a Riemannian globally symmetric space of type 4 in Cartan classification. In a well known fashion one can attach to the Riemannian metric a second order differential operator, called Laplace–Beltrami operator, which is invariant in the sense that its action (on C^∞-functions $\mathbf{X} \to \mathbf{C}$) commute with the action of $\mathrm{SL}(3, \mathbf{C})$ on $\mathbf{X}$, and which is uniquely determined (up to a constant factor) by this condition.

Actually, we have an algebra, let us denote it by $\mathbf{D}(\mathbf{X})$, of invariant differential operators on $\mathbf{X}$, which is generated by the Laplace–Beltrami operator and by one third order operator. $\mathbf{D}(\mathbf{X})$ is a commutative $\mathbf{C}$-algebra isomorphic to the algebra of polynomials of 2 variables over $\mathbf{C}$. We shall descuss this in more details in 1.3.9.

From now on we save the notations (1.3.1).

1.3.2 Involution ι. First, let us recall we defined the involutions of the group $\mathrm{SL}(2, \mathbf{C})$ and the space $\mathbf{H}$ in 0.3.9, and then the involution of the group $\mathrm{SL}(3, \mathbf{C})$

in 1.1.5. Since SU(3) is invariant under the involution ι of $\mathrm{SL}(3, \mathbf{C})$ in 1.1.5, we get, by passing to quotient, an involution of $\mathbf{X}$, which we shall denote again by ι. For all $\gamma \in \mathrm{SL}(3, \mathbf{C})$ and $w \in \mathbf{X}$ as in (1.3.1) we have ${}^\iota\gamma\, {}^\iota w = {}^\iota(\gamma w)$, and ${}^\iota w = (-z_3, -z_2 + z_1 z_3, -z_1, v, u)$, because of ${}^\iota n(w) = n(-z_3, -z_2 + z_1 z_3, -z_1)$, ${}^\iota a(w) = \mathrm{diag}(u^{2/3}v^{1/3}, u^{-1/3}v^{1/3}, u^{-1/3}v^{-2/3})$.

To avoid any possible misunderstandings let us summarize:

$$
\begin{aligned}
&{}^\iota\gamma = \sigma\, {}^t\gamma^{-1}\sigma^{-1} \text{ for } \gamma \in \mathrm{SL}(2, \mathbf{C}),\ \sigma = \begin{pmatrix} 0 & 1 \\ 1 & 0 \end{pmatrix};\\
&{}^\iota w = (-z, v) \text{ for } w = (z, v) \in \mathbf{H};\\
&{}^\iota\gamma = \sigma\, {}^t\gamma^{-1}\sigma^{-1} \text{ for } \gamma \in \mathrm{SL}(3, \mathbf{C}),\ \sigma \text{ being defined in (1.1.5)};\\
&{}^\iota w = (-z_3, -z_2 + z_1 z_3, -z_1, v, u) \text{ for } w = (z_1, z_2, z_3, u, v) \in \mathbf{X};\\
&{}^\iota f(w) = f({}^\iota w) \text{ for any } f\colon \mathbf{H} \to \mathbf{C},\ w \in \mathbf{H}, \text{ and for any } f\colon \mathbf{X} \to \mathbf{C},\ w \in \mathbf{X};
\end{aligned}
$$

1.3.3 Cubic metaplectic forms. Let Γ be a discrete subgroup of $\mathrm{SL}(3, \mathbf{C})$. Let $\tilde{\psi}\colon \Gamma \to \mathbf{C}^*$ be an unitary character, that is a group homomorphism with values in the unit circle. We say $F\colon \mathbf{X} \to \mathbf{C}$ is an automorphic form under Γ with multiplier system $\tilde{\psi}$ if F is a C^∞-function satisfying the conditions

(a) $F(\gamma w) = \tilde{\psi}(\gamma) F(w)$ for $\gamma \in \Gamma$, $w \in \mathbf{X}$;

(b) F is an eigenfunction of all the invariant differential operators on $\mathbf{X}$;

(c) There is $c \in \mathbf{R}$ such that $|F(w)| < \mathit{Norm}\big(n(w)a(w)\big)^c$ for all $w \in \mathbf{X}$.

In the point (c), by $\mathit{Norm}(\delta)$ for $\delta \in \mathrm{SL}(3, \mathbf{C})$ we mean $\mathit{trace}(\delta\, {}^t\bar{\delta})$ (just like in [32]). The point (b) of this definition require for any $D \in \mathbf{D}(\mathbf{X})$ one has $DF = \lambda_F(D)F$ with some $\lambda_F(D) \in \mathbf{C}$. It is clear then the mapping $D \mapsto \lambda_F(D)$ is a $\mathbf{C}$-algebras homomorphism, i. e., λ_F is a character of $\mathbf{D}(\mathbf{X})$. This case one say F is a joint eigenfunction of $\mathbf{D}(\mathbf{X})$ attached to λ_F. One get more general and more natural notion of automorphic forms by replacing (b) in the definition above by

(b′) One has $DF = 0$ for all $D \in I_F$, where I_F is an ideal of finite codimention of $\mathbf{D}(\mathbf{X})$.

However, all automorphic forms we shall deal in Part 1 satisfy (b), and we shall not have an occasion to refer to (b′).

Let $\psi = \bar{\kappa}_3\colon \tilde{\Gamma}^{(3)}_{\mathrm{princ}}(3) \to \mathbf{C}^*$ be the Bass–Milnor–Serre homomorphism, see 1.2.7, and let Γ be a subgroup of finite index in $\tilde{\Gamma}^{(3)}_{\mathrm{princ}}(3)$. We say $F\colon \mathbf{X} \to \mathbf{C}$ is a cubic metaplectic form under Γ if F is an automorphic form under Γ with multiplier system ψ.

In what follows we restrict ourselves mainly by consideration of metaplectic forms under the groups $\Gamma(q)$, $q \subset (3)$.

Notice that the isomorphism of analytic manifolds

$$(z_1, z_2, z_3) \mapsto n(z_1, z_2, z_3), \tag{1.3.2}$$

carries the Lebesgue measure on $\mathbf{C}^3$ to a Haar measure on N. Let a Lebesgue measurable set $\mathcal{F}_q \subset \mathbf{C}^3$ be such that its image under (1.3.2) contains a unique representative of each coset in $\Delta\backslash N$; $\Delta = \Gamma(q) \cap N$. One can take, for example, $\mathcal{F}_q = (\mathbf{C}/q)^3$ where $\mathbf{C}/q$ is a fundamental domain of the lattice q. We write $\mathrm{vol}(\mathcal{F}_q)$ for the volume of $\mathcal{F}_q$ and set $c = \mathrm{vol}(\mathcal{F}_q)^{-1}$. It is obvious that $\mathrm{vol}(\mathcal{F}_q)$ does not depend on the particular choice of $\mathcal{F}_q$. Now let $F: \mathbf{X} \to \mathbf{C}$ be an automorphic form under $\Gamma(q)$ with multiplier system $\tilde{\psi}$. Assume that Δ is contained in the kernel of the homomorphim $\tilde{\psi}$. Given $\mu, \nu \in q^*$, let

$$c_{\mu,\nu}(u, v; F) = c \iiint_{\mathcal{F}_q} F(w)\overline{e(\mu z_1 + \nu z_3)}\, dz_1\, dz_2\, dz_3 \tag{1.3.3}$$

with w as in (1.3.1). Since Δ is contained in both the kernel of $\tilde{\psi}$ and the kernel of the homomorphism (1.1.7), we have that the integral (1.3.3) does not depend on the particular choice of $\mathcal{F}_q$. We call $c_{\mu,\nu}(u, v; F)$ the μ, ν^{th} Fourier coefficients of the form F. In view of Proposition 1.2.8 (a), this definition is valid for cubic metaplectic forms F.

If F is an automorphic form under the group $\Gamma(q)$ and $\lambda \in \mathrm{SL}(3, \mathbf{C})$ is any matrix satisfying

$$\lambda\Gamma(q)\lambda^{-1} = \Gamma(q) \quad \text{and} \quad \tilde{\psi}(\lambda\gamma\lambda^{-1}) = \tilde{\psi}(\gamma) \quad \text{for all} \quad \gamma \in \Gamma(q),$$

then the function $w \mapsto F(\lambda w)$, $w \in \mathbf{X}$, also is an automorphic form under the group $\Gamma(q)$ with just the same multiplier system $\tilde{\psi}$ as F is. This can be easily checked. In particular, if F is a cubic metaplectic form under $\Gamma(q)$ and $\lambda \in \mathrm{SL}(3, \mathbf{Z})$, then $w \mapsto F(\lambda w)$, $w \in \mathbf{X}$, is a cubic metaplectic form under $\Gamma(q)$ too, see Theorem 0.2.2 (f).

Although the given above definition of cubic metaplectic forms is sufficient for our aims in the present notes, let us pointed out that one can give slightly different definitions, which may be even more convenient at least at some cases. For example, let we are given homomorphism $\chi: \Gamma(q) \to \mathbf{C}^*$, and let us assume that $\chi(\gamma) = 1$ for all $\gamma \in \Gamma^{(3)}_{\text{princ}}(q)$. Such type χ can be defined as

$$\chi(\gamma) = \chi_1(a_1)\chi_2(b_2)\chi_3(c_3) \quad \text{for} \quad \gamma = \begin{pmatrix} a_1 & a_2 & a_3 \\ b_1 & b_2 & b_3 \\ c_1 & c_2 & c_3 \end{pmatrix} \in \Gamma(q),$$

where χ_j are Dirichlet characters $\bmod q$, $j = 1, 2, 3$. Then, let us define homomorphism $\tilde{\psi}: \Gamma(q) \to \mathbf{C}^*$ by $\tilde{\psi}(\gamma) = \psi(\gamma)\chi(\gamma)$, $\gamma \in \Gamma(q)$, with Bass–Milnor–Serre's ψ. Now, let F be an automorphic form under $\Gamma(q)$ with multiplier system $\tilde{\psi}$. In accordance with our definition, we have that F as a cubic metaplectic form under $\Gamma^{(3)}_{\text{princ}}(q)$, just because of $\tilde{\psi}$ agrees with ψ on $\Gamma^{(3)}_{\text{princ}}(q)$. On the other hand, it may occurs more convenient to slightly extend the definition to treate F as a cubic metaplectic form under $\Gamma(q)$ with the above defined multiplier system $\tilde{\psi}$.

1.3.4 The action of upper triangular matrices.

Proposition 1.3.1 (a) *If* $\gamma = n(l_1, l_2, l_3)$, $l_1, l_2, l_3 \in \mathbf{C}$, *then*

$$Z_j(\gamma w) = z_j + l_j + \begin{cases} 0 & \text{for } j = 1,3, \\ l_3 z_1 & \text{for } j = 2, \end{cases} \qquad \begin{aligned} U(\gamma w) &= u, \\ V(\gamma w) &= v, \end{aligned}$$

$$\widetilde{Q}(\gamma w) = (z_1 + l_1, u), \quad Q(\gamma w) = (z_3 + l_3, v).$$

(b) *If* $\gamma = \mathrm{diag}(t_1, t_2, t_1^{-1} t_2^{-1})$, $t_1, t_2 \in \mathbf{C}^*$, *then*

$$Z_j(\gamma w) = z_j \begin{cases} t_1 t_2^2 & \text{if } j = 1, \\ t_1^2 t_2 & \text{if } j = 2, \\ t_1 t_2^{-1} & \text{if } j = 3, \end{cases} \qquad \begin{aligned} U(\gamma w) &= |t_1 t_2^2| u, \\ V(\gamma w) &= |t_1 t_2^{-1}| v, \end{aligned}$$

$$\widetilde{Q}(\gamma w) = (t_1 t_2^2 z_1, |t_1 t_2^2| u), \quad Q(\gamma w) = (t_1 t_2^{-1} z_3, |t_1 t_2^{-1}| v). \qquad \square$$

Proof. (a) By definition of the $\mathrm{SL}(3,\mathbf{C})$ action on $\mathbf{X}$ one has $n(\gamma w) a(\gamma w) = \gamma n(w) a(w) k$ with some $k \in \mathrm{SU}(3)$. Since $\gamma \in N$, this implies[†] $n(\gamma w) = \gamma n(w)$, $a(\gamma w) = a(w)$ and we receive the desired results applying the formulae in 1.1.6.

(b) Let $w_\gamma = (t_1 t_2^2 z_1, t_1^2 t_2 z_2, t_1 t_2^{-1} z_3, |t_1 t_2^2| u, |t_1 t_2^{-1}| v)$ and $\varepsilon_1 = t_1/|t_1|$, $\varepsilon_2 = t_2/|t_2|$. We only have to prove $w_\gamma = \gamma w$. One can check easily that $\gamma n(w) = n(w_\gamma)\gamma$, $\gamma a(w) = a(w_\gamma) k$, with $k = \mathrm{diag}(\varepsilon_1, \varepsilon_2, \varepsilon_1^{-1} \varepsilon_2^{-1}) \in \mathrm{SU}(3)$. This yields $\gamma n(w) a(w) = n(w_\gamma) a(w_\gamma) k$ with the same k, and thus $w_\gamma = \gamma w$. ∎

1.3.5 The action of p and $\tilde{p}$. The next proposition describes the action on the space $\mathbf{X}$ of the matrices p and $\tilde{p}$ defined in 1.1.4.

Proposition 1.3.2 *For* $w = (z_1, z_2, z_3, u, v) \in \mathbf{X}$ *we have*

$$\text{(a)} \quad \begin{aligned} &Z_3(pw) = -\bar{z}_3 L^{-1}, \quad && Z_2(pw) = -z_1, \quad && Z_1(pw) = z_2, \\ &V(pw) = v L^{-1}, && U(pw) = u L^{1/2}, && \text{where } L = v^2 + |z_3|^2; \end{aligned}$$

$$\text{(b)} \quad \begin{aligned} &Z_3(\tilde{p}w) = z_2 - z_1 z_3, \quad && (Z_2 - Z_1 Z_3)(\tilde{p}w) = -z_3, \quad && Z_1(\tilde{p}w) = -\bar{z}_1 L^{-1}, \\ &V(\tilde{p}w) = v L^{1/2}, && U(\tilde{p}w) = u L^{-1}, && \text{where } L = u^2 + |z_1|^2. \end{aligned} \qquad \square$$

Proof. To prove (a) we set

$$\begin{aligned} u_p = U(pw), \quad v_p = V(pw), \quad s_j = Z_j(pw), \\ \text{so that} \quad pw = (s_1, s_2, s_3, u_p, v_p). \end{aligned} \tag{1.3.4}$$

According to the definition of the $\mathrm{SL}(3,\mathbf{C})$ action on $\mathbf{X}$ we have

$$n(pw) a(pw) = p n(w) a(w) k \qquad \text{with some} \quad k \in \mathrm{SU}(3). \tag{1.3.5}$$

[†] Due to the uniqueness in the Iwasawa decomposition 1.1.2.

Since $k{}^t\bar{k}$ is the identity and $a(pw)$, $a(w)$ are real symmetric matrices, (1.3.5) implies

$$n(pw)a(pw)^2\,{}^t\overline{n(pw)} = (pn(w))a(w)^2\,{}^t\overline{(pn(w))}. \qquad (1.3.6)$$

Then by routine computation we find that

$$n(pw)a(pw)^2\,{}^t\overline{n(pw)} = (m_{ij}), \quad (pn(w))a(w)^2\,{}^t\overline{(pn(w))} = (l_{ij}) \qquad (1.3.7)$$

with

$$\begin{aligned}
m_{12} &= s_3u_p^{2/3}v_p^{-2/3} + \bar{s}_1s_2u_p^{-4/3}v_p^{-2/3},\\
m_{22} &= u_p^{2/3}v_p^{-2/3} + |s_1|^2u_p^{-4/3}v_p^{-2/3},\\
m_{13} &= s_2u_p^{-4/3}v_p^{-2/3},\\
m_{23} &= s_1u_p^{-4/3}v_p^{-2/3},\\
m_{33} &= u_p^{-4/3}v_p^{-2/3},\\
l_{12} &= -\bar{z}_3u^{2/3}v^{-2/3} - z_1\bar{z}_2u^{-4/3}v^{-2/3},\\
l_{22} &= u^{2/3}v^{4/3} + |z_3|^2u^{2/3}v^{-2/3} + |z_2|^2u^{-4/3}v^{-2/3},\\
l_{13} &= -z_1u^{-4/3}v^{-2/3},\\
l_{23} &= z_2u^{-4/3}v^{-2/3},\\
l_{33} &= u^{-4/3}v^{-2/3}.
\end{aligned}$$

It follows from (1.3.6), (1.3.7) that $m_{ij} = l_{ij}$ for all i, j. So we have 5 equations which provide the expressions for s_1, s_2, s_3, u_p, v_p in terms of z_1, z_2, z_3, u, v. This expressions are stated in part (a). To prove part (b) one can replace p by $\tilde{p}$ in all the formulae (1.3.4),...,(1.3.7), and thus to get equations determining $\tilde{p}w$ similar to that for pw. However, it is simpler to notice that $\tilde{p} = {}^{\iota}p$, and thus to get $\tilde{p}w = {}^{\iota}pw = {}^{\iota}(p^{\iota}w)$. One can find $p^{\iota}w$ by means of the formulae of part (a) applyied to ${}^{\iota}w$ (given in 1.3.2) instead of w. Then, applying ι to $p^{\iota}w$ one can get the formulae of part (b). ∎

1.3.6 The mappings Q and $\tilde{Q}$. The mappings $Q\colon \mathbf{X} \to \mathbf{H}$ and $\tilde{Q}\colon \mathbf{X} \to \mathbf{H}$ are defined in 1.3.1 by the formulae

$$Q(w) = (z_3, v), \qquad \tilde{Q}(w) = (z_1, u).$$

We have that $\mathrm{SL}(2, \mathbf{C})$ acts on $\mathbf{H}$, and $\mathrm{SL}(3, \mathbf{C})$ acts on $\mathbf{X}$. These actions are related one with another as it is described in the following propositions.

Proposition 1.3.3 *Let* $\gamma = h_3(\alpha)$, $\alpha = \begin{pmatrix} a & b \\ c & d \end{pmatrix} \in \mathrm{SL}(2, \mathbf{C})$. *Then we have* $(U^2V)(\gamma w) = u^2v$ *and*

$$Q(\gamma w) = \begin{pmatrix} a & b \\ c & d \end{pmatrix} Q(w), \quad \begin{pmatrix} Z_2(\gamma w) \\ Z_1(\gamma w) \end{pmatrix} = \begin{pmatrix} a & b \\ c & d \end{pmatrix} \begin{pmatrix} z_2 \\ z_1 \end{pmatrix}. \qquad \square$$

Proposition 1.3.4 *Let* $\gamma = h_1(\alpha)$, $\alpha = \begin{pmatrix} a & b \\ c & d \end{pmatrix} \in \mathrm{SL}(2,\mathbf{C})$. *Then we have* $(UV^2)(\gamma w) = uv^2$ *and*

$$\widetilde{Q}(\gamma w) = \begin{pmatrix} a & b \\ c & d \end{pmatrix} \widetilde{Q}(w), \quad \begin{pmatrix} (-Z_2 + Z_1 Z_3)(\gamma w) \\ Z_3(\gamma w) \end{pmatrix} = \begin{pmatrix} a & b \\ c & d \end{pmatrix} \begin{pmatrix} -z_2 + z_1 z_3 \\ z_3 \end{pmatrix}. \quad \square$$

Proof (of Propositions 1.3.3, 1.3.4). If $c \neq 0$, express α as the product $\delta_1\delta_2\delta_3\delta_4$ with

$$\delta_1 = \begin{pmatrix} 1 & a/c \\ 0 & 1 \end{pmatrix}, \quad \delta_2 = \begin{pmatrix} 1/c & 0 \\ 0 & c \end{pmatrix}, \quad \delta_3 = \begin{pmatrix} 0 & -1 \\ 1 & 0 \end{pmatrix}, \quad \delta_4 = \begin{pmatrix} 1 & d/c \\ 0 & 1 \end{pmatrix}.$$

After this we have: $\gamma = \varepsilon_1\varepsilon_2\varepsilon_3\varepsilon_4$ with $\varepsilon_i = h_j(\delta_i)$, $j = 3$ for the case of Proposition 1.3.3 and $j = 1$ for the case of Proposition 1.3.4. An action of the matrices $\varepsilon_1, \varepsilon_2, \varepsilon_4$ on $\mathbf{X}$ is described in Proposition 1.3.1. Now notice that $\varepsilon_3 = p$ for the case of Proposition 1.3.3 and that $\varepsilon_3 = \tilde{p}\,\mathrm{diag}(1,-1,-1)$ for the case of Proposition 1.3.4. An action of the matrices p and $\tilde{p}$ is described in Proposition 1.3.2. Applying successively the formulae of these propositions we obtain the desired formulae. If $c = 0$, express α as the product

$$\alpha = \begin{pmatrix} 1 & b/d \\ 0 & 1 \end{pmatrix} \begin{pmatrix} a & 0 \\ 0 & d \end{pmatrix}$$

and do as in previous case. ∎

1.3.7 Eisenstein series. Let $\tilde{\psi}\colon \Gamma(q) \to \mathbf{C}^*$ be an unitary character. Let us assume that its kernel is a subgroup of finite index in $\Gamma(q)$ containing $\Delta = \Gamma(q) \cap N$. Then, let us define characters $\tilde{\kappa}'\colon \Gamma_2(q) \to \mathbf{C}^*$ and $\tilde{\kappa}''\colon \Gamma_2(q) \to \mathbf{C}^*$ setting

$$\tilde{\kappa}'(\gamma) = \tilde{\psi}(h_3(\gamma)), \quad \tilde{\kappa}''(\gamma) = \tilde{\psi}(h_1(\gamma)) \quad \text{for any } \gamma \in \Gamma_2(q).$$

Given an automorphic form $f\colon \mathbf{H} \to \mathbf{C}$ under $\Gamma_2(q)$ with multiplier system $\tilde{\kappa}'$ or $\tilde{\kappa}''$, one can attach to f the series

$$\mathrm{E}(w,s;f) = \sum_{\gamma} \bar{\tilde{\psi}}(\gamma) U(\gamma w)^{2s} V(\gamma w)^s f(Q(\gamma w)), \tag{1.3.8}$$

or, respectively,

$$\widetilde{\mathrm{E}}(w,s;f) = \sum_{\gamma} \bar{\tilde{\psi}}(\gamma) U(\gamma w)^{s} V(\gamma w)^{2s} f(\widetilde{Q}(\gamma w)), \tag{1.3.9}$$

$w \in \mathbf{X}$, $s \in \mathbf{C}$. The summation is carried out over a set of matrices $\gamma \in \Gamma(q)$ that contains a unique representative of each coset in $\Delta_\times\backslash\Gamma(q)$ or $\widetilde{\Delta}_\times\backslash\Gamma(q)$, according to the case (1.3.8) or (1.3.9). In both cases the terms do not depend on the choice of representatives. To prove, notice first that $\Delta_\times = \Delta h_3(\Gamma_2(q))$ and $\widetilde{\Delta}_\times = \Delta h_1(\Gamma_2(q))$, see 1.2.2. Recall that Δ consists of upper triangular unipotent matrices only. So we have:

$(U^2V)(\delta w) = (U^2V)(w)$ for $\delta \in \Delta$ (by Proposition 1.3.1), for $\delta \in h_3(\Gamma_2(q))$ (by Proposition 1.3.3), and thus for all $\delta \in \Delta_\times$;

$(UV^2)(\delta w) = (UV^2)(w)$ for $\delta \in \Delta$ (by Proposition 1.3.1), for $\delta \in h_1(\Gamma_2(q))$ (by Proposition 1.3.4), and thus for all $\delta \in \widetilde{\Delta}_\times$.

At the same time:

For (1.3.8), expressing $\delta \in \Delta_\times$ as the product $\delta' h_3(\lambda)$ with $\delta' = n(*,*,0) \in \Delta$, $\lambda \in \Gamma_2(q)$ (see 1.2.2), we find $\tilde{\psi}(\delta) = \tilde{\kappa}'(\lambda)$, and then $f(Q(\delta w)) = f(\lambda Q(w)) = \tilde{\kappa}'(\lambda) f(Q(w))$ (see Proposition 1.3.1, Proposition 1.3.3), so we have $\overline{\tilde{\psi}}(\delta) f(Q(\delta w)) = f(Q(w))$ for all $\delta \in \Delta_\times$;

For (1.3.9), expressing $\delta \in \widetilde{\Delta}_\times$ as the product $\delta' h_1(\lambda)$ with $\delta' = n(0,*,*) \in \Delta$, $\lambda \in \Gamma_2(q)$ (see 1.2.2), we find $\tilde{\psi}(\delta) = \tilde{\kappa}''(\lambda)$, and then $f(\widetilde{Q}(\delta w)) = f(\lambda \widetilde{Q}(w)) = \tilde{\kappa}''(\lambda) f(\widetilde{Q}(w))$ (see Proposition 1.3.1, Proposition 1.3.4), so we have $\overline{\tilde{\psi}}(\delta) f(\widetilde{Q}(\delta w)) = f(\widetilde{Q}(w))$ for all $\delta \in \widetilde{\Delta}_\times$.

Thus it is proved that the series (1.3.8), (1.3.9) are defined correctly. These series are known in Selberg–Langlands theory as the maximal parabolic Eisenstein series attached to the form f and to the maximal parabolic subgroups P and $\widetilde{P}$. Let us now define one more series

$$\mathrm{E}_{\min}(w; s,t) = \sum_{\gamma} \overline{\tilde{\psi}}(\gamma) U(\gamma w)^{2s+t} V(\gamma w)^{2t+s}, \quad w \in \mathbf{X}, \quad s \in \mathbf{C}, \qquad (1.3.10)$$

known as the minimal parabolic Eisenstein series. Here the summation is carried out over a set of matrices $\gamma \in \Gamma(q)$ that contains a unique representative of each coset in $\Delta \backslash \Gamma(q)$. Just like as above, one can prove the terms of the series do not depend on the choice of representatives.

The next theorem is known from the general Eisenstein series theory [59].

Theorem 1.3.5 (a) *The minimal parabolic Eisenstein series* (1.3.10) *converges absolutely and locally uniformly in* $\mathbf{X} \times \{(s,t) \in \mathbf{C}^2 \mid \Re(s) > \tau,\ \Re(t) > \tau\}$ *for some* $\tau \in \mathbf{R}$. *For each* $w \in \mathbf{X}$ *the function* $\mathrm{E}_{\min}(w;\cdot,\cdot)$ *is regular in* $\{(s,t) \in \mathbf{C}^2 \mid \Re(s) > \tau,\ \Re(t) > \tau\}$ *and can be extended meromorphically to* $\mathbf{C}^2$.

(b) *The maximal parabolic Eisenstein series* (1.3.8), (1.3.9) *converge absolutely and locally uniformly in* $\mathbf{X} \times \{s \in \mathbf{C} \mid \Re(s) > \tau\}$ *for some* $\tau \in \mathbf{R}$. *Given* $w \in \mathbf{X}$, *the functions* $\mathrm{E}(w,\cdot;f)$, $\widetilde{\mathrm{E}}(w,\cdot;f)$ *are regular in* $\{s \in \mathbf{C} \mid \Re(s) > \tau\}$ *and can be extended meromorphically to* $\mathbf{C}$.

(c) *All the functions* $\mathrm{E}(\cdot,s;f)$, $\widetilde{\mathrm{E}}(\cdot,s;f)$ *and* $\mathrm{E}_{\min}(\cdot;s,t)$ *are automorphic forms on* $\mathbf{X}$ *under* $\Gamma(q)$ *with multiplier system* $\tilde{\psi}$, *whenever* s *and* (s,t) *are regular points.* ∎

It is seen from Proposition 1.2.8 (a)† that one can take $\tilde{\psi}$ to be the Bass–

† Notice also that the kernel of $\psi|_{\Gamma(q)}$ is a subgroup of finite index ($= 3$) in $\Gamma(q)$.

Milnor–Serre ψ, and thus to get the well defined cubic metaplectic minimal parabolic Eisenstein series (1.3.10). With such a choice of $\tilde{\psi}$, we get that both $\tilde{\kappa}'$ and $\tilde{\kappa}''$ are nothing but Kubota's κ, see Proposition 1.2.9. So, given a cubic metaplectic form $f\colon \mathbf{H} \to \mathbf{C}$ under $\Gamma_2(q)$, we have two well defined maximal parabolic cubic metaplectic Eisenstein series (1.3.8), (1.3.9) attached to f. Some elementary relations satisfied by the cubic metaplectic Eisenstein series are collected in the next theorem.

Theorem 1.3.6 *Let $f\colon \mathbf{H} \to \mathbf{C}$ be a cubic metaplectic form under the group $\Gamma_2(q)$, and let $\mathrm{E}_*(\cdot,t)\colon \mathbf{H} \to \mathbf{C}$ be the Eisenstein series defined in* 0.3.5, *under $\Gamma_2(q)$ too. Let $\mathrm{E}(\cdot,s;f)$, $\widetilde{\mathrm{E}}(\cdot,s;f)$ and $\mathrm{E}_{\min}(\cdot;t,s)$ be the above defined cubic metaplectic Eisenstein series on $\mathbf{X}$, all under the group $\Gamma(q)$. Let ι denotes the involutions defined in* 0.3.9, 1.1.5, 1.3.2. *Then the relations*

(a) $\mathrm{E}_{\min}({}^{\iota}w;s,t) = \mathrm{E}_{\min}(w;t,s)$,

(b) $\widetilde{\mathrm{E}}({}^{\iota}w,s;f) = \mathrm{E}(w,s;{}^{\iota}f)$,

(c) $\mathrm{E}\big(w,s;\mathrm{E}_*(\cdot,t)\big) = \mathrm{E}_{\min}(w;s-t/3,2t/3)$,

(d) $\widetilde{\mathrm{E}}\big(w,s;\mathrm{E}_*(\cdot,t)\big) = \mathrm{E}_{\min}(w;2t/3,s-t/3)$

are valid for all $w \in \mathbf{X}$, and for all $s,t \in \mathbf{C}$ whenever the series are absolutely convergent. □

Proof. (a) By definition (1.3.10) we have

$$\mathrm{E}_{\min}({}^{\iota}w;s,t) = \sum_{\gamma\in\nabla} \overline{\psi}(\gamma)U(\gamma\,{}^{\iota}w)^{2s+t}V(\gamma\,{}^{\iota}w)^{2t+s}, \tag{1.3.11}$$

where ∇ is any subset of $\Gamma(q)$ which contains a unique representative of each coset in $\Delta\backslash\Gamma(q)$. For such $\nabla \subset \Gamma(q)$, the set ${}^{\iota}\nabla = \{\,{}^{\iota}\gamma \mid \gamma \in \nabla\,\}$ also contains a unique representative of each coset in $\Delta\backslash\Gamma(q)$. In fact, two matrices $\gamma,\varepsilon \in \Gamma(q)$ lie in the same coset in $\Delta\backslash\Gamma(q)$ if and only if $\gamma = \delta\varepsilon$ with some $\delta \in \Delta$; hence if and only if ${}^{\iota}\gamma = {}^{\iota}\delta\,{}^{\iota}\varepsilon$ with some $\delta \in \Delta$, and it remains to notice that Δ is invariant under ι. This allows us to rewrite (1.3.11) as

$$\mathrm{E}_{\min}({}^{\iota}w;s,t) = \sum_{\gamma\in\nabla} \overline{\psi}({}^{\iota}\gamma)U({}^{\iota}\gamma\,{}^{\iota}w)^{2s+t}V({}^{\iota}\gamma\,{}^{\iota}w)^{2t+s}. \tag{1.3.12}$$

By Proposition 1.2.8 (c) we have $\psi({}^{\iota}\gamma) = \psi(\gamma)$. Also, see 1.3.2, we have ${}^{\iota}\gamma\,{}^{\iota}w = {}^{\iota}(\gamma w)$ and $U\big({}^{\iota}(\gamma w)\big) = V(\gamma w)$, $V\big({}^{\iota}(\gamma w)\big) = U(\gamma w)$. So, (1.3.12) yields

$$\mathrm{E}_{\min}({}^{\iota}w;s,t) = \sum_{\gamma\in\nabla} \overline{\psi}(\gamma)V(\gamma w)^{2s+t}U(\gamma w)^{2t+s} = \mathrm{E}_{\min}(w;t,s).$$

(b) By definition (1.3.9) we have

$$\widetilde{\mathrm{E}}({}^{\iota}w,s;f) = \sum_{\gamma\in\nabla} \overline{\psi}(\gamma)(UV^2)(\gamma\,{}^{\iota}w)^{s}f(\widetilde{Q}(\gamma\,{}^{\iota}w)), \tag{1.3.13}$$

where ∇ is any subset of $\Gamma(q)$ which contains a unique representative of each coset in $\widetilde{\Delta}_\times\backslash\Gamma(q)$. Obviously, ${}^\iota\gamma \in \nabla$ if and only if $\gamma \in {}^\iota\nabla$; hence one can rewrite (1.3.13) as

$$\widetilde{\mathrm{E}}({}^\iota w, s; f) = \sum_{\gamma\in{}^\iota\nabla} \overline{\psi}({}^\iota\gamma)(UV^2)({}^\iota\gamma{}^\iota w)^s f(\widetilde{Q}({}^\iota\gamma{}^\iota w)). \qquad (1.3.14)$$

Two matrices $\gamma, \varepsilon \in \Gamma(q)$ lie in the same coset in $\widetilde{\Delta}_\times\backslash\Gamma(q)$ if and only if $\gamma = \delta\varepsilon$ with some $\delta \in \widetilde{\Delta}_\times$; hence, if and only if ${}^\iota\gamma = {}^\iota\delta{}^\iota\varepsilon$ with some $\delta \in \widetilde{\Delta}_\times$. Since ${}^\iota\widetilde{\Delta}_\times = \Delta_\times$, this means that ${}^\iota\nabla$ contains a unique representative of each coset in $\Delta_\times\backslash\Gamma(q)$. Then, as in (a), we have $\psi({}^\iota\gamma) = \psi(\gamma)$, $(UV^2)({}^\iota\gamma{}^\iota w) = (U^2V)(\gamma w)$. Also, see 1.3.2 and 1.3.6, we have

$$\begin{aligned}\widetilde{Q}({}^\iota\gamma{}^\iota w) = \widetilde{Q}({}^\iota(\gamma w)) &= \big(Z_1({}^\iota(\gamma w)), U({}^\iota(\gamma w))\big) \\ &= \big(-Z_3(\gamma w), V(\gamma w)\big) = {}^\iota\big(Z_3(\gamma w), V(\gamma w)\big) = {}^\iota Q(\gamma w),\end{aligned}$$

and thus $f(\widetilde{Q}({}^\iota\gamma{}^\iota w)) = f({}^\iota Q(\gamma w)) = {}^\iota f(Q(\gamma w))$. So, (1.3.14) implies

$$\widetilde{\mathrm{E}}({}^\iota w, s; f) = \sum_{\gamma\in{}^\iota\nabla} \overline{\psi}(\gamma)(U^2V)(\gamma w)^s\, {}^\iota f(Q(\gamma w)) = \mathrm{E}(w, s; {}^\iota f).$$

(c) Let Λ_* be a subset of $\Gamma_2(q)$ which contains a unique representative of each coset in $\Delta_*\backslash\Gamma_2(q)$, where Δ_* is the subgroup of upper triangular matrices in $\Gamma_2(q)$. It is easy to find, see Proposition 1.2.1, that the set $\Lambda = h_3(\Lambda_*)$ contains a unique representative of each coset in $\Delta\backslash\Delta_\times$. So, we have

$$\sum_{\lambda\in\Lambda} \overline{\psi}(\lambda)V(\lambda w)^t = \sum_{\delta\in\Lambda_*} \overline{\kappa}(\delta)V\big(h_3(\delta)w\big)^t = \mathrm{E}_*\big(Q(w), t\big),$$

by definition (see 0.3.4) of the Eisenstein series E_*, Proposition 1.2.8 (b) and Proposition 1.3.3. Now, let Λ' be a subset of $\Gamma(q)$ which contains a unique representative of each coset in $\Delta_\times\backslash\Gamma(q)$. Set $\nabla = \{\,\lambda\lambda' \mid \lambda \in \Lambda,\ \lambda' \in \Lambda'\}$ and notice that a mapping $\Lambda \times \Lambda' \to \nabla$ defined by $(\lambda, \lambda') \mapsto \lambda\lambda'$ is bijective and that ∇ contains a unique representative of each coset in $\Delta\backslash\Gamma(q)$. Also notice that, by Proposition 1.3.3, $(U^2V)(\lambda\lambda' w) = (U^2V)(\lambda' w)$. These notices yield

$$\begin{aligned}\mathrm{E}_{\min}(w; s - t/3, 2t/3) &= \sum_{\lambda'\in\Lambda',\, \lambda\in\Lambda} \overline{\psi}(\lambda\lambda')U(\lambda\lambda' w)^{2s}V(\lambda\lambda' w)^{s+t} \\ &= \sum_{\lambda'\in\Lambda'} \overline{\psi}(\lambda')(U^2V)(\lambda' w)^s \sum_{\lambda\in\Lambda} \overline{\psi}(\lambda)V(\lambda\lambda' w)^t \\ &= \sum_{\lambda'\in\Lambda'} \overline{\psi}(\lambda')(U^2V)(\lambda' w)^s \mathrm{E}_*\big(Q(\lambda' w), t\big) = \mathrm{E}(w, s; \mathrm{E}_*(\cdot, t)).\end{aligned}$$

(d) This can be deduced from (c) by using parts (a) and (b). In fact, for

$f = \mathrm{E}_*(\cdot, t)$ one has ${}^{\iota}f = f$, by Proposition 0.3.11 and thus

$$\begin{aligned}\widetilde{\mathrm{E}}(w,s;f) &= \mathrm{E}({}^{\iota}w,s;{}^{\iota}f) = \mathrm{E}({}^{\iota}w,s;f)\\ &= \mathrm{E}_{\mathrm{min}}({}^{\iota}w;s-t/3,2t/3) = \mathrm{E}_{\mathrm{min}}(w;2t/3,s-t/3). \qquad \blacksquare\end{aligned}$$

The analytic continuation principle yields immediately parts (a) and (b) of Theorem 1.3.6 are valid without any restriction on s and t. Also, parts (c) and (d) are usefull to prove part (a) of Theorem 1.3.5. We should say the significance of the Eisenstein series is based mainly on their role in the description of the spectrum of the invariant differential operators. Also, we should say the Eisenstein series satisfy some functional equations. However, we do not need these deep and significant points, because our interests are concentrated only on some very specific functions which are residues of the Eisenstein series at their poles. We refer to Selberg [87], [88], Langlands [59], Harish-Chandra [32] for the Eisenstein series theory.

We are interested in one more elementary observation.

Proposition 1.3.7 *Let $f\colon \mathbf{H} \to \mathbf{C}$ be a cubic metaplectic form under $\Gamma_2(3)$. Suppose further that f is invariant under some $\lambda \in \mathrm{SL}(2,\mathbf{Z})$, that is $f(\lambda(z,v)) = f(z,v)$ for all $(z,v) \in \mathbf{H}$. Then for the cubic metaplectic Eisenstein series attached to f and $\Gamma(3)$ we have*

(a) $\mathrm{E}(\delta w,s;f) = \mathrm{E}(w,s;f)$ *for* $\delta = h_3(\lambda)$;

(b) $\widetilde{\mathrm{E}}(\delta w,s;f) = \widetilde{\mathrm{E}}(w,s;f)$ *for* $\delta = h_1(\lambda)$. □

Proof. (a) By analytic continuation principle it is sufficient to prove the desired formula assuming $\Re(s)$ is so large that the Eisenstein series is absolutely convergent. We have $\Gamma(3) = \Gamma^{(3)}_{\mathrm{princ}}(3)$ and $\Gamma_2(3) = \Gamma^{(2)}_{\mathrm{princ}}(3)$, that are normal subgroups of $\mathrm{SL}(3,\mathcal{O})$ and $\mathrm{SL}(2,\mathcal{O})$, respectively. This yields further $\delta\Gamma(3)\delta^{-1} = \Gamma(3)$ and $\delta\Delta_\times\delta^{-1} = \Delta_\times$. Now, by definition (1.3.8), assuming $\Re(s)$ is sufficiently large we have

$$\mathrm{E}(\delta w,s;f) = \sum_{\gamma\in\Omega} \overline{\psi}(\gamma)(U^2V)(\gamma\delta w)^s f(Q(\gamma\delta w)),$$

where $\Omega \subset \Gamma(3)$ contains a unique representative of each coset in $\Delta_\times\backslash\Gamma(3)$. For such Ω, the set $\delta\Omega\delta^{-1}$ also contains a unique representative of each coset in $\Delta_\times\backslash\Gamma(3)$, just because of $\delta\Gamma(3)\delta^{-1} = \Gamma(3)$ and $\delta\Delta_\times\delta^{-1} = \Delta_\times$. So, one can replace γ by $\delta\gamma\delta^{-1}$ in the right-hand side to obtain

$$\mathrm{E}(\delta w,s;f) = \sum_{\gamma\in\Omega} \overline{\psi}(\delta\gamma\delta^{-1})(U^2V)(\delta\gamma w)^s f(Q(\delta\gamma w)), \qquad \text{where}$$

$\psi(\delta\gamma\delta^{-1}) = \psi(\gamma)$, by Theorem 0.2.2 (f);

$(U^2V)(\delta\gamma w) = (U^2V)(\gamma w)$, by Proposition 1.3.3;

$f(Q(\delta\gamma w)) = f(Q(\gamma w))$, by Proposition 1.3.3 and the assumptions.

So, the right-hand side is equal to

$$\sum_{\gamma\in\Omega} \overline{\psi}(\gamma)(U^2V)(\gamma w)^s f(Q(\gamma w)) = \mathrm{E}(w,s;f), \quad \text{as required.}$$

(b) This can be proved by quite similar arguments, or can be deduced from (a) by means of Theorem 1.3.6 (b), let us omit the details. ∎

1.3.8 Fourier expansion according to Piatetski-Shapiro and Shalika. In this subsection we shall specialize and adapt to the case of the metaplectic forms on $\mathbf{X}$ the construction of the Fourier expansions, which was proposed originally by Piatetski-Shapiro [75] and Shalika [90] in context of adelic automorphic forms on general linear groups.

The main statement will be given in Theorem 1.3.9 below. Befor we need some notations and preliminary considerations. Let, as usual, $q \subset (3)$ be an ideal of $\mathcal{O}$, e_2 denotes 2×2 identity matrix, $w = (z_1, z_2, z_3, u, v) \in \mathbf{X}$. Let

$$\Delta' = \text{the group of upper triangular matrices } \begin{pmatrix} * & * \\ 0 & * \end{pmatrix} \in \mathrm{SL}(2,\mathcal{O}),$$

$$\widetilde{\nabla} \subset \mathrm{SL}(2,\mathcal{O}) \text{ contains a unique representative of each coset in } \mathrm{SL}(2,\mathcal{O})/\Delta',$$

$$\widehat{\delta} = \begin{pmatrix} \delta & & 0 \\ & & 0 \\ 0 & 0 & 1 \end{pmatrix} \quad \text{for any } \delta \in \mathrm{SL}(2,\mathbf{C}).$$

The next theorem gives us preliminary Fourier expansions, instructive and interesting in their own right.

Theorem 1.3.8 *Let $F\colon \mathbf{X} \to \mathbf{C}$ be an infinitely differentiable function satisfying*

$$F(\lambda w) = F(w) \quad \textit{for all } \lambda = \begin{pmatrix} 1 & 0 & l_2 \\ 0 & 1 & l_1 \\ 0 & 0 & 1 \end{pmatrix}, \quad l_1, l_2 \in q, \tag{1.3.15}$$

and let $\delta \in \mathrm{SL}(2,\mathcal{O})$. Then one has

(a) $F(\widehat{\delta} w) = \sum_{m,n\in q^*} \widetilde{C}^{\delta}_{m,n}(z_3, v; u^2 v) e(mz_2 + nz_1)$, *where*†

$$\widetilde{C}^{\delta}_{m,n}(z_3, v; u^2v) = \mathrm{vol}(\mathbf{C}/q)^{-2} \int_{\mathbf{C}/q}\int_{\mathbf{C}/q} F(\widehat{\delta} w)\overline{e(mz_2 + nz_1)}\, dz_1\, dz_2; \tag{1.3.16}$$

(b) $\widetilde{C}^{\delta}_{m,n}(\xi(z_3, v); u^2 v) = \widetilde{C}^{\delta\xi}_{(m,n)\xi}(z_3, v; u^2 v)$ *for all* $m, n \in q^*$, $\xi \in \mathrm{SL}(2,\mathcal{O})$;

† The coefficients are the functions of z_3, u, v. But, for further convenience, we take as variable $u^2 v$ instead of u, and we consider the pair (z_3, v) as a point in $\mathbf{H}$. Notice, see Proposition 1.3.3, that the function $w \mapsto u^2 v$ is invariant under $\widehat{\delta}$ for any $\delta \in \mathrm{SL}(2,\mathbf{C})$.

(c) $F(w) = \widetilde{C}^{e_2}_{0,0}(z_3, v; u^2v) + \sum_{\xi\in\widetilde{\nabla}} \sum_{\mu\in q^*\setminus\{0\}} \widetilde{C}^{\xi}_{0,\mu}(\xi^{-1}(z_3,v); u^2v)\, e\big(\mu Z_1(\widehat{\xi}^{-1}w)\big).$

□

Proof. For any λ in (1.3.15) we have $F(\widehat{\delta}\lambda w) = F((\widehat{\delta}\lambda\widehat{\delta}^{-1})\widehat{\delta}w) = F(\widehat{\delta}w)$, because of

$$\widehat{\delta}\lambda\widehat{\delta}^{-1} = \begin{pmatrix} 1 & 0 & l_2' \\ 0 & 1 & l_1' \\ 0 & 0 & 1 \end{pmatrix} \quad \text{with some} \quad l_1', l_2' \in q,$$

We have also $\lambda w = (z_1 + l_1, z_2 + l_2, z_3, u, v)$, see Proposition 1.3.1 (a). These observations show $(z_1, z_2) \mapsto F(\widehat{\delta}w)$ is a periodic function with the lattice of periods $q \times q$, and thus we have (a).

To prove (b), we take the expansion (a) first with $\delta\xi$ instead of δ, and then with $\tilde{w} = \widehat{\xi}w$ instead of w. Thus we get

$$F\big((\widehat{\delta\xi})w\big) = \sum_{m,n\in q^*} \widetilde{C}^{\delta\xi}_{m,n}(z_3, v; u^2v)\, e(mz_2 + nz_1), \tag{1.3.17}$$

$$F(\widehat{\delta}\tilde{w}) = \sum_{m,n\in q^*} \widetilde{C}^{\delta}_{m,n}\big(Z_3(\tilde{w}), V(\tilde{w}); (U^2V)(\tilde{w})\big)\, e\big(mZ_2(\tilde{w}) + nZ_1(\tilde{w})\big). \tag{1.3.18}$$

By Proposition 1.3.3 we have

$$Z_2(\tilde{w}) = az_2 + bz_1, \quad (Z_3(\tilde{w}), V(\tilde{w})) = \xi(z_3, v),$$
$$Z_1(\tilde{w}) = cz_2 + dz_1, \quad (U^2V)(\tilde{w}) = u^2v, \quad \text{if} \quad \xi = \begin{pmatrix} a & b \\ c & d \end{pmatrix},$$

and thus we can rewrite (1.3.18) as

$$F(\widehat{\delta}(\widehat{\xi}w)) = \sum_{m,n\in q^*} \widetilde{C}^{\delta}_{m,n}(\xi(z_3, v); u^2v)\, e\big((am + nc)z_2 + (bm + dn)z_1\big). \tag{1.3.19}$$

Since $(\widehat{\delta\xi})w = \widehat{\delta}(\widehat{\xi}w)$, we have two Fourier expansions (1.3.17) and (1.3.19) for just one function. Equating the coefficients we get (b).

To get (c) we take the expansion (a) with $\delta = e_2$, i.e.

$$F(w) = \widetilde{C}^{e_2}_{0,0}(z_3, v; u^2v) + \sum_{\substack{m,n\in q^* \\ (m,n)\neq(0,0)}} \widetilde{C}^{e_2}_{m,n}(z_3, v; u^2v) e(mz_2 + nz_1), \tag{1.3.20}$$

and then we apply (b) and Proposition 1.3.3 to find that

$$\widetilde{C}^{e_2}_{m,n}(z_3, v; u^2v) = \widetilde{C}^{\xi}_{0,\mu}(\xi^{-1}(z_3, v); u^2v) \quad \text{and} \quad mz_2 + nz_1 = \mu Z_1(\widehat{\xi}^{-1}w)$$

if $\mu \in q^*$ and $\xi \in \mathrm{SL}(2, \mathcal{O})$ are so that $(0, \mu) = (m, n)\xi$. Then we should notice only that the mapping $(\mu, \xi) \mapsto (m, n) = (0, \mu)\xi^{-1}$ is a bijection

$$\{(\mu,\xi) \mid \mu \in q^*,\ \mu \neq 0,\ \xi \in \widetilde{\nabla}\} \to \{(m,n) \mid m,n \in q^*,\ (m,n) \neq (0,0)\}. \quad \blacksquare$$

For any $\delta \in \mathrm{SL}(2,\mathcal{O})$ and ideal $q \subset (3)$ let us choose $h_{\delta,q} \in \mathbf{Q}(\sqrt{-3})$ such that

$$\kappa(\delta \begin{pmatrix} 1 & k \\ 0 & 1 \end{pmatrix} \delta^{-1}) = e(h_{\delta,q}k) \qquad \text{for all} \quad k \in q$$

(compair with 0.3.4). For a congruence subgroup Γ of $\mathrm{SL}(2,\mathcal{O})$ we choose $\nabla_\Gamma \subset \mathrm{SL}(2,\mathcal{O})$, and $\Omega_{\Gamma,\alpha} \subset \Gamma$, for each $\alpha \in \nabla_\Gamma$, in such a way that[†]:

∇_Γ contains a unique representative of each double coset in $\Gamma\backslash\mathrm{SL}(2,\mathcal{O})/\Delta'$; in other words, each cusp of Γ is equivalent to $\alpha\infty$ for unique $\alpha \in \nabla_\Gamma$;

$\Omega_{\Gamma,\alpha}$ contains a unique representaive of each coset in $\Gamma_{\alpha\infty}\backslash\Gamma$, where we write $\Gamma_{\alpha\infty}$ for the stabilizer in Γ of the cusps $\alpha\infty$, so $\Gamma_{\alpha\infty} = \Gamma \cap \alpha\Delta'\alpha^{-1}$.

One can easily find one has the disjoint unions

$$\mathrm{SL}(2,\mathcal{O}) = \bigcup_{\alpha\in\nabla_\Gamma,\ \gamma\in\Omega_{\Gamma,\alpha}} \gamma^{-1}\alpha\Delta' = \bigcup_{\alpha\in\nabla_\Gamma,\ \gamma\in\Omega_{\Gamma,\alpha}} \Delta'\alpha^{-1}\gamma. \tag{1.3.21}$$

Like in 1.3.3, let $\mathcal{F}_q \subset \mathbf{C}^3$ be any Lebesgue measurable set such that its image under (1.3.2) contains a unique representative of each coset in $(\Gamma^{(3)}_{\mathrm{princ}}(q) \cap N)\backslash N$.

Theorem 1.3.9 *Let $F\colon \mathbf{X} \to \mathbf{C}$ be a C^∞-function satisfying*

$$F(\lambda w) = \psi(\lambda)F(w) \qquad \textit{for all} \quad \lambda = \begin{pmatrix} * & * & * \\ * & * & * \\ 0 & 0 & 1 \end{pmatrix} \in \Gamma^{(3)}_{\mathrm{princ}}(q). \tag{1.3.22}$$

Then, for any $\delta \in \mathrm{SL}(2,\mathcal{O})$ and $\mu \in q^*$, $\nu \in h_{\delta,q} + q^*$, the integrals

$$C^{\delta}_{\mu,\nu}(u,v) = \mathrm{vol}(\mathbf{C}/q)^{-3} \iiint_{\mathcal{F}_q} F(\widehat{\delta} w)\overline{e(\mu z_1 + \nu z_3)}\, dz_1 dz_2 dz_3 \tag{1.3.23}$$

do not depend on the particular choice of $\mathcal{F}_q$. Suppose further that

$$F(\lambda w) = \psi(\lambda)F(w) \qquad \textit{for all} \quad \lambda = \begin{pmatrix} \gamma & & 0 \\ & & 0 \\ 0 & 0 & 1 \end{pmatrix}, \qquad \gamma \in \Gamma, \tag{1.3.24}$$

where Γ is a subgroup of $\widetilde{\Gamma}^{(2)}_{\mathrm{princ}}(3)$ such that $\Gamma^{(2)}_{\mathrm{princ}}(q) \subset \Gamma$. Then one has the expansion

$$F(w) = F_*(w) + \sum_{\alpha\in\nabla_\Gamma} \sum_{\gamma\in\Omega_{\Gamma,\alpha}} \overline{\psi}(\widehat{\gamma}) F_\alpha(\widehat{\alpha}^{-1}\widehat{\gamma} w), \tag{1.3.25}$$

where

$$F_*(w) = \sum_{\nu\in q^*} C^{e_2}_{0,\nu}(u,v)e(\nu z_3),$$

$$F_\alpha(w) = \sum_{\substack{\mu\in q^*\backslash\{0\} \\ \nu\in h_{\alpha,q}+q^*}} C^{\alpha}_{\mu,\nu}(u,v)e(\mu z_1 + \nu z_3). \qquad \blacksquare \tag{1.3.26}$$

---

† For needed definitions see 0.3.1.

Proof. First, let us examine the integrands in (1.3.23). For $\delta \in \mathrm{SL}(2,\mathcal{O})$ and

$$\lambda = \begin{pmatrix} 1 & l_3 & l_2 \\ 0 & 1 & l_1 \\ 0 & 0 & 1 \end{pmatrix}, \qquad l_1, l_2, l_3 \in q, \tag{1.3.27}$$

one has

$$\widehat{\delta}\lambda\widehat{\delta}^{-1} = \begin{pmatrix} \lambda' & & * \\ & & * \\ 0 & 0 & 1 \end{pmatrix} \in \Gamma^{(3)}_{\text{princ}}(q) \quad \text{with} \quad \lambda' = \delta \begin{pmatrix} 1 & l_3 \\ 0 & 1 \end{pmatrix} \delta^{-1},$$

and consequently† $\psi(\widehat{\delta}\lambda\widehat{\delta}^{-1}) = \kappa(\lambda') = e(h_{\delta,q} l_3)$. One more remark is $e(\nu l_3) = e(h_{\delta,q} l_3)$ for $\nu \in h_{\delta,q} + q^*$, $l_3 \in q$. After all, and in view of Proposition 1.3.1 (a), we get

$$\begin{aligned} F(\widehat{\delta}\lambda w) = F((\widehat{\delta}\lambda\widehat{\delta}^{-1})\widehat{\delta}w) &= \psi(\widehat{\delta}\lambda\widehat{\delta}^{-1})F(\widehat{\delta}w) \\ &= e(h_{\delta,q} l_3)F(\widehat{\delta}w), \\ e\left(\mu Z_1(\lambda w) + \nu Z_3(\lambda w)\right) &= e\left(\mu(z_1 + l_1) + \nu(z_3 + l_3)\right) \\ &= e(h_{\delta,q} l_3)e(\mu z_1 + \nu z_3). \end{aligned} \tag{1.3.28}$$

for any λ in (1.3.27), $\delta \in \mathrm{SL}(2,\mathcal{O})$, $\mu \in q^*$, $\nu \in h_{\delta,q} + q^*$. In its turn, (1.3.28) yields the invariance of the integrands in (1.3.23) under the matrix in (1.3.27), and thus the independance of the integrals (1.3.23) on the choice of $\mathcal{F}_q$.

To get the expansion (1.3.25) we first notice that our present assumptions (1.3.22) imply (1.3.15), because of $\psi(\lambda) = 1$ for λ in (1.3.15) (see Proposition 1.2.8 (a)), and thus we have for F the expansion of Theorem 1.3.8 (c). It is clear from (1.3.21), we can take $\widetilde{\nabla}$ consisting of the matrices $\xi = \gamma^{-1}\alpha$ with $\alpha \in \nabla_\Gamma$, $\gamma \in \Omega_{\Gamma,\alpha}$, and then to rewrite the expansion of Theorem 1.3.8 (c) as

$$\begin{aligned} F(w) &= \widetilde{C}^{e_2}_{0,0}(z_3, v; u^2 v) \\ &+ \sum_{\alpha \in \nabla_\Gamma} \sum_{\gamma \in \Omega_{\Gamma,\alpha}} \sum_{\mu \in q^* \setminus \{0\}} \widetilde{C}^{\gamma^{-1}\alpha}_{0,\mu}(\alpha^{-1}\gamma(z_3, v); u^2 v)\, e(\mu Z_1(\widehat{\alpha}^{-1}\widehat{\gamma}w)). \end{aligned} \tag{1.3.29}$$

Next, just from the definition (1.3.16) and from (1.3.22) it follows

$$\widetilde{C}^{\gamma^{-1}\alpha}_{0,\mu} = \overline{\psi}(\widehat{\gamma})\widetilde{C}^{\alpha}_{0,\mu} \tag{1.3.30}$$

for any $\gamma \in \Gamma$, $\alpha \in \mathrm{SL}(2,\mathcal{O})$, $\mu \in q^*$. Thus we can rewrite (1.3.29) as in (1.3.25) with

$$\begin{aligned} F_*(w) &= \widetilde{C}^{e_2}_{0,0}(z_3, v; u^2 v), \\ F_\alpha(w) &= \sum_{\mu \in q^* \setminus \{0\}} \widetilde{C}^{\alpha}_{0,\mu}(z_3, v; u^2 v) e(\mu z_1), \end{aligned} \tag{1.3.31}$$

† See also Proposition 1.2.8 (b) or Theorem 0.2.2 (b).

and it remains only to transform the right-hand sides of (1.3.31) to the form stated in (1.3.26). For this let us look at the functions $z_3 \mapsto \widetilde{C}^{\alpha}_{0,\mu}(z_3, v; u^2 v)$, with any $\alpha \in \mathrm{SL}(2,\mathcal{O})$, $\mu \in q^*$. (The case $\alpha = e_2$, $\mu = 0$ is not excluded. One can take $h_{e_2,q} = 0$.) Let

$$\xi = \begin{pmatrix} 1 & l \\ 0 & 1 \end{pmatrix}, \quad l \in q, \tag{1.3.32}$$

Applying (0.3.2), and then Theorem 1.3.8 (b) with $m = 0$, $n = \mu$, $\delta = \alpha$, and with ξ in (1.3.32) we find

$$\widetilde{C}^{\alpha}_{0,\mu}(z_3 + l, v; u^2 v) = \widetilde{C}^{\alpha}_{0,\mu}(\xi(z_3, v); u^2 v) = \widetilde{C}^{\alpha\xi}_{(0,\mu)\xi}(z_3, v; u^2 v).$$

Next, with ξ in (1.3.32) we have $(0,\mu)\xi = (0,\mu)$ and $\alpha\xi\alpha^{-1} \in \Gamma^{(2)}_{\mathrm{princ}}(q) \subset \Gamma$. We can apply (1.3.30) with $\gamma = \alpha\xi^{-1}\alpha^{-1}$ to get

$$\widetilde{C}^{\alpha\xi}_{(0,\mu)\xi} = \widetilde{C}^{\alpha\xi}_{0,\mu} = \widetilde{C}^{(\alpha\xi\alpha^{-1})\alpha}_{0,\mu} = \kappa(\alpha\xi\alpha^{-1})\widetilde{C}^{\alpha}_{0,\mu} = e(h_{\alpha,q} l)\widetilde{C}^{\alpha}_{0,\mu}.$$

Thus we have

$$\widetilde{C}^{\alpha}_{0,\mu}(z_3 + l, v; u^2 v) = e(h_{\alpha,q} l)\widetilde{C}^{\alpha}_{0,\mu}(z_3, v; u^2 v)$$

for all $l \in q$, $\alpha \in \mathrm{SL}(2,\mathcal{O})$, $\mu \in q^*$. This yields immediately Fourier expansions

$$\widetilde{C}^{\alpha}_{0,\mu}(z_3, v; u^2 v) = \sum_{\nu \in h_{\alpha,q} + q^*} C^{\alpha}_{\mu,\nu}(u, v) e(\nu z_3) \tag{1.3.33}$$

with coefficients

$$C^{\alpha}_{\mu,\nu}(u, v) = \mathrm{vol}(\mathbf{C}/q)^{-1} \int_{\mathbf{C}/q} \widetilde{C}^{\alpha}_{0,\mu}(z_3, v; u^2 v) \overline{e(\nu z_3)}\, dz_3. \tag{1.3.34}$$

Substituting (1.3.33) into (1.3.31) we get the expansions (1.3.26) with coefficients (1.3.34). In its turn, the expressions (1.3.34) agree with (1.3.23) in view of (1.3.16). ∎

Let $F\colon \mathbf{X} \to \mathbf{C}$ be a cubic metaplectic form under some congruence subgroup. By point (a) of the definition in 1.3.3 we have that F satisfies the conditions (1.3.22) and (1.3.24) with appropriately choosen q and Γ. It is naturally to choose minimal possible q in (1.3.22) and maximal possible Γ in (1.3.24). However, if we prefer to have $h_{\alpha,q} = 0$ in (1.3.26), then we can take $q \subset (9)$ and $\Gamma \subset \Gamma^{(2)}_{\mathrm{princ}}(9)$, ignoring the form F maybe satisfies the conditions (1.3.22), (1.3.24) not only for $\lambda \in \Gamma^{(3)}_{\mathrm{princ}}(9)$.

Let $F\colon \mathbf{X} \to \mathbf{C}$ be a cubic metaplectic form under $\Gamma(q)$. Its Fourier coefficients $c_{\mu,\nu}(u, v; F)$ defined by (1.3.3) are nothing but $C^{e_2}_{\mu,\nu}(u, v)$ in (1.3.23). We see that to have the Fourier expansion of F we need not only the Fourier coefficients (1.3.3) of F itself, but also the Fourier coefficients of the forms $F_\delta\colon \mathbf{X} \to \mathbf{C}$, defined as $F_\delta(w) = F(\delta w)$, $w \in \mathbf{X}$. If $q \subset (9)$, the so defined F_δ

is a cubic metaplectic form under $\Gamma(q)$, and $C^{\delta}_{\mu,\nu}(u,v)$ in (1.3.23) is just the Fourier coefficient $c_{\mu,\nu}(u,v;F_\delta)$ in (1.3.3). The case $q \not\subset (9)$ is slightly more complicated.

It should be clear, one can replace Bass–Milnor–Serre's ψ in Theorem 1.3.9 by some other homomorphisms (for instance, by the trivial one; in this case we have $h_{\alpha,q} = 0$).

Again, let $F\colon \mathbf{X} \to \mathbf{C}$ be a cubic metaplectic form under some congruence subgroup. Then the Fourier coefficients in (1.3.23) as well as in (1.3.3) satisfy some differential equations, which one may get applying the invariant differential operators to F, and taking into account that F is their eigenfunction (see the definition in 1.3.3, and compare with 0.3.4). This leads to the concept of Whittaker functions, which we shall introduce in 1.3.10.

1.3.9 Invariant differential operators. It was already pointed out that the invariant differential operators algebra $\mathbf{D}(\mathbf{X})$ is isomorphic to the algebra of polynomials of 2 variabels over $\mathbf{C}$. This means, in particular, the algebra is generated by 2 operators. Also, one knows the product of the degres of the generators of $\mathbf{D}(\mathbf{X})$ is equal to the order of the Weyl group, i.e., 6. We have to take Lalace–Beltrami operator on $\mathbf{X}$ as one of generators, thus an another one should be of degree 3. Well known methods†, based on the consideration of universal enveloping algebra and related objects, allow to find explicitly the generates of $\mathbf{D}(\mathbf{X})$. However, the computation needed is very massive.

Let us turn to 0.3.1 and to notice that the Laplace–Beltrami operator on $\mathbf{H}$ can be written as

$$\mathrm{D}_{\text{L-B}} = -v^2\Big(4\frac{\partial^2}{\partial z\,\partial \bar z} + \frac{\partial^2}{\partial v^2}\Big) + v\frac{\partial}{\partial v},$$

if we adopt the commonly used notations

$$\frac{\partial}{\partial z} = \frac{1}{2}\Big(\frac{\partial}{\partial x} - i\frac{\partial}{\partial y}\Big), \qquad \frac{\partial}{\partial \bar z} = \frac{1}{2}\Big(\frac{\partial}{\partial x} + i\frac{\partial}{\partial y}\Big),$$

$x = \Re(z)$, $y = \Im(z)$. With these notations, let us further set

$$\mathrm{D}^{(1)}_{\text{L-B}} = -u^2\Big(4\frac{\partial^2}{\partial z_1\partial \bar z_1} + \frac{\partial^2}{\partial u^2}\Big) + u\frac{\partial}{\partial u},$$

$$\mathrm{D}^{(3)}_{\text{L-B}} = -v^2\Big(4\frac{\partial^2}{\partial z_3\partial \bar z_3} + \frac{\partial^2}{\partial v^2}\Big) + v\frac{\partial}{\partial v}.$$

We consider $\mathrm{D}^{(1)}_{\text{L-B}}$ and $\mathrm{D}^{(3)}_{\text{L-B}}$ as operators acting on C^∞–functions $\mathbf{X} \to \mathbf{C}$.

† See, for example, Baily [2], Bourbaki [13], Varadarajan [98].

Theorem 1.3.10 *The algebra* $\mathbf{D}(\mathbf{X})$ *is generated by the following two operators*

$$\mathrm{D}_2 = \mathrm{D}^{(1)}_{\text{L-B}} + \mathrm{D}^{(3)}_{\text{L-B}} + uv\frac{\partial^2}{\partial u\,\partial v} - 4u^2(v^2+|z_3|^2)\frac{\partial^2}{\partial z_2\partial \bar z_2} - 4u^2\Big(z_3\frac{\partial^2}{\partial z_1\partial z_2} + \bar z_3\frac{\partial^2}{\partial \bar z_1\partial \bar z_2}\Big),$$

$$\begin{aligned}\mathrm{D}_3 = &-\Big(1-\frac{1}{2}v\frac{\partial}{\partial v}\Big)\mathrm{D}^{(1)}_{\text{L-B}} + \Big(1-\frac{1}{2}u\frac{\partial}{\partial u}\Big)\mathrm{D}^{(3)}_{\text{L-B}} \\ &+ 4u^2v^2\Big(\frac{\partial^3}{\partial z_1\partial \bar z_2\partial z_3} + \frac{\partial^3}{\partial \bar z_1\partial z_2\partial \bar z_3}\Big) \\ &+ 4u^2v^2\frac{\partial^2}{\partial z_2\partial \bar z_2}\Big(z_3\frac{\partial}{\partial z_3} + \bar z_3\frac{\partial}{\partial \bar z_3}\Big) + 2u^2v(v^2-|z_3|^2)\frac{\partial^3}{\partial z_2\partial \bar z_2\partial v} \\ &- 2u^3v^2\frac{\partial^3}{\partial z_2\partial \bar z_2\partial u} - 2u^2v\Big(z_3\frac{\partial^2}{\partial \bar z_1\partial z_2} + \bar z_3\frac{\partial^2}{\partial z_1\partial \bar z_2}\Big)\frac{\partial}{\partial v} \\ &+ 4u^2(v^2+|z_3|^2)\frac{\partial^2}{\partial z_2\partial \bar z_2} + 4u^2\Big(z_3\frac{\partial^2}{\partial z_1\partial z_2} + \bar z_3\frac{\partial^2}{\partial \bar z_1\partial \bar z_2}\Big). \quad \blacksquare\end{aligned}$$

This is shown by Kapitanski and the author in the work in progress [45].

Let $s, t \in \mathbf{C}$ and let $f\colon \mathbf{H} \to \mathbf{C}$ be an eigenfunction of the Laplace–Beltrami operator, say $\mathrm{D}_{\text{L-B}} f = r(2-r)$ with some $r \in \mathbf{C}$. For $w = (z_1, z_2, z_3, u, v) \in \mathbf{X}$ we set

$$\begin{aligned}\nabla_{s,t}(w) &= u^{2s+t}v^{s+2t}, \\ \nabla_s(w; f) &= u^{2s}v^s f(z_3, v), \qquad (1.3.35) \\ \tilde\nabla_s(w; f) &= u^s v^{2s} f(z_1, u).\end{aligned}$$

It can be easily checked $\nabla_{s,t}$ is a joint eigenfunction of D_2 and D_3, and so of all the invariant differential operators. This allows us to define a character $\chi_{s,t}\colon \mathbf{D}(\mathbf{X}) \to \mathbf{C}$ setting $D\nabla_{s,t} = \chi_{s,t}(D)\nabla_{s,t}$ for $D \in \mathbf{D}(\mathbf{X})$. It is known, $\chi_{s,t}$, $s, t \in \mathbf{C}$, are all of characters of $\mathbf{D}(\mathbf{X})$. One has

$$\begin{aligned}\chi_{s,t}(\mathrm{D}_2) &= -3(s^2+t^2+st) + 6(s+t), \\ \chi_{s,t}(\mathrm{D}_3) &= -(s-t)(s^2+t^2+5st/2-3s-3t+2).\end{aligned} \qquad (1.3.36)$$

Also, one can check the functions $\nabla_s(\cdot; f)$ and $\tilde\nabla_s(\cdot; f)$ are joint eigenfunctions of $\mathbf{D}(\mathbf{X})$ attached to the characters $\chi_{s-r/3,2r/3}$ and $\chi_{2r/3,s-r/3}$, respectively.

Our observations on the functions (1.3.35) are known in the general Eisenstein series theory. In fact the translations of the functions in (1.3.35) occur as the terms in the Eisenstein series in 1.3.7, and the Eisenstein series are eigenfunctions of all the invariant differential operators just because the functions (1.3.35) are. To be more precise, we have that the Eisenstein series $\mathrm{E}_{\min}(\cdot; s, t)$,

$\mathrm{E}(\cdot, s; f)$ and $\widetilde{\mathrm{E}}(\cdot, s; f)$ are joint eigenfunctions of $\mathbf{D}(\mathbf{X})$ attached, respectively, to the characters $\chi_{s,t}$, $\chi_{s-r/3,2r/3}$ and $\chi_{2r/3,s-r/3}$.

1.3.10 Whittaker functions. Let $\chi_{s,t}$ be the character of $\mathbf{D}(\mathbf{X})$ defined in 1.3.9. Let $\mu, \nu \in \mathbf{C}$, and let $\phi_{\mu,\nu}\colon N \to \mathbf{C}^*$ be the unitary character of the group N defined as in 1.1.6. So we have $\phi_{\mu,\nu}(n(z_1, z_2, z_3)) = e(\mu z_1 + \nu z_3)$ for all $n(z_1, z_2, z_3) \in N$. We say an C^∞-function $W\colon \mathbf{X} \to \mathbf{C}$ is an Whittaker function attached to $\chi_{s,t}$ and $\phi_{\mu,\nu}$ if

(a) $W(\delta w) = \phi_{\mu,\nu}(\delta) W(w)$ for all $\delta \in N$, $w \in \mathbf{X}$;

(b) $DW = \chi_{s,t}(D) W$ for all $D \in \mathbf{D}(\mathbf{X})$;

(c) $W(z_1, z_2, z_3, u, v)$ is of at most polynomial growth as $u, v \to \infty$.

It follows immediately from (a) that W can be written uniquely as the product

$$W(w) = R(u, v) e(\mu z_1 + \nu z_3), \tag{1.3.37}$$

where $R\colon \mathbf{R}_+^* \times \mathbf{R}_+^* \to \mathbf{C}$ is a C^∞-function called the radial part of W, and, as usually, $w = (z_1, z_2, z_3, u, v) \in \mathbf{X}$.

To clarify the origin and the role of Whittaker functions, let $c_{\mu,\nu}(u, v; F)$ be the μ, ν^{th} Fourier coefficient of the form F attached to the character $\chi_{s,t}$, as it was defined in (1.3.3). Then one can show

$$w \mapsto c_{\mu,\nu}(u, v; F) e(\mu z_1 + \nu z_3) \tag{1.3.38}$$

is an Whittaker function attached to the characters $\chi_{s,t}$ and $\phi_{\mu,\nu}$. Indeed, (1.3.38) satisfies (a), as one can see from Proposition 1.3.1 (a). For (b) let us look at function

$$w \mapsto c \iiint_{\mathcal{F}_q} F(n(l) w) \overline{e(\mu l_1 + \nu l_3)}\, dl_1\, dl_2\, dl_3, \tag{1.3.39}$$

where $n(l)$ as in (1.1.1) and all other notations as in (1.3.3). Given any $D \in \mathbf{D}(\mathbf{X})$, we have D commute with the integration in (1.3.39) and with the action of $n(l)$, and thus we have the function (1.3.39) is eigenfunction of D with just the same eigenvalue as F itself. In the meantime, we have

$$\begin{aligned} n(l) w &= n(l) n(z)(0, 0, 0, u, v) \\ &= n(l_1 + z_1, l_2 + z_2 + l_3 z_1, l_3 + z_3)(0, 0, 0, u, v), \end{aligned}$$

by Proposition 1.3.1 (a). Thus, changing varibles[†] in (1.3.39) and comparing with (1.3.3) we find (1.3.39) is just the same function as (1.3.38). At last, (1.3.38) satisfies the growth condition (c) just because F satisfies the growth condition (c) in the definition given in 1.3.10.

In particular, one can take F to be the Eisenstein series to get some integral representation for Whittaker functions. This will be treated in Section 1.4.

[†] $z_1 \mapsto z_1 - l_1$, $z_2 \mapsto z_2 - l_2 - l_3 z_1$, $z_3 \mapsto z_3 - l_3$.

The simplest Whittaker function is the function $\nabla_{s,t}(w)$ defined in (1.3.35). It is attached to the characters $\chi_{s,t}$ and $\phi_{0,0}$.

The Whittaker function W is said to be non-degenerate if it is attached to the non-degenerate character $\phi_{\mu,\nu}$, i.e. if $\mu\nu \neq 0$. Otherwise, W is said to be degenerate. It is known, the radial parts of degenerate Whittaker functions attached to the characters $\phi_{\mu,0}$ and $\phi_{0,\nu}$ with $\mu\nu \neq 0$ are, in essense, the Bessel–MacDonald functions.

Let us look more closely on non-degenerate Whittaker functions. Let $R_{s,t}$ be the radial part of an Whittaker function $W_{s,t}^{1,1}$ attached to the characters $\chi_{s,t}$ and $\phi_{1,1}$. We then have the function $W_{s,t}^{\mu,\nu}$ defined as

$$W_{s,t}^{\mu,\nu}(w) = R_{s,t}(|\mu|u, |\nu|v)e(\mu z_1 + \nu z_3), \tag{1.3.40}$$

is an Whittaker function attached to the characters $\chi_{s,t}$ and $\phi_{\mu,\nu}$. To explain, let us choose diagonal matrix, say $\delta_{\mu,\nu}$, in such a way that

$$\delta_{\mu,\nu} w = (\mu z_1, \mu\nu z_2, \nu z_3, |\mu|u, |\nu|v).$$

Then one can easily check that the function $w \mapsto W_{s,t}^{1,1}(\delta_{\mu,\nu} w)$ satisfies (a), (b) and (c) of the definition above, and that this function is nothing but (1.3.40).

In the meantime, one knows (multiplicity one theorem) the conditions (a), (b) and (c) determine non-degenerate Whittaker functions uniquely up to constant factors. Thus, the study of non-degenerate Whittaker functions can be reduced to study of $R_{s,t}$ only. Also, one knows the factors above can be choosen in dependence of s and t in such a way, that, given any $u, v \in \mathbf{R}_+^*$, function $(s,t) \mapsto R_{2/3+s,2/3+t}(u,v)$ is regular on $\mathbf{C} \times \mathbf{C}$ and is invariant under the action of the Weyl group on $\mathbf{C} \times \mathbf{C}$. To describe the action we mean let us turn to 1.1.3, and let $\Phi_{s,t}\colon A \to \mathbf{C}^*$ be character defined by

$$\Phi_{s,t}(\mathrm{diag}(u^{1/3}v^{2/3}, u^{1/3}v^{-1/3}, u^{-2/3}v^{-1/3})) = u^{2s+t}v^{s+2t}.$$

Given ε in the Weyl group, we have ${}^{\varepsilon}\Phi_{s,t} = \Phi_{s',t'}$ with some $(s',t') \in \mathbf{C} \times \mathbf{C}$ uniquely determined by ε and (s,t). Setting (s',t') to be the image of (s,t) under ε we get the action of the Weyl group on $\mathbf{C} \times \mathbf{C}$. One can easily find

$$\begin{gathered} (s,t) \stackrel{p}{\longmapsto} (s+t,-t), \qquad (s,t) \stackrel{\tilde{p}}{\longmapsto} (-s, s+t), \\ (s,t) \stackrel{p\tilde{p}}{\longmapsto} (t, -s-t), \qquad (s,t) \stackrel{\tilde{p}p}{\longmapsto} (-s-t, s), \\ (s,t) \stackrel{\sigma}{\longmapsto} (-t,-s), \end{gathered} \tag{1.3.41}$$

with the notations in 1.1.4. We have also that $\chi_{2/3+s,2/3+t}$ is invariant under the action (1.3.41) on (s,t), and (compare with (1.3.36))

$$\begin{aligned} \chi_{2/3+s,2/3+t}(\mathrm{D}_2) &= -3(s^2+t^2+st)+4, \\ \chi_{2/3+s,2/3+t}(\mathrm{D}_3) &= -(s-t)(s^2+t^2+5st/2). \end{aligned}$$

Let W be an Whittaker function attached to $\chi_{s,t}$ and $\phi_{\mu,\nu}$. Then the condition

(b) in the definition is equivalent to

$$\begin{aligned} \widetilde{\mathrm{D}}_2 W &= \lambda_2 W \quad \text{with } \lambda_2 = \chi_{s,t}(\mathrm{D}_2), \\ \widetilde{\mathrm{D}}_3 W &= \lambda_3 W \quad \text{with } \lambda_3 = \chi_{s,t}(\mathrm{D}_3), \end{aligned} \tag{1.3.42}$$

where we write $\widetilde{\mathrm{D}}_2$ and $\widetilde{\mathrm{D}}_3$ for the operators obtained from D_2 and D_3 by omitting all monomes containing differentiation over z_2 or $\overline{z}_2$ in the expressions given by Theorem 1.3.10. So,

$$\widetilde{\mathrm{D}}_2 = \mathrm{D}^{(1)}_{\text{L-B}} + \mathrm{D}^{(3)}_{\text{L-B}} + uv\frac{\partial^2}{\partial u \partial v}$$

$$\widetilde{\mathrm{D}}_3 = -\Big(1 - \frac{1}{2}v\frac{\partial}{\partial v}\Big)\mathrm{D}^{(1)}_{\text{L-B}} + \Big(1 - \frac{1}{2}u\frac{\partial}{\partial u}\Big)\mathrm{D}^{(3)}_{\text{L-B}}.$$

Writing W as in (1.3.37) and substituting to (1.3.42) one can get differential equations satisfyed by the radial part R of W. These are

$$\widetilde{\mathrm{D}}^{\mu,\nu}_2 R = \lambda_2 R, \qquad \widetilde{\mathrm{D}}^{\mu,\nu}_3 R = \lambda_3 R, \tag{1.3.43}$$

where λ_2, λ_3 as in (1.3.42), (1.3.36), and

$$\begin{aligned} \widetilde{\mathrm{D}}^{\mu,\nu}_2 &= -u^2\frac{\partial^2}{\partial u^2} - v^2\frac{\partial^2}{\partial v^2} + uv\frac{\partial^2}{\partial u \partial v} + u\frac{\partial}{\partial u} + v\frac{\partial}{\partial v} \\ &\quad + 16\pi^2|\mu|^2u^2 + 16\pi^2|\nu|^2v^2, \\ \widetilde{\mathrm{D}}^{\mu,\nu}_3 &= -\Big(1 - \frac{1}{2}v\frac{\partial}{\partial v}\Big)\Big(-u^2\frac{\partial^2}{\partial u^2} + u\frac{\partial}{\partial u} + 16\pi^2|\mu|^2u^2\Big) \\ &\quad + \Big(1 - \frac{1}{2}u\frac{\partial}{\partial u}\Big)\Big(-v^2\frac{\partial^2}{\partial v^2} + v\frac{\partial}{\partial v} + 16\pi^2|\nu|^2v^2\Big). \end{aligned} \tag{1.3.44}$$

For the general Whittaker functions theory see Jacquet [43], Hashizume [33], Shalika [90]. We shall continue our discussion in subsection 1.4.7.

1.4 Eisenstein series Fourier coefficients

1.4.1 Basic theorem. Let $\tilde{\kappa}\colon \Gamma_2(q) \to \mathbf{C}^*$ and $\tilde{\psi}\colon \Gamma(q) \to \mathbf{C}^*$ $(q \subset (3))$ be unitary characters satisfying the conditions:

the kernel of $\tilde{\kappa}$ is a subgroup of finite index in $\Gamma_2(q)$,

the kernel of $\tilde{\psi}$ is a subgroup of finite index in $\Gamma(q)$,

$\tilde{\kappa}(\gamma) = \tilde{\psi}(h_3(\gamma))$ for $\gamma \in \Gamma_2(q)$,

$\tilde{\kappa}(\gamma) = 1$ for $\gamma \in \Delta_*$ and for $\gamma \in {}^t\Delta_*$,

$$\tilde{\psi}(\gamma) = 1 \text{ for } \gamma \in \Delta \text{ and for } \gamma \in {}^t\Delta,$$
$$\tilde{\psi}(\sigma\gamma\sigma^{-1}) = \tilde{\psi}(\gamma) \text{ for all } \gamma \in \Gamma(q).$$

Recall, Δ_* and Δ are the subgroups of upper triangular unipotent matrices in $\Gamma_2(q)$ and in $\Gamma(q)$, and

$$\sigma = -\begin{pmatrix} 0 & 0 & 1 \\ 0 & 1 & 0 \\ 1 & 0 & 0 \end{pmatrix} \quad \text{(as in 1.1.4)}.$$

Let $f\colon \mathbf{H} \to \mathbf{C}$ be an automorphic form under the group $\Gamma_2(q)$ with multiplier system $\tilde{\kappa}$, and let $\mathrm{E}(\cdot, s; f)$ be the maximal parabolic Eisenstein series attached to f and $\tilde{\psi}$, as it was defined in 1.3.7. More precisely, we mean that $\mathrm{E}(\cdot, s; f)$ is the series (1.3.8), so it is an automorphic form under $\Gamma(q)$ with multiplier system $\tilde{\psi}$. Then we have that

$$w \mapsto \mathrm{E}(\sigma w, s; f), \qquad w \in \mathbf{X}, \tag{1.4.1}$$

is an automorphic form under the group $\Gamma(q)$ with multiplier system $\tilde{\psi}$, as well as the Eisenstein series $\mathrm{E}(\cdot, s; f)$ itself. This is just because of we have $\sigma\Gamma(q)\sigma^{-1} = \Gamma(q)$ and $\tilde{\psi}(\sigma\gamma\sigma^{-1}) = \tilde{\psi}(\gamma)$ for all $\gamma \in \Gamma(q)$, see remark in 1.3.3. We shall evaluate the Fourier coefficiets

$$c_{\mu,\nu}(u, v; \mathrm{E}(\sigma(\cdot), s; f)) \tag{1.4.2}$$

of the form (1.4.1). The result is stated below in Theorem 1.4.1. We assume that $\mu, \nu \in q^*$. Only for such μ and ν are the coefficients (1.4.2) defined. Also, in our computations we assume $\Re(s)$ to be sufficiently large for the integrals and the series involved were absolutely convergent. However, we have that $s \mapsto c_{\mu,\nu}(u, v; \mathrm{E}(\sigma w, s; f))$ is a meromorphic function on $\mathbf{C}$, because the Eisenstein series is. Also, the functions $(s,t) \mapsto J_{\mu,\nu}(u, v, s, t)$ in the formulae (1.4.11), (1.4.12) in Theorem 1.4.1 are meromorphic functions on $\mathbf{C}^2$. This is known from the Whittaker functions theory, and this will be seen from our formulae in 1.4.5. So, we have the formulae (1.4.6), (1.4.7) in Theorem 1.4.1 still valid without any restrictions on s, and we have the Dirihlet series $\mathcal{D}_{\mu,\nu}$ and $\mathcal{D}_{0,\nu}(\cdot, t)$ in (1.4.9), (1.4.8) are meromorphic functions on $\mathbf{C}$.

By Theorem 0.2.2 (c), (f) and Proposition 1.2.8 (a), (b), our conditions on $\tilde{\kappa}$ and $\tilde{\psi}$ are satisfied by Kubota's κ and Bass–Milnor–Serre's ψ. Therefore, Theorem 1.4.1 is valid for the cubic metaplectic Eisenstein series, and this is just the case we are interested in.

For the sake of brevity, let us write from now on $c_{\mu,\nu}$ instead of (1.4.2). Immediately from the definitions we get

$$\begin{aligned} c_{\mu,\nu} &= c \iiint_{\mathcal{F}_q} \mathrm{E}(\sigma w, s; f)\overline{e(\mu z_1 + \nu z_3)}\, dz_1\, dz_2\, dz_3 \\ &= c \sum_{\gamma} \bar{\tilde{\psi}}(\gamma) \iiint_{\mathcal{F}_q} \nabla(\gamma\sigma w)\overline{e(\mu z_1 + \nu z_3)}\, dz_1\, dz_2\, dz_3. \end{aligned} \tag{1.4.3}$$

Here the summation is carried out over some set of matrices γ that contains a unique representative of each coset in $\Delta_\times\backslash\Gamma(q)$; $\mathcal{F}_q$ denotes $(\mathbf{C}/q)^3$ or any other set indicated in 1.3.3, $c = \mathrm{vol}(\mathcal{F}_q)^{-1}$, and

$$\nabla(w) = U(w)^{2s} V(w)^s f(Q(w)), \qquad w = (z_1, z_2, z_3, u, v) \in \mathbf{X}. \tag{1.4.4}$$

We shall use the notations of Corollary 0.3.3 for the form $f\colon \mathbf{H} \to \mathbf{C}$. So, $\mathrm{D_{L\text{-}B}} f = r(2-r)$ with $r \in \mathbf{C}$, and $\rho(\nu)$ means the νth Fourier coefficient of f about the cusp 0. In addition, let

$$m_+ = m\rho_+, \qquad m_- = m\rho_-, \qquad m = \mathrm{vol}(\mathbf{C}/q)^{-2}. \tag{1.4.5}$$

The notations introduced here are remained till the end of this section.

Theorem 1.4.1 *For the Eisenstein series* (1.4.1) *Fourier coefficients* $c_{\mu,\nu}$ *the following formulae hold. First, we have either*

$$c_{0,\nu} = m_+ \mathcal{D}_{0,\nu}(s,r) J_{0,\nu}(u,v,s,r) + m_- \mathcal{D}_{0,\nu}(s,2-r) J_{0,\nu}(u,v,s,2-r)$$

or (1.4.6)

$$c_{0,\nu} = m_+ \frac{\partial}{\partial t} \mathcal{D}_{0,\nu}(s,t) J_{0,\nu}(u,v,s,t)\Big|_{t=1} + m_- \mathcal{D}_{0,\nu}(s,1) J_{0,\nu}(u,v,s,1),$$

according to $r \neq 1$ *or* $r = 1$. *Then, for any* r, *we have*

$$c_{\mu,\nu} = m\mathcal{D}_{\mu,\nu}(s) J_{\mu,\nu}(u,v,s,r). \tag{1.4.7}$$

Here $\mu, \nu \in q^*$, $\mu \neq 0$, $s \in \mathbf{C}$ *and*

$$\mathcal{D}_{0,\nu}(s,t) = \sum_{\gamma} \bar{\bar{\psi}}(\gamma) e(\nu c_2/c_3) |c_3|^{2-t-3s} |k_\gamma|^{2t} |l_\gamma|^{-2}, \tag{1.4.8}$$

$$\mathcal{D}_{\mu,\nu}(s) = \sum_{\gamma} \bar{\bar{\psi}}(\gamma) e(\mu n_\gamma/k_\gamma + \nu c_2/c_3)\, \rho(-\mu c_3/k_\gamma^2) |c_3|^{1-3s} |k_\gamma/l_\gamma|^2. \tag{1.4.9}$$

The summation in (1.4.8), (1.4.9) *is over some set of matrices* $\gamma \in \Gamma(q)$ *that contains a unique representative of each double coset in* $\Delta_\times\backslash\Gamma(q)/{}^t\Delta$. *This set of representatives is chosen so that*

$$a_2c_3 - a_3c_2 = 0, \qquad k_\gamma^2/l_\gamma \in \mathcal{O}, \tag{1.4.10}$$

and arbitrarily otherwise. We mean by a_j, b_j, c_j *the entries of the first, second and third rows of the matrix* γ, *as in* (1.1.9), (1.2.6);

$$n_\gamma = b_1c_3 - b_3c_1, \quad k_\gamma = b_2c_3 - b_3c_2, \quad l_\gamma = \gcd(c_2, c_3).$$

The coefficients $\rho(\lambda)$, $\lambda \in q^*$, *and* r *are defined with respect to the form* f *as in Corollary* 0.3.3. *In addition, it is assumed that* $\rho(\lambda) = 0$ *if* $\lambda \notin q^*$.

$$J_{\mu,\nu}(u,v,s,t) = \iint\limits_{\mathbf{C}\,\mathbf{C}} (U^{2s}V^{s+t})(\widehat{\sigma w}) \overline{e(\nu z_3)}\, dz_2 dz_3 \quad \text{if } \mu = 0, \text{ and} \tag{1.4.11}$$

$$= \iint\limits_{\mathbf{C}\,\mathbf{C}} (U^{2s}V^{s+1})(\widehat{\sigma w}) K_{t-1}(4\pi|\mu| V(\widehat{\sigma w})) \overline{e(\mu Z_3(\widehat{\sigma w}) + \nu z_3)}\, dz_2 dz_3 \tag{1.4.12}$$

otherwise. In both formulae,

$$\widehat{\sigma} = -\begin{pmatrix} 0 & 1 & 0 \\ 0 & 0 & -1 \\ 1 & 0 & 0 \end{pmatrix}, \qquad \widehat{w} = (0, z_2, z_3, u, v). \tag{1.4.13}$$

The series (1.4.8), (1.4.9) *and the integrals* (1.4.11), (1.4.12) *are absolutely convergent whenever* $\Re(s)$ *is sufficiently large.* □

The existence of a representative system satisfying (1.4.10) is clear from Theorem 1.2.10 and Proposition 1.2.5.

We shall prove Theorem 1.4.1 in 1.4.3, and then, in 1.4.4 and 1.4.5, we shall find more manageable formulae for the series $\mathcal{D}_{\mu,\nu}$ and the integrals $J_{\mu,\nu}$.

1.4.2 An auxiliary proposition. Let p, σ and $\widehat{\sigma}$ be the matrices defined by (1.1.5), (1.4.13). It is easy to check that

$$\sigma = p\widehat{\sigma}. \tag{1.4.14}$$

Proposition 1.4.2 *Let* $a, b \in \mathbf{C}^*$ *and*

$$\widehat{w} = (0, z_2, z_3, u, v) \in \mathbf{X}. \tag{1.4.15}$$

Then

(a) $Q(\sigma w) = \begin{pmatrix} 0 & -1 \\ 1 & 0 \end{pmatrix} (-z_1 + Z_3(\widehat{\sigma}\widehat{w}), V(\widehat{\sigma}\widehat{w}))$;

(b) $\left(\frac{a}{b} Z_3(\sigma w), \left|\frac{a}{b}\right| V(\sigma w)\right) = \begin{pmatrix} 0 & -1 \\ 1 & 0 \end{pmatrix} \left(\frac{b}{a}(-z_1 + Z_3(\widehat{\sigma}\widehat{w})), \left|\frac{b}{a}\right| V(\widehat{\sigma}\widehat{w})\right)$;

(c) $(U^2V)(\sigma w) = (U^2V)(\widehat{\sigma}\widehat{w})$. □

Proof. (a) By Proposition 1.3.3 and (1.4.14),

$$Q(\sigma w) = Q(p\widehat{\sigma} w) = \begin{pmatrix} 0 & -1 \\ 1 & 0 \end{pmatrix} Q(\widehat{\sigma} w). \tag{1.4.16}$$

One has $a(w) = a(\widehat{w})$, $n(w) = \delta n(\widehat{w})$ with $\delta = n(z_1, 0, 0)$ and $\widehat{\sigma}\delta = \delta'\widehat{\sigma}$ with $\delta' = n(0, 0, -z_1)$. Consequently, $\widehat{\sigma} n(w) a(w) = \widehat{\sigma}\delta n(\widehat{w}) a(\widehat{w}) = \delta'\widehat{\sigma} n(\widehat{w}) a(\widehat{w})$ and, so,

$$\widehat{\sigma} w = \delta'\widehat{\sigma}\widehat{w}. \tag{1.4.17}$$

Since

$$\delta' = h_3\left(\begin{pmatrix} 1 & -z_1 \\ 0 & 1 \end{pmatrix}\right),$$

from (1.4.17), Proposition 1.3.3 and (0.3.2) follows

$$Q(\widehat{\sigma} w) = Q(\delta'\widehat{\sigma}\widehat{w}) = \begin{pmatrix} 1 & -z_1 \\ 0 & 1 \end{pmatrix} Q(\widehat{\sigma}\widehat{w}) = (-z_1 + Z_3(\widehat{\sigma}\widehat{w}), V(\widehat{\sigma}\widehat{w})),$$

which, in conjunction with (1.4.16), yields (a).

(b) Let us take $t \in \mathbf{C}$ such that $t^2 = a/b$. Then, by means of (0.3.2), we find that

the left-hand side of (b) equals $\begin{pmatrix} t & 0 \\ 0 & 1/t \end{pmatrix} (Z_3(\sigma w), V(\sigma w))$,

the right-hand side of (b) equals $\begin{pmatrix} 0 & -1 \\ 1 & 0 \end{pmatrix} \begin{pmatrix} 1/t & 0 \\ 0 & t \end{pmatrix} (-z_1 + Z_3(\widehat{\sigma}\widehat{w}), V(\widehat{\sigma}\widehat{w}))$.

Now to prove (b) we notice that

$$\begin{pmatrix} t & 0 \\ 0 & 1/t \end{pmatrix} = \begin{pmatrix} 0 & -1 \\ 1 & 0 \end{pmatrix} \begin{pmatrix} 1/t & 0 \\ 0 & t \end{pmatrix} \begin{pmatrix} 0 & -1 \\ 1 & 0 \end{pmatrix}^{-1}, \quad \text{and we apply (a).}$$

(c) This follows from Proposition 1.3.3 and (1.4.14), (1.4.17) —

$$\begin{aligned} (U^2V)(\sigma w) &= (U^2V)(p\widehat{\sigma} w) = (U^2V)(\widehat{\sigma} w) \\ &= (U^2V)(\delta'\widehat{\sigma}\widehat{w}) = (U^2V)(\widehat{\sigma}\widehat{w}). \quad \blacksquare \end{aligned}$$

1.4.3 Proof of Theorem 1.4.1. From the formula (1.4.3) and Proposition 1.2.7 (a) we get

$$c_{\mu,\nu} = c \sum_{\gamma} \overline{\tilde{\psi}}(\gamma) \sum_{h} \overline{\tilde{\psi}}(h) \iiint_{\mathcal{F}_q} \nabla(\gamma h \sigma w) \overline{e(\mu z_1 + \nu z_3)}\, dz_1 dz_2 dz_3.$$

Here w is as in (1.4.4) and the summation is over $\gamma \in \Omega$, $h \in \Lambda_\gamma$, with the notations of Proposition 1.2.7. Then, since $\Lambda_\gamma \subset {}^t\Delta$, our assumptions on $\tilde{\psi}$ yield $\tilde{\psi}(h) = 1$ for all $h \in \Lambda_\gamma$. Changing $h \in \Lambda_\gamma$ by $\sigma h \sigma^{-1}$ in the formula above we get

$$c_{\mu,\nu} = c \sum_{\gamma} \overline{\tilde{\psi}}(\gamma) \sum_{h} \iiint_{\mathcal{F}_q} \nabla(\gamma \sigma h w) \overline{e(\mu z_1 + \nu z_3)}\, dz_1 dz_2 dz_3, \tag{1.4.18}$$

where the summation is over $\gamma \in \Omega$, $h \in \sigma^{-1}\Lambda_\gamma\sigma$. Then, by means of Proposition 1.2.7 (b), (c), we reduce (1.4.18) to the following form:

$$c_{\mu,\nu} = c \sum_{\gamma} \overline{\tilde{\psi}}(\gamma) \iiint_{\mathcal{F}_\gamma} \nabla(\gamma \sigma w) \overline{e(\mu z_1 + \nu z_3)}\, dz_1 dz_2 dz_3; \tag{1.4.19}$$

here the summation is over $\gamma \in \Omega$ and

$$\mathcal{F}_\gamma = \{(z_1, z_2, z_3) \mid z_1 \in \mathbf{C}/\vartheta q,\ z_2, z_3 \in \mathbf{C}\}, \qquad \vartheta = c_3/l_\gamma. \tag{1.4.20}$$

Now express $\gamma\sigma$ in (1.4.19) by the formulae (1.1.11), (1.1.12) of Proposition 1.1.3. (The needed conditions are hold, since $c_3 \equiv b_2c_3 - b_3c_2 \equiv 1 \pmod 3$ for

$\gamma \in \Gamma(q)$.) Choose the representative system Ω in such a way that (1.4.10) take place. Then for $\gamma \in \Omega$ one has $l_3 = 0$ in the decomposition (1.1.11). By using Proposition 1.3.1 we find

$$\begin{aligned}(U^2V)(\gamma\sigma w) &= (U^2V)(\tilde{a}\sigma\tilde{n}w) = |t_1t_2|^3(U^2V)(\sigma\tilde{n}w),\\ Q(\gamma\sigma w) &= Q(\tilde{a}\sigma\tilde{n}w) = \Big(\frac{t_1}{t_2}Z_3(\sigma\tilde{n}w), \Big|\frac{t_1}{t_2}\Big|V(\sigma\tilde{n}w)\Big);\end{aligned} \tag{1.4.21}$$

here $\tilde{a}$ and $\tilde{n}$ are respectively the second and the last matrices in the decomposition (1.1.11), and t_1, t_2 are as in (1.1.12), so that

$$t_1t_2 = c_3^{-1}, \qquad t_1/t_2 = c_3/k_\gamma^2. \tag{1.4.22}$$

From (1.4.4), (1.4.21), (1.4.22) we obtain

$$\nabla(\gamma\sigma w) = \frac{1}{|c_3|^{3s}}(U^2V)(\sigma\tilde{n}w)^s f\Big(\frac{t_1}{t_2}Z_1(\sigma n w), \Big|\frac{t_1}{t_2}\Big|V(\sigma n w)\Big). \tag{1.4.23}$$

We now substitute the right-hand side of (1.4.23) into (1.4.19) and, using Proposition 1.3.1 (a), we make a change of variable in the integral (1.4.19) so that the integrand becomes independent of n. As a result, we obtain

$$\begin{aligned}c_{\mu,\nu} &= c\sum_\gamma \overline{\tilde{\psi}}(\gamma)e(\mu n_\gamma/k_\gamma + \nu c_2/c_3)|c_3|^{-3s}I,\\ I &= \iiint_{\mathcal{F}_\gamma}(U^2V)(\sigma w)^s f\Big(\frac{t_1}{t_2}Z_3(\sigma w), \Big|\frac{t_1}{t_2}\Big|V(\sigma w)\Big)\overline{e(\mu z_1 + \nu z_3)}\,dz_1dz_2dz_3.\end{aligned} \tag{1.4.24}$$

By means of Proposition 1.4.2 we find

$$I = \iint_{\mathbf{C}\,\mathbf{C}}(U^2V)(\widehat{\sigma}\widehat{w})^s\,\overline{e(\nu z_3)}\,\widehat{I}\,dz_2dz_3, \tag{1.4.25}$$

$$\widehat{I} = \int_{\mathbf{C}/\vartheta q} f\Big(\begin{pmatrix}0 & -1\\ 1 & 0\end{pmatrix}\Big)\Big(\frac{t_2}{t_1}(-z_1 + Z_3(\widehat{\sigma}\widehat{w})), \Big|\frac{t_2}{t_1}\Big|V(\widehat{\sigma}\widehat{w})\Big)\overline{e(\mu z_1)}\,dz_1 \tag{1.4.26}$$

with $\widehat{\sigma}$ as in (1.4.13) and $\widehat{w}$ as in (1.4.15). To evaluate integral $\widehat{I}$ we notice, first, that in accordance with our stipulation about Ω we have $\vartheta t_2/t_1 \in \mathcal{O}$. And, second,

$$\int_{\mathbf{C}/\vartheta q} e(\lambda z_3)\,dz_3 = \begin{cases}\mathrm{vol}(\mathbf{C}/q)|\vartheta|^2 & \text{if } \lambda = 0,\\ 0 & \text{if } \lambda\vartheta \in q^*,\ \lambda \neq 0.\end{cases}$$

Now, substituting into the right-hand side of (1.4.26) the Fourier expansion of the form f from Corrolary 0.3.3 and integrating term-by-term, we obtain

$$\widehat{I} = \mathrm{vol}(\mathbf{C}/q)|\vartheta|^2 \cdot \tag{1.4.27}$$
$$\cdot \begin{cases} \rho_+ \left|\frac{t_2}{t_1}\right|^r V(\widehat{\sigma}\widehat{w})^r + \rho_- \left|\frac{t_2}{t_1}\right|^{2-r} V(\widehat{\sigma}\widehat{w})^{2-r} & \text{if } \mu = 0,\ r \neq 1, \\ \rho_+ \left|\frac{t_2}{t_1}\right| V(\widehat{\sigma}\widehat{w}) \log\left(\left|\frac{t_2}{t_1}\right| V(\widehat{\sigma}\widehat{w})\right) + \rho_- \left|\frac{t_2}{t_1}\right| V(\widehat{\sigma}\widehat{w}) & \text{if } \mu = 0,\ r = 1, \\ \rho(-\frac{t_1}{t_2}\mu)\left|\frac{t_2}{t_1}\right| V(\widehat{\sigma}\widehat{w}) K_{r-1}(4\pi|\mu| V(\widehat{\sigma}\widehat{w}))\, \overline{e(\mu Z_3(\widehat{\sigma}\widehat{w}))} & \text{if } \mu \neq 0. \end{cases}$$

The formulae (1.4.6)[†] , ..., (1.4.9) and (1.4.11), (1.4.12) follow from (1.4.24), ..., (1.4.27) and (1.4.22), (1.4.5). ■

1.4.4 On the series $\mathcal{D}_{\mu,\nu}$. We can apply Theorem 1.2.10 to give more manageable form for the series $\mathcal{D}_{\mu,\nu}$ that arose in the formulae for the Eisenstein series Fourier coefficients given by Theorem 1.4.1. Indeed, by means of Theorem 1.2.10 we can rewrite (1.4.8), (1.4.9) as the sums over n_1, n_2 and

$$m_1 \in \mathbf{m}_1(n), \quad m_2 \in \mathbf{m}_2(n; m_1, l_1), \tag{1.4.28}$$

where $n = (n_1, n_2)$. Then, by means of Proposition 1.2.5 we can rewrite the terms in (1.4.8) and (1.4.9) in terms of n_1, n_2, m_1, m_2, l_1. This leads to the following formulae:

$$\mathcal{D}_{0,\nu}(s,t) = \sum_{\substack{n_1, n_2 \equiv 1(3) \\ \gcd(n_1 n_2, q) = 1}} \frac{\mathcal{E}(0,\nu;n)}{|n_1|^{3s+t-2}|n_2|^{3s-t}}, \tag{1.4.29}$$

$$\mathcal{D}_{\mu,\nu}(s) = \sum_{\substack{n_1, n_2 \equiv 1(3) \\ \gcd(n_1 n_2, q) = 1,\ \nu n_1/n_2 \in q^*}} \frac{\rho(-\mu n_1/n_2)\mathcal{E}(\mu,\nu;n)}{|n_1 n_2|^{3s-1}}, \quad \mu \neq 0, \tag{1.4.30}$$

where

$$\mathcal{E}(\mu,\nu;n) = \sum_{\substack{m_1, m_2 \\ \text{as in (1.4.28)}}} \bar{\tilde{\psi}}\left(\begin{pmatrix} k_2 & 0 & l_2 \\ 0 & 1 & 0 \\ m_2 & 0 & n_2 \end{pmatrix}\begin{pmatrix} 1 & 0 & 0 \\ 0 & k_1 & l_1 \\ 0 & m_1 & n_1 \end{pmatrix}\right) e\left(-\mu\frac{l_1 m_2}{n_2} + \nu\frac{m_1}{n_1}\right)$$

$$= \sum_{m_1} \bar{\tilde{\psi}}\left(\begin{pmatrix} 1 & 0 & 0 \\ 0 & k_1 & l_1 \\ 0 & m_1 & n_1 \end{pmatrix}\right) e\left(\nu\frac{m_1}{n_1}\right) \sum_{m_2} \bar{\tilde{\psi}}\left(\begin{pmatrix} k_2 & 0 & l_2 \\ 0 & 1 & 0 \\ m_2 & 0 & n_2 \end{pmatrix}\right) e\left(-\mu\frac{l_1 m_2}{n_2}\right). \tag{1.4.31}$$

Let us now restrict our attention by the case of cubic metaplectic Eisenstein series only, so by the case $\tilde{\psi}$ is Bass–Milnor–Serre's ψ, leaving the general case for the interested reader.

---

† For the exceptional case $\mu = 0$, $r = 1$ notice also that $v \log v = \frac{\partial}{\partial t} v^t \big|_{t=1}$.

Theorem 1.4.3 *For the cubic metaplectic Eisenstein series* $\mathrm{E}(\sigma(\cdot), s; f)$, *the series* $\mathcal{D}_{0,\nu}$ *in* (1.4.8) *can be expressed as*

$$\mathcal{D}_{0,\nu}(s,t) = \sum_{n_1,n_2} \frac{S(\nu,n_1)\,S(0,n_2)}{|n_1|^{3s+t-2}|n_2|^{3s-t}}.$$

Here: $\nu \in q^*$, $s,t \in \mathbf{C}$; *the summation is carried out over* $n_1, n_2 \in \mathcal{O}$, $n_1, n_2 \equiv 1 \pmod 3$, $\gcd(n_1 n_2, q) = 1$; $\Re(s)$ *is assumed to be sufficiently large for the absolute convergence of the series in the right-hand side.* □

Theorem 1.4.4 *For the cubic metaplectic Eisenstein series* $\mathrm{E}(\sigma(\cdot), s; f)$, *the series* $\mathcal{D}_{\mu,\nu}$ *in* (1.4.9) *can be expressed as*

$$\mathcal{D}_{\mu,\nu}(s) = \sum_{n_1,n_2} \frac{\rho(-\mu n_1/n_2)S(\mu,n_2)S(\nu,n_1,n_2)}{|n_1 n_2|^{3s-1}}.$$

Here: $\mu,\nu \in q^*$, $\mu \neq 0$, $s \in \mathbf{C}$; *the summation is carried out over* $n_1, n_2 \in \mathcal{O}$, $n_1, n_2 \equiv 1 \pmod 3$, $\gcd(n_1 n_2, q) = 1$, *subject to the conditions*

$$\mu n_1/n_2 \in q^*, \qquad S(\mu, n_2) \neq 0; \tag{1.4.32}$$

$\Re(s)$ *is assumed to be sufficiently large for the absolute convergence of the series in the right-hand side.* □

One remark has to be done in connection with Theorem 1.4.4. The Gauß sum $S(\nu, n_1, n_2)$ is defined (see 0.1.4) only if n_2 can be factored as ld^3 with $l \mid n_1^\infty$, $l, d \in \mathcal{O}$. Let us show that this take place under the condition (1.4.32). Let $p \in \mathcal{O}$ be a prime, $p \mid n_2$ and $p \nmid n_1$. From $\mu n_1/n_2 \in q^*$ it follows that $\mathrm{ord}_p\, \mu \geq \mathrm{ord}_p\, n_2$ and thus either $S(\mu, n_2) = 0$ or $\mathrm{ord}_p\, n_2 \equiv 0 \pmod 3$, as required.

Proof (of Theorem 1.4.3 and Theorem 1.4.4). In view of Proposition 1.2.9 we can rewrite (1.4.31) as

$$\mathcal{E}(\mu,\nu;n) = \sum_{\substack{m_1,m_2 \\ \text{as in (1.4.28)}}} \left(\frac{m_1}{n_1}\right)\left(\frac{m_2}{n_2}\right) \mathrm{e}\Big(-\mu\frac{l_1 m_2}{n_2} + \nu\frac{m_1}{n_1}\Big). \tag{1.4.33}$$

For $\mu = 0$, this yields

$$\mathcal{E}(0,\nu;n) = S(\nu,n_1)S(0,n_2). \tag{1.4.34}$$

Substituting (1.4.34) into (1.4.29) we obtain Theorem 1.4.3. In view of (1.4.30), to prove Theorem 1.4.4 we have only to show that, under the conditions $\mu \neq 0$, $\mu n_1/n_2 \in q^*$, one has

$$\mathcal{E}(\mu,\nu;n) = \begin{cases} 0 & \text{if } S(\mu,n_2) = 0, \\ S(\mu,n_2)S(\nu,n_1,n_2) & \text{otherwise.} \end{cases}$$

To this end choose $\mathfrak{l}_1(n)$ in 1.2.8 satisfying $\gcd(l_1, n_2) = 1$ for all $l_1 \in \mathfrak{l}_1(n)$.

This is possible for all n. Recall that l_1 in (1.4.33) is determined modn_1 by $m_1 l_1 \equiv -1 \pmod{n_1}$. From (1.4.33) we now obtain

$$\begin{aligned}
\mathcal{E}(\mu,\nu;n) &= \sum_{m_1} \left(\frac{m_1}{n_1}\right) e\left(\nu\frac{m_1}{n_1}\right) \sum_{m_2} \left(\frac{m_2}{n_2}\right) e\left(-\mu\frac{l_1 m_2}{n_2}\right) \\
&= \sum_{m_1} \left(\frac{m_1}{n_1}\right) \left(\frac{l_1}{n_2}\right)^{-1} e\left(\nu\frac{m_1}{n_1}\right) \sum_{m_2} \left(\frac{l_1 m_2}{n_2}\right) e\left(-\mu\frac{l_1 m_2}{n_2}\right) \\
&= S(\mu,n_2) \sum_{m_1} \left(\frac{m_1}{n_1}\right) \left(\frac{l_1}{n_2}\right)^{-1} e\left(\nu\frac{m_1}{n_1}\right) \\
&= \begin{cases} 0 \quad \text{if} \quad S(\mu,n_2)=0, \\ S(\mu,n_2) \sum_{m_1} \left(\frac{m_1}{n_1}\right) \left(\frac{m_1}{n_2}\right) e\left(\nu\frac{m_1}{n_1}\right) = S(\mu,n_2) S(\nu,n_1,n_2) \quad \text{otherwise,} \end{cases}
\end{aligned}$$

as required. ∎

1.4.5 Evaluation of integrals. Our aim is to examine the functions $J_{\mu,\nu}$ that arose in Theorem 1.4.1. The formulae we give for $J_{\mu,\nu}$ with $\mu\nu = 0$, i.e. for degenerate case, are in agreement with those given in [33] in a more general context. The formula for $J_{\mu,\nu}$ with $\mu\nu \neq 0$ is new and it is SL(3, **C**)-analoque of the formula given by Vinogradov and Tahtajan [100] for SL(3, **R**)-case.

Recall that, as usually, Γ and K denote the gamma-function and the Bessel–MacDonald function, see 0.1.6. We need one lemma. Let $p, \tilde{p}$ and σ be the matrices definned in (1.1.5). Then for $\widehat{\sigma}$ in (1.4.13) one has $\widehat{\sigma} = \tilde{p}p$.

Lemma 1.4.5 *If* $\widehat{w} = (0, z_2, z_3, u, v) \in \mathbf{X}$, *then*

$$\begin{aligned}
p\widehat{w} &= (z_2, 0, -\bar{z}_3 L^{-1}, uL^{1/2}, vL^{-1}), \\
\widehat{\sigma}\widehat{w} &= \tilde{p}p\widehat{w} = (*, *, z_2\bar{z}_3 L^{-1}, uL^{1/2}R^{-2}, vL^{-1}R),
\end{aligned}$$

where $L = v^2 + |z_3|^2$ *and* $R = (u^2 L + |z_2|^2)^{1/2}$. □

Proof. Apply successively parts (a) and (b) of Proposition 1.3.2. ∎

Now we can evaluate the integrals $J_{\mu,\nu}$ with $\mu = 0$.

Theorem 1.4.6 *For the integral* $J_{0,0}$ *from* (1.4.11) *we have*

$$J_{0,0}(u,v,s,t) = \frac{2^2\pi^2 u^{2-s+t} v^{4-2s}}{(3s-t-2)(3s+t-4)}. \quad \square$$

Theorem 1.4.7 *For the integrals* $J_{0,\nu}$ *from* (1.4.11) *we have*

$$J_{0,\nu}(u,v,s,t) = \frac{(2\pi)^{l+2} |\nu|^l u^{2-s+t} v^{2-s/2+t/2}}{(3s-t-2)\Gamma(l+1)} K_l(4\pi|\nu|v);$$

here $l = 3s/2 + t/2 - 2$ *and* $\nu \in \mathbf{C}$, $\nu \neq 0$. □

Proof (of Theorem 1.4.6 and Theorem 1.4.7). By means of Lemma 1.4.5 one can rewrite the integrals $J_{0,\nu}$ as

$$J_{0,\nu}(u,v,s,t) = u^{2s}v^{s+t} \iint\limits_{\mathbf{C}\,\mathbf{C}} L^{-t} R^{t-3s} \overline{e(\nu z_3)}\, dz_2 dz_3.$$

Notice that L does not depend on z_2 and that

$$\int\limits_{\mathbf{C}} R^{t-3s}\, dz_2 = \frac{2\pi}{3s-t-2}(u^2 L)^{1+t/2-3s/2},$$

by (0.1.33). Thus we have

$$J_{0,\nu}(u,v,s,t) = \frac{2\pi u^{2-s+t} v^{s+t}}{3s-t-2} \int\limits_{\mathbf{C}} L^{-r} \overline{e(\nu z_3)}\, dz_3$$

with $r = 3s/2+t/2-1$. Applying (0.1.33) to the last integral we obtain Theorem 1.4.6 if $\nu = 0$, and Theorem 1.4.7 if $\nu \neq 0$. ■

It is not so easy to evaluate the integrals $J_{\mu,\nu}$ with $\mu \neq 0$. Applying Lemma 1.4.5 we find the expression

$$J_{\mu,\nu}(u,v,s,t) = u^{2s}v^{s+1} \iint\limits_{\mathbf{C}\,\mathbf{C}} K_{t-1}\Big(4\pi|\mu|v\frac{R}{L}\Big) e\Big(-\mu\frac{z_2\bar{z}_3}{L} - \nu z_3\Big) \frac{dz_2 dz_3}{R^{3s-1}L}.$$

Now let us make the substitution $z_2 = uL^{1/2}y_2$, $z_3 = vy_3$, and let us set $Y_j = (1+|y_j|^2)^{1/2}$, to get

$$J_{\mu,\nu}(u,v,s,t) = u^{3-s}v^{4-2s}\tilde{J}(\alpha,\beta,\gamma,\delta), \tag{1.4.35}$$

where $\alpha = \mu u$, $\beta = \nu v$, $\gamma = 3s/2 - 1/2$, $\delta = t/2 - 1/2$, and

$$\tilde{J}(\alpha,\beta,\gamma,\delta) = \iint\limits_{\mathbf{C}\,\mathbf{C}} K_{2\delta}\Big(4\pi|\alpha|\frac{Y_2}{Y_3}\Big) e\Big(-\alpha\frac{y_2\bar{y}_3}{Y_3} - \beta y_3\Big) \frac{dy_2 dy_3}{Y_2^{2\gamma}Y_3^{2\gamma}}. \tag{1.4.36}$$

Before continuing, we would like to state the final results.

Theorem 1.4.8 *For the integrals $J_{\mu,0}$ from (1.4.12) we have*

$$J_{\mu,0}(u,v,s,t) = \frac{2^2\pi^2 u^{3-s}v^{4-2s}}{(3s-t-2)(3s+t-4)} K_{t-1}(4\pi|\mu|u);$$

here $\mu \in \mathbf{C}$, $\mu \neq 0$. □

Theorem 1.4.9 *For the integrals $J_{\mu,\nu}$ from* (1.4.12) *we have*

$$J_{\mu,\nu}(u,v,s,t)=\frac{2^{3s-2}\pi^{3s-1}|\mu|^{s-3}|\nu|^{2s-4}}{\Gamma\left(\frac{3s+t}{2}-1\right)\Gamma\left(\frac{3s-t}{2}\right)}R_{s-t/3,2t/3}(|\mu|u,|\nu|v),$$

where

$$R_{s-t/3,2t/3}(u,v)=u^{2+m}v^{2-m}\int_0^\infty K_l\left(4\pi u\sqrt{1+x}\right)K_l\left(4\pi v\sqrt{1+1/x}\right)x^{3m/2-1}\,dx$$

(is the radial part of an Whittaker function attached to the characters $\phi_{1,1}$ and $\chi_{s-t/3,2t/3}$, see 1.3.9, 1.3.10*), $l=3s/2+t/2-2$, $m=s/2-t/2$, and $\mu,\nu\in\mathbf{C}$, $\mu\nu\neq 0$.* □

Proof (of Theorem 1.4.8 and Theorem 1.4.9). We only have to evaluate the integrals (1.4.36). We can and do assume $\Re(\gamma)$ so large that the integrals (1.4.36) and all the integrals in the computations below converge absolutely. Immediately from the definitions given in 0.1.6 we find that

$$K_{2\delta}(ct)=\frac{c^{2\delta}}{2}\int_0^\infty \exp\left(-(a+c^2a^{-1})\frac{t}{2}\right)\frac{da}{a^{2\delta+1}},\tag{1.4.37}$$

$$c^{-\gamma+\delta}=\frac{1}{\Gamma(\gamma-\delta)}\int_0^\infty \exp(-cb)b^{\gamma-\delta-1}\,db,\tag{1.4.38}$$

at least for $c,t\in\mathbf{R}_+^*$. Replacing $K_{2\delta}(\ldots)$ in (1.4.36) by the right-hand side of (1.4.37) with $t=4\pi|\alpha|Y_3^{-1}$, $c=Y_2$ and applying then (1.4.38) with $c=Y_2^2Y_3^{-1}$ we find

$$\begin{aligned}&2\Gamma(\gamma-\delta)\widetilde{J}(\alpha,\beta,\gamma,\delta)\\&=\int_{\mathbf{C}}\int_0^\infty\int_0^\infty \nabla\exp\left(-\frac{2\pi|\alpha|(a+a^{-1})+b}{Y_3}\right)\frac{\overline{e(\beta y_3)}\,da\,db\,dy_3}{a^{2\delta+1}b^{1-\gamma+\delta}Y_3^{3\gamma-\delta}},\end{aligned}\tag{1.4.39}$$

with

$$\nabla=\int_{\mathbf{C}}\exp\left(-(2\pi|\alpha|a^{-1}+b)\frac{|y_2|^2}{Y_3}\right)e\left(-\alpha\frac{y_2\overline{y}_3}{Y_3}\right)dy_2.$$

According to (0.1.32),

$$\nabla=\pi Y_3h^{-1}\exp(-4\pi^2|\alpha|^2|y_3|^2h^{-1}Y_3^{-1})\quad\text{with}\quad h=2\pi|\alpha|a^{-1}+b.$$

Substituting the last expression for ∇ into (1.4.39) we obtain

$$\begin{aligned}&2\pi^{-1}\Gamma(\gamma-\delta)\widetilde{J}(\alpha,\beta,\gamma,\delta)\\&=\int_{\mathbf{C}}\int_0^\infty\int_0^\infty \exp\left(-\frac{2\pi|\alpha|a+h+4\pi^2|\alpha|^2|y_3|^2h^{-1}}{Y_3}\right)\frac{\overline{e(\beta y_3)}\,da\,db\,dy_3}{ha^{2\delta+1}b^{1-\gamma+\delta}Y_3^{3\gamma-\delta-1}}.\end{aligned}\tag{1.4.40}$$

Let us now assume $h = 2\pi|\alpha|a^{-1} + b$ and $x = (2\pi|\alpha|Y_3^2)^{-1}ab$ as new variables instead of a and b. The old variables can be expressed as

$$a = 2\pi|\alpha|(1 + Y_3^2 x)h^{-1}, \quad b = hxY_3^2(1 + Y_3^2 x)^{-1}.$$

Jacobian is equal to $2\pi|\alpha|Y_3^2 h^{-1}$. As a result, we have

$$\widetilde{J}(\alpha, \beta, \gamma, \delta) = 2^{-2\delta-1}\pi^{-2\delta+1}|\alpha|^{-2\delta}\Gamma(\gamma-\delta)^{-1} \cdot \\ \cdot \int_0^\infty \int_{\mathbf{C}} \nabla \frac{\overline{e(\beta y_3)}\, x^{\gamma-\delta-1}\, dy_3\, dx}{(1 + Y_3^2 x)^{\gamma+\delta} Y_3^{\gamma+\delta-1}}, \tag{1.4.41}$$

$$\nabla = \int_0^\infty \exp\left(-4\pi^2|\alpha|^2(1+x)Y_3 h^{-1} - Y_3^{-1}h\right) h^{\gamma+\delta-2}\, dh.$$

Applying (1.4.37) with $(1-\gamma-\delta)/2$ instead of δ and with

$$c = 2\pi|\alpha|\sqrt{1+x}\,Y_3, \quad t = 2Y_3^{-1}$$

we get

$$\nabla = 2\left(2\pi|\alpha|\sqrt{1+x}\,Y_3\right)^{\gamma+\delta-1} K_{\gamma+\delta-1}\left(4\pi|\alpha|\sqrt{1+x}\right).$$

Substituting the last expression for ∇ into (1.4.41) we obtain

$$\widetilde{J}(\alpha, \beta, \gamma, \delta) = \frac{(2|\alpha|)^{\gamma-\delta-1}\pi^{\gamma-\delta}}{\Gamma(\gamma-\delta)} \cdot \\ \cdot \int_0^\infty \widehat{\nabla}_\beta K_{\gamma+\delta-1}\left(4\pi|\alpha|\sqrt{1+x}\right) \frac{x^{\gamma-\delta-1}\, dx}{(1+x)^{1/2-\gamma/2-\delta/2}}, \tag{1.4.42}$$

where $\widehat{\nabla}_\beta = \int_{\mathbf{C}} \frac{e(-\beta y_3)\, dy_3}{(1 + Y_3^2 x)^{\gamma+\delta}}$, and, by (0.1.33), we have

$$\widehat{\nabla}_0 = \frac{\pi}{\gamma+\delta-1}(1 + x^{-1})^{1-\gamma-\delta} x^{-\gamma-\delta},$$

$$\widehat{\nabla}_\beta = \frac{(2\pi)^{\gamma+\delta}|\beta|^{\gamma+\delta-1}}{\Gamma(\gamma+\delta)}(1 + x^{-1})^{1/2-\gamma/2-\delta/2} x^{-\gamma-\delta} K_{\gamma+\delta-1}\left(4\pi|\beta|\sqrt{1 + 1/x}\right)$$

for $\beta \neq 0$. Replacing $\widehat{\nabla}_\beta$ in (1.4.42) by the right-hand sides of the last two formulae we find

$$\widetilde{J}(\alpha, 0; \gamma, \delta) = \frac{(2|\alpha|)^{\gamma-\delta-1}\pi^{\gamma-\delta+1}}{(\gamma+\delta-1)\Gamma(\gamma-\delta)} \cdot \\ \cdot \int_0^\infty K_{\gamma+\delta-1}\left(4\pi|\alpha|\sqrt{1+x}\right)(1+x)^{1/2-\gamma/2-\delta/2} x^{\gamma-\delta-2}\, dx, \tag{1.4.43}$$

$$\widetilde{J}(\alpha,\beta,\gamma,\delta) = \frac{2^{2\gamma-1}\pi^{2\gamma}|\alpha|^{\gamma-\delta-1}|\beta|^{\gamma+\delta-1}}{\Gamma(\gamma-\delta)\Gamma(\gamma+\delta)} \cdot \int_0^\infty K_{\gamma+\delta-1}(4\pi|\alpha|\sqrt{1+x})K_{\gamma+\delta-1}(4\pi|\beta|\sqrt{1+1/x})\,x^{\gamma/2-3\delta/2-3/2}\,dx \tag{1.4.44}$$

for $\beta \neq 0$. Theorem 1.4.9 follows from (1.4.44), (1.4.35). Theorem 1.4.8 follows from (1.4.35), (1.4.43) and (0.1.22).

1.4.6 Mellin transform. We would like now to complete our exposition by deducing explicit formula for the Mellin transform of $J_{\mu,\nu}(\cdot,\cdot,s,t)$ i.e. for

$$M_{\mu,\nu}(a,b,s,t) = \int_0^\infty\int_0^\infty J_{\mu,\nu}(u,v,s,t)\,u^{a-1}v^{b-1}\,du\,dv \tag{1.4.45}$$

where $\mu,\nu,s,t,a,b \in \mathbf{C}$, $\mu\nu \neq 0$, and we assume that $\Re(a)$ and $\Re(b)$ are sufficiently large for absolute convergence of this integral.

Corollary 1.4.10 *For the integrals $M_{\mu,\nu}(a,b,s,t)$ defined by (1.4.45) with $\mu,\nu \in \mathbf{C}$, $\mu\nu \neq 0$ one has the expression*

$$M_{\mu,\nu}(a,b,s,t) = 2^{3s-10-a-b}\pi^{3s-5-a-b}|\mu|^{s-3-a}|\nu|^{2s-4-b}\cdot$$

$$\cdot\frac{\Gamma(\frac{a}{2}+s)\Gamma(\frac{b}{2}+2-s)\Gamma(\frac{a-s-t}{2}+2)\Gamma(\frac{b+s+t}{2})\Gamma(\frac{a-s+t}{2}+1)\Gamma(\frac{b+s-t}{2}+1)}{\Gamma(\frac{3s+t}{2}-1)\Gamma(\frac{3s-t}{2})\Gamma(\frac{a+b}{2}+2)}$$

which is valid for all $a,b,s,t \in \mathbf{C}$, so gives a meromorphic continuation of $M_{\mu,\nu}(\cdot,\cdot,s,t)$ to $\mathbf{C}\times\mathbf{C}$. In particular, if $s=t=4/3$, then

$$M_{\mu,\nu}(a,b,s,t) = h_{\mu,\nu}(a,b)\frac{\Gamma(\frac{3a}{2}+2)\Gamma(\frac{3b}{2}+2)}{\Gamma(\frac{a+b}{2}+2)}$$

with $h_{\mu,\nu}(a,b) = 2^{-6-a-b}3^{-3a/2-3b/2-1/2}\pi^{-a-b}|\mu|^{-5/3-a}|\nu|^{-4/3-b}$. □

Proof. Let us substitute the expression given by Theorem 1.4.9 into (1.4.45) and change the order of integration to get

$$M_{\mu,\nu}(a,b,s,t) = \frac{2^{3s-2}\pi^{3s-1}|\mu|^{(3s-t)/2-1}|\nu|^{(3s+t)/2-2}}{\Gamma(\frac{3s+t}{2}-1)\Gamma(\frac{3s-t}{2})} \cdot \int_0^\infty A(x)B(x)x^{3(s-t)/4-1}\,dx, \tag{1.4.46}$$

where

$$A(x) = \int_0^\infty K_{(3s+t)/2-2}\big(4\pi|\mu|u\sqrt{1+x}\big)\, u^{1+a+(s-t)/2}\, du,$$

$$B(x) = \int_0^\infty K_{(3s+t)/2-2}\big(4\pi|\nu|v\sqrt{1+1/x}\big)\, v^{1+b-(s-t)/2}\, dv$$

are the Mellin transforms of the Bessel–MacDonald function. By (0.1.22),

$$A(x) = 2^{-4-a-s/2+t/2}\big(\pi|\mu|\sqrt{1+x}\big)^{-2-a-(s-t)/2}\Gamma\Big(\frac{a}{2}+s\Big)\Gamma\Big(\frac{a-s-t}{2}+2\Big),$$

$$B(x) = 2^{-4-b+s/2-t/2}\big(\pi|\nu|\sqrt{1+1/x}\big)^{-2-b+(s-t)/2}\Gamma\Big(\frac{b+s+t}{2}\Big)\Gamma\Big(\frac{b}{2}+2-s\Big).$$

Substituting the last expressions into (1.4.46) we get

$$M_{\mu,\nu}(a,b,s,t) = 2^{3s-10-a-b}\pi^{3s-5-a-b}|\mu|^{s-3-a}|\nu|^{2s-4-b}\cdot$$

$$\cdot\frac{\Gamma\big(\frac{a}{2}+s\big)\Gamma\big(\frac{a-s-t}{2}+2\big)\Gamma\big(\frac{b+s+t}{2}\big)\Gamma\big(\frac{b}{2}+2-s\big)}{\Gamma\big(\frac{3s+t}{2}-1\big)\Gamma\big(\frac{3s-t}{2}\big)}\int_0^\infty \frac{x^{b/2+s/2-t/2}\,dx}{(1+x)^{2+a/2+b/2}}.$$

Now it is sufficient to apply Euler's formula (0.1.20) to express the last integral in terms of the gamma-function and, thus, to obtain the desired formula. The simplification for $s = t = 4/3$ achieved by means of the triplication formula —

$$\Gamma(a/2+2/3)\Gamma(a/2+1)\Gamma(a/2+4/3) = 2\pi 3^{-3a/2-3/2}\Gamma(3a/2+2),$$
$$\Gamma(b/2+2/3)\Gamma(b/2+1)\Gamma(b/2+4/3) = 2\pi 3^{-3b/2-3/2}\Gamma(3b/2+2),$$
$$\Gamma(4/3)\Gamma(5/3) = 2^2 3^{-5/2}\pi. \quad\blacksquare$$

The formulae for the Mellin transform in Corollary 1.4.10 were proved in [15] by means of an entirely different method. As a consequence, it is noticed in [15], that, up to a constant factor which is irrelevent in the moment, $J_{1,1}(u,v,4/3,4/3)$ is equal to[†]

$$(uv)^{4/3}\sqrt{u^{2/3}+v^{2/3}}\,K_{1/3}\big(4\pi(u^{2/3}+v^{2/3})^{3/2}\big).^{\dagger}$$

In view of the relationship between the Bessel–MacDonald function $K_{1/3}$ with the Airy function (0.1.26), this formula is in agreement with the formula (1.4.52) we give in Theorem 1.4.13 below.

[†] One can take the formula 4.635 (1)+(2) in [30] with $n = 2$, $p_1 = a + 4/3$, $p_2 = b + 4/3$, $\alpha_1 = \alpha_2 = 2/3$, $q_1 = q_2 = 1$, $f(x) = xK_{1/3}(4\pi x^{3/2})$ to find that the double Mellin transform of this function coincides (up to a constant factor) with $M_{1,1}(a,b,4/3,4/3)$.

1.4.7 More on Whittaker functions. We would like to study in more detail the non-degenerate Whittaker functions.

Theorem 1.4.11 *Given $s,t \in \mathbf{C}$, the function $R_{s,t}\colon \mathbf{R}_+^* \times \mathbf{R}_+^* \to \mathbf{C}$ defined as*

$$R_{s,t}(u,v) = u^{2+m}v^{2-m} \int_0^\infty K_l\big(4\pi u\sqrt{1+x}\big)K_l\big(4\pi v\sqrt{1+1/x}\big)\,x^{3m/2-1}\,dx$$

with $l = 3(s+t)/2-2$, $m = s/2-t/2$ is the radial part of an Whittaker function attached to the characters $\phi_{1,1}$, $\chi_{s,t}$. One can realize it also as the convolution

$$R_{s,t}(u,v) = u^{2+m}v^{2-m} \int_0^1 K_l\Big(\frac{4\pi u}{\sqrt{x}}\Big)K_l\Big(\frac{4\pi v}{\sqrt{1-x}}\Big)\frac{dx}{x^{1+3m/2}(1-x)^{1-3m/2}}$$

with l, m as above. □

Proof. According to Theorem 1.4.9 we have

$$J_{1,1}(u,v,s+t/2,3t/2) = \frac{2^{3s+3t/2-2}\pi^{3s+3t/2-1}}{\Gamma(3s/2+3t/2-1)\Gamma(3s/2)}R_{s,t}(u,v). \qquad (1.4.47)$$

Thus, $R_{s,t}$ is the radial part of Whittaker function attached to the same characters $\phi_{1,1}$, $\chi_{s,t}$ as $J_{1,1}(\cdot,\cdot,s+t/2,3t/2)$ is. Changing varible $x \mapsto -1+1/x$ we get the second expression for $R_{s,t}$. ■

Recall, our formulae for $R_{s,t}$ are quite similar to that given in 1978 by Vinogradov and Tahtajan [100] for the radial parts of non-degenerate Whittaker functions for the group $\mathrm{SL}(3,\mathbf{R})$. Such type of formulae occur as very convenient tools for further study of Whiitaker functions. One evidence was given in the preciding subsection. One can apply these formulae also to eximine asymptotic behaviour of $R_{s,t}(u,v)$ as $u \to \infty$ and/or $v \to \infty$. For this one has to replace the Bessel–MacDonald functions under the integral sign by the well known asymptotic expansions and then integrate term-by-term.

All facts concerning $R_{s,t}$, which are known from the genreal Whittaker functions theory, follow easily from the expressions given in Theorem 1.4.9 and Theorem 1.4.11. In particular, we can prove the functional equations mentioned in 1.3.9. One of them, that is $R_{s,t} = R_{4/3-t,4/3-s}$, follows immediately from Theorem 1.4.11 just because $(s,t) \mapsto (4/3-t, 4/3-s)$ transforms $m \mapsto m$ and $l \mapsto -l$, and $K_{-l} = K_l$. To get other equations one can follow procedure proposed in [100], but we find it is more elegant to deduce all the functional equations from Corollary 1.4.10.

Theorem 1.4.12 *Given any $u,v \in \mathbf{R}_+^*$, the function $(s,t) \mapsto R_{s,t}(u,v)$, defined in Theorem 1.4.11 and in Theorem 1.4.9, is entire function on $\mathbf{C}\times\mathbf{C}$ satisfying the functional equations*

$$\begin{aligned} R_{s,t} &= R_{-2/3+s+t,4/3-t} = R_{4/3-s,-2/3+s+t} \\ &= R_{t,2-s-t} = R_{2-s-t,s} = R_{4/3-t,4/3-s}. \end{aligned} \qquad \square$$

Proof. The fact that $(s,t) \mapsto R_{s,t}(u,v)$ is entire function follows just from the integral expressions in Theorem 1.4.11, because of the Bessel–MacDonald function is entire function of index and satisfies the conditions we need to differentiate under integral sign. To prove the functional equations let us set

$$\nabla(c;s,t) = \Gamma\Big(\frac{c+t}{2}+s\Big)\Gamma\Big(\frac{c-s+t}{2}+1\Big)\Gamma\Big(\frac{c-s}{2}-t+2\Big), \qquad (1.4.48)$$

and then let us look at the Mellin transform

$$M(a,b,s,t) = \int_0^\infty\!\!\int_0^\infty R_{s,t}(u,v)u^{a-1}v^{b-1}\,du\,dv.$$

In view of Corollary 1.4.10 and (1.4.47) we have

$$2^{8+a+b}\pi^{4+a+b}\Gamma\Big(\frac{a+b}{2}+2\Big)M(a,b,s,t) = \nabla(a;s,t)\nabla(b;s,t),$$

We see that $\nabla(c;s,t)$, and thus $M(a,b,s,t)$ also, are invariant under the linear transformations

$$(s,t) \mapsto (-2/3+s+t, 4/3-t), \ldots, \; (s,t) \mapsto (4/3-t, 4/3-s),$$

because these transformations change only the order of the factors in (1.4.48). Taking inverse Mellin transform, we get $R_{s,t}$ is invariant under the same linear transformations. ∎

The expressions for the radial parts of Whittaker functions given in Theorem 1.4.11 can be simplified further for some particular cases. If $t = s$, then we have $m = 0$ and $R_{s,s}(u,v) = R_{s,s}(v,u)$ for all $s \in \mathbf{C}$, $u, v \in \mathbf{R}_+^*$. If s, t are choosen in such a way that $s+t = (2n+1)/3$ with $n \in \mathbf{Z}$, then l is half of odd integer and $K_l(z)$ can be expressed as $z^{-1/2}\exp(-z)$ times polynomial of z^{-1}, see [101]. We will need in one more particular case described in the theorem below.

Theorem 1.4.13 *If $s = t = 8/9$ or $s = 10/9$, $t = 4/9$, then for the function $R_{s,t}\colon \mathbf{R}_+^* \times \mathbf{R}_+^* \to \mathbf{C}$ defined in Theorem 1.4.11 one has*

$$R_{s,t}(u,v) = h(uv)^{4/3}\widetilde{R}\big((6\pi u)^{2/3}, (6\pi v)^{2/3}\big), \qquad (1.4.49)$$

where $h = 2^{-4/3}3^{-1/3}\pi^{2/3}$ and

$$\widetilde{R}(a,b) = \int_0^1 Ai'\Big(\frac{a}{x^{1/3}}\Big)Ai'\Big(\frac{b}{(1-x)^{1/3}}\Big)\frac{dx}{x^{2/3}(1-x)^{2/3}} \qquad (1.4.50)$$

$$= \frac{ab}{a+b}\int_0^1 Ai\Big(\frac{a}{x^{1/3}}\Big)Ai\Big(\frac{b}{(1-x)^{1/3}}\Big)\frac{dx}{x^{4/3}(1-x)^{4/3}} \qquad (1.4.51)$$

$$= Ai(a+b). \quad \square \qquad (1.4.52)$$

Proof. By Theorem 1.4.12 we have that $R_{8/9,8/9} = R_{10/9,4/9}$. To obtain (1.4.49) with $\widetilde{R}$ as in (1.4.50) we take $s = t = 8/9$ in Theorem 1.4.11 and apply (0.1.26) to replace $K_{2/3}$ by the derivative of the Airy function. For the expression (1.4.51) we begin with Theorem 1.4.11 with $s = 10/9$, $t = 4/9$, and we apply then (0.1.26) to get

$$\widetilde{R}(a,b)/a = \int_0^1 Ai\Big(\frac{a}{x^{1/3}}\Big)\, Ai\Big(\frac{b}{(1-x)^{1/3}}\Big) \frac{dx}{x^{4/3}(1-x)^{1/3}}. \tag{1.4.53}$$

It is clear just from Theorem 1.4.11 and also form (1.4.50) that $\widetilde{R}(a,b) = \widetilde{R}(b,a)$. Thus, besides (1.4.53), we have

$$\widetilde{R}(a,b)/b = \int_0^1 Ai\Big(\frac{a}{x^{1/3}}\Big)\, Ai\Big(\frac{b}{(1-x)^{1/3}}\Big) \frac{dx}{x^{1/3}(1-x)^{4/3}}. \tag{1.4.54}$$

More detailed, to get (1.4.54) we change a by b and b by a, and we change variable x by $1-x$ in (1.4.53). Summing (1.4.53) and (1.4.54) we get (1.4.51).

For the last expression (1.4.52) for $\widetilde{R}$ let us look at the integrals in the right-hand sides of (1.4.50) and (1.4.53). Recall, Ai is entire function, and $Ai(z)$, as well as derivatives $Ai^{(k)}(z)$ of any order k, decay exponentially as $z \to \infty$, $z \in \mathbf{R}$. This yields the integrals in the right-hand sides of (1.4.50) and (1.4.53) represent real analytic functions of two real positive parameters a and b, and that the differentiation over parameters under integral sign is admissiable in (1.4.50) and (1.4.53). Now, let us differentiate (1.4.50) over a, and let us express Ai'' according to (0.1.24). Then, let us differentiate (1.4.53) over b. Comparing results we get

$$\frac{\partial \widetilde{R}}{\partial a} = \frac{\partial \widetilde{R}}{\partial b}.$$

It is well known, this yields $\widetilde{R}(a,b) = P(a+b)$ with some real analytic function P. To determine P we first notice that $P(x)$ decay exponentially as $x \to \infty$, $x \in \mathbf{R}_+^*$. Then, let us differentiate (1.4.50) over a and over b, and let us express Ai'' according to (0.1.24). Comparing the result with (1.4.51) we get

$$\frac{\partial^2 \widetilde{R}}{\partial a\, \partial b} = (a+b)\widetilde{R}.$$

Therefore, $P''(x) = xP(x)$, $x \in \mathbf{R}_+^*$, i.e. P satisfies just the same differential equation as the Airy function does. We know the Airy function is, up to a constant factor, the only solution of this equation which decays exponentially at infinity. Thus, with some $c \in \mathbf{C}$, we have $P(x) = c\,Ai(x)$ for any $x \in \mathbf{R}_+^*$. Now, it is clear, the integral (1.4.50), being equal $c\,Ai(a+b)$ for $a, b \in \mathbf{R}_+^*$, still valid and equal $c\,Ai(0)$ for $a = b = 0$. This notice allows us to find c by means of the formulae in 0.1.6, namely

$$c\,Ai(0) = \int_0^1 Ai'(0)Ai'(0)\frac{dx}{x^{2/3}(1-x)^{2/3}}$$

$$= \left(Ai'(0)\right)^2 B(1/3,1/3) = 3^{-2/3}\Gamma(2/3)^{-1} = Ai(0), \quad \text{so } c = 1. \quad \blacksquare$$

It is interesting to notice that the theorem above gives rise to 'integral addition formulae' for the Airy function. More definitely, one can treate the integrals in the right-hand sides in (1.4.50), (1.4.51) as integral expressions for $Ai(a+b)$, without any reference to Whittaker functions. From this point of view it is not reasonable to deal with real positive a, b only. By elementary considerations (see the asimptotics for Ai and Ai' in 0.1.6) we find the integrals in the right-hand sides in (1.4.50), (1.4.51), as well as the integrals obtained from those by differentiation over a or/and b, are absolutely and locally uniformly convergent in the sector

$$|\arg a| < \pi/3, \quad |\arg b| < \pi/3, \quad a, b \in \mathbf{C},$$

and, by analitic continuation principle, our expressions for $Ai(a+b)$ still valid for all a, b in this sector. See [85].

We can give an alternative proof for (1.4.52), which is instructive, thought it is not so simple as the one given above. For this let us turn to the differential equations (1.3.43) with $\lambda_2 = 32/9$, $\lambda_3 = 0$, $\mu = \nu = 1$ satisfied by the function (1.4.49). By standard computation we find that these equations are equivalent to the following

$$a^2\frac{\partial^2 \widetilde{R}}{\partial a^2} - ab\frac{\partial^2 \widetilde{R}}{\partial a\,\partial b} + b^2\frac{\partial^2 \widetilde{R}}{\partial b^2} = (a^3+b^3)\widetilde{R}, \tag{1.4.55}$$

$$\begin{aligned} a^2 b\frac{\partial^3 \widetilde{R}}{\partial a^2 \partial b} - ab^2\frac{\partial^3 \widetilde{R}}{\partial a\,\partial b^2} - a^2\frac{\partial^2 \widetilde{R}}{\partial a^2} + b^2\frac{\partial^2 \widetilde{R}}{\partial b^2} \\ + ab^3\frac{\partial \widetilde{R}}{\partial a} - a^3 b\frac{\partial \widetilde{R}}{\partial b} + (a^3-b^3)\widetilde{R} = 0. \end{aligned} \tag{1.4.56}$$

Since $a^3+b^3 = (a^2-ab+b^2)(a+b)$, it is just clear that if any function $\widetilde{R}$ satisfies

$$\frac{\partial^2 \widetilde{R}}{\partial a^2} = \frac{\partial^2 \widetilde{R}}{\partial a\,\partial b} = \frac{\partial^2 \widetilde{R}}{\partial b^2} = (a+b)\widetilde{R}, \tag{1.4.57}$$

then it satisfies (1.4.55). Also, one can easily find, if $\widetilde{R}$ satisfies (1.4.57) and

$$\frac{\partial \widetilde{R}}{\partial a} = \frac{\partial \widetilde{R}}{\partial b}, \tag{1.4.58}$$

then it satisfies (1.4.56). In the meantime it is obvious the system (1.4.57), (1.4.58) has the solution $(a,b) \mapsto Ai(a+b)$. In view of the Shalika multiplicity one theorem, we have that $\widetilde{R}(a,b) = c\,Ai(a+b)$ with some $c \in \mathbf{C}$. It remains to show $c = 1$, and this can be done as in the proof of theorem above.

1.5 Eisenstein series $E(\sigma(\cdot), s; \Theta_{\text{K-P}})$ and cubic theta function

1.5.1 Preliminary remarks. The first goal of this section is to study the Fourier coefficients of the Eisenstein series

$$E(\sigma(\cdot), s; \Theta_{\text{K-P}})\colon \mathbf{X} \to \mathbf{C} \tag{1.5.1}$$

attached to the group $\Gamma(3)$ and to the Kubota–Patterson cubic theta function $\Theta_{\text{K-P}}$ defined in 0.3.8. Then, in 1.5.5, we find that the Eisenstein series (1.5.1) has a simple pole at the point $s = 4/3$. In 1.5.6, we define the cubic theta function $\Theta\colon \mathbf{X} \to \mathbf{C}$ as the residue of the Eisenstein series at this point and we give explicit expressions in terms of Patterson's τ-function for its Fourier coefficients. The main result, stated in Theorem 1.5.6, is the one given in [78], except that the integrals defining Whittaker functions were not evaluated in [78].

Let us write $c_{\mu,\nu}$ for the Fourier coefficients of the Eisenstein series (1.5.1), see (1.4.2), (1.4.3). To evaluate $c_{\mu,\nu}$ one can apply the general formulae obtained in Section 1.4. It happens that the general formulae simplify dramatically in this particular case. It is especially important to notice that the series $\mathcal{D}_{\mu,\nu}(s)$ with $\mu\nu \neq 0$ have Euler products. Up to a finite number of Euler factors, $\mathcal{D}_{\mu,\nu}(s)$ is $L_{\mu,\nu}(3s/2-1)/\zeta_*(9s/2-3)$, where $L_{\mu,\nu}$ is the cubic Hecke series defined in 0.1.5. As in [78], this fact has an essential role in the proof of Theorem 1.5.6.

We shall use the results obtained in Section 1.4 taking $q = (3)$ and $f = \Theta_{\text{K-P}}$. With such a choice we have $q^* = (\sqrt{-3})^{-3}\mathcal{O}$, $\text{vol}(\mathbf{C}/q) = 2^{-1}3^{5/2}$. Also, it follows from Theorem 0.3.8 (a), (c) that, in the notations of Corollary 0.3.3, for $f = \Theta_{\text{K-P}}$ we have $\rho_+ = 0$, $\rho_- = 2^{-1}3^{-1/2}$, $\kappa = 4/3$. Thus, in Theorem 1.4.1 for (1.5.1) we have

$$m_+ = 0, \quad m_- = 2/3^{11/2}, \quad m = 2^2/3^5, \quad r = 4/3. \tag{1.5.2}$$

Throughout this section

$$u, v \in \mathbf{R}_+^*, \quad \zeta_*(s) = (1-3^{-s})\zeta_{\mathbf{Q}(\sqrt{-3})}(s), \quad s \in \mathbf{C}, \quad \mu, \nu \in (\sqrt{-3})^{-3}\mathcal{O},$$
$$p, p' \equiv 1 \pmod 3 \text{ are prime}, \quad n_1, n_2 \in \mathcal{O}, \quad n_1, n_2 \equiv 1 \pmod 3.$$

1.5.2 Fourier coefficients $c_{0,\nu}$.

Theorem 1.5.1 *For the Eisenstein series $E(\sigma(\cdot), s; \Theta_{\text{K-P}})$ Fourier coefficient $c_{0,0}$ we have*

$$c_{0,0} = h\frac{\zeta_*(9s/2-5)}{\zeta_*(9s/2-3)} I_{0,0}(u, v, s),$$

where $h = 2^3\pi^2/3^{11/2}$ and

$$I_{0,0}(u, v, s) = \frac{u^{8/3-s}v^{4-2s}}{(3s-8/3)(3s-10/3)}. \qquad \square$$

Proof. From Theorem 1.4.1, (1.5.2) and Theorem 1.4.6, it follows that

$$c_{0,0} = 2^1 3^{-11/2} \mathcal{D}_{0,0}(s, 2/3) J_{0,0}(u, v, s, 2/3)$$

with $J_{0,0}(u, v, s, 2/3) = 2^2\pi^2 I_{0,0}(u, v, s)$. Thus we get

$$c_{0,0} = h\mathcal{D}_{0,0}(s, 2/3) I_{0,0}(u, v, s),$$

and it remains to apply Theorem 1.4.3 and (0.1.17) to find that $\mathcal{D}_{0,0}(s, 2/3) = \widetilde{\mathcal{D}}(3s - 4/3)\widetilde{\mathcal{D}}(3s - 2/3)$ with

$$\widetilde{\mathcal{D}}(t) = \sum_{n\equiv 1(3)} \frac{S(0,n)}{|n|^t} = \frac{\zeta_*(3t/2-3)}{\zeta_*(3t/2-2)}. \quad \blacksquare$$

Theorem 1.5.2 *For the Eisenstein series* $\mathrm{E}(\sigma(\cdot), s; \Theta_{\text{K-P}})$ *Fourier coefficients* $c_{0,\nu}$, $\nu \neq 0$, *we have*

$$c_{0,\nu} = h\frac{\zeta_*(9s/2-4)}{\zeta_*(9s/2-3)}\Psi(3s/2 - 2/3, \nu) I_{0,\nu}(u, v, s),$$

where $h = 2^3\pi^2/3^{11/2}$,

$$I_{0,\nu}(u, v, s) = \frac{u^{8/3-s}v^{4-2s}}{(3s-8/3)\Gamma(3s/2-2/3)}(2\pi|\nu|v)^{3s/2-5/3}K_{3s/2-5/3}(4\pi|\nu|v),$$

and Ψ *is the series* (0.3.11) (*with* $q = (3)$). □

Proof. As in the proof of the preceding theorem, we apply Theorem 1.4.1, Theorem 1.4.7 and (1.5.2) to get $c_{0,\nu} = h\mathcal{D}_{0,\nu}(s, 2/3) I_{0,\nu}(u, v, s)$. Then, for the request form for $\mathcal{D}_{0,\nu}(s, 2/3)$ apply (0.1.17), (0.3.11) and Theorem 1.4.3. ■

1.5.3 Fourier coefficients $c_{\mu,0}$ with $\mu \neq 0$.

Theorem 1.5.3 *For the Eisenstein series* $\mathrm{E}(\sigma(\cdot), s; \Theta_{\text{K-P}})$ *Fourier coefficients* $c_{\mu,0}$, $\mu \neq 0$, *we have*

$$c_{\mu,0} = h\tau(\mu)\frac{\zeta_*(9s/2-5)}{\zeta_*(9s/2-3)} I_{\mu,0}(u, v, s),$$

where $h = 2^4\pi^2/3^5$, τ *is the function defined in* (0.3.15) *and*

$$I_{\mu,0}(u, v, s) = \frac{u^{3-s}v^{4-2s}}{(3s-8/3)(3s-10/3)} K_{1/3}(4\pi|\mu|u). \quad \square$$

Proof. From Theorem 1.4.1, (1.5.2) and Theorem 1.4.8, it follows that

$$c_{\mu,0} = 2^2 3^{-5} \mathcal{D}_{\mu,0}(s) J_{\mu,0}(u, v, s, 4/3)$$

and $J_{\mu,0}(u,v,s,4/3) = 2^2\pi^2 I_{\mu,0}(u,v,s)$. This yields $c_{\mu,0} = h\mathcal{D}_{\mu,0}(s) I_{\mu,0}(u,v,s)$, and it remains to show that

$$\mathcal{D}_{\mu,0}(s) = \tau(\mu)\frac{\zeta_*(9s/2-5)}{\zeta_*(9s/2-3)}. \tag{1.5.3}$$

For this we can apply Theorem 1.4.4. First, let us set

$$\mu' = \prod_p p^{\vartheta_p} \quad \text{with} \quad \vartheta_p = \operatorname{ord}_p \mu \quad \text{for each prime} \quad p \equiv 1(3).$$

Then, for $n_1 \equiv n_2 \equiv 1 \pmod 3$ and prime $p \equiv 1 \pmod 3$, we set

$$\alpha_p = \operatorname{ord}_p n_1, \quad \beta_p = \operatorname{ord}_p n_2, \quad l_p = \vartheta_p + \alpha_p - \beta_p.$$

It follows easily from the formulae given in 0.1.4 that

$$\begin{aligned} S(0,n_1,n_2) &= \begin{cases} \tilde{\varphi}(n_1) = \prod_p \tilde{\varphi}(p^{\alpha_p}) & \text{if } n_1 n_2 \text{ is a cube,} \\ 0 \quad \text{for other } n_1,\, n_2 \text{ under (1.4.32),} \end{cases} \\ S(\mu,n_2) &= \Big\{ \prod_{p\neq p'} \left(\frac{p'}{p}\right)^{\beta_p\beta_{p'}} \Big\} \Big\{ \prod_p \left(\frac{\mu p^{-\vartheta_p}}{p}\right)^{-\beta_p} S(p^{\vartheta_p}, p^{\beta_p}) \Big\}. \end{aligned} \tag{1.5.4}$$

Here and in what follows all products are taken either over prime $p \equiv 1 \pmod 3$ or over prime p and p' subject to the conditions $p \equiv p' \equiv 1 \pmod 3$, $p \neq p'$. Corollary 0.3.10 provides the factoriztion of τ:

$$\tau(-\mu n_1/n_2) = \tau(\mu\mu'^{-1})\Big\{ \prod_{p\neq p'} \left(\frac{p'}{p}\right)^{-l_p l_{p'}} \Big\} \Big\{ \prod_p \left(\frac{\mu\mu'^{-1}}{p}\right)^{l_p} \tau(p^{l_p}) \Big\}. \tag{1.5.5}$$

If $n_1 n_2$ is a cube, then $\alpha_p + \beta_p \equiv 0 \pmod 3$ for all p and, multiplying (1.5.5), (1.5.4), we find that almost all cubic residue symbols eliminated and

$$\tau(-\mu n_1/n_2) S(\mu,n_2)\tilde{\varphi}(n_1) = k(\mu) \prod_p \tau(p^{\vartheta_p+\alpha_p-\beta_p}) S(p^{\vartheta_p}, p^{\beta_p}) \tilde{\varphi}(p^{\alpha_p}) \tag{1.5.6}$$

with $k(\mu)$ as in Corollary 0.3.10. From Theorem 1.4.4 (with $\rho = \tau$) and (1.5.6), we obtain the decomposition

$$\mathcal{D}_{\mu,0}(s) = k(\mu) \prod_p Q^{(p)}(\vartheta_p; s) \tag{1.5.7}$$

with

$$Q^{(p)}(\delta; s) = \sum_{\alpha,\beta} \frac{\tau(p^{\delta+\alpha-\beta}) S(p^\delta, p^\beta)\tilde{\varphi}(p^\alpha)}{\|p^{\alpha+\beta}\|^{(3s-1)/2}}, \tag{1.5.8}$$

where the summation is carried out over non-negative rational integers α, β subject to the conditions $\delta + \alpha - \beta \geq 0$, $\alpha + \beta \equiv 0 \pmod 3$.

Let us now evaluate the sum in (1.5.8). The term associated with $\alpha = \beta = 0$ is just $\tau(p^\delta)$. Then, let: $\mathcal{A}$ be the sum of all terms in (1.5.8) with $\alpha = 0$, $\beta \geq 1$; $\mathcal{B}$ be the sum of all terms in (1.5.8) with $\alpha \geq 1$, $\delta \geq \beta \geq 0$; $\mathcal{C}$ be the sum of all terms in (1.5.8) with $\alpha \geq 1$, $\beta = \delta + 1$. Recall,

$$\tilde{\varphi}(1) = 1, \quad \tilde{\varphi}(p^l) = \|p\|^l\Big(1 - \frac{1}{\|p\|}\Big) \text{ if } l \geq 1; \quad S(p^\delta, p^\beta) = 0 \text{ if } \beta \geq \delta + 2;$$

$$S(p^\delta, p^\beta) = \tilde{\varphi}(p^\beta) \text{ if } \beta \leq \delta, \ \beta \equiv 0 \ (\mathrm{mod}\ 3);$$

see 0.1.4. So, we have

$$Q^{(p)}(\delta; s) = \tau(p^\delta) + \mathcal{A} + \mathcal{B} + \mathcal{C} \tag{1.5.9}$$

with

$$\mathcal{A} = \Big(1 - \frac{1}{\|p\|}\Big) \sum_{\substack{1 \leq \beta \leq \delta \\ \beta \equiv 0(3)}} \frac{\tau(p^{\delta-\beta})}{\|p\|^{(3s/2-3/2)\beta}},$$

$$\mathcal{B} = \Big(1 - \frac{1}{\|p\|}\Big) \sum_{\substack{\alpha \geq 1,\, \beta \geq 0 \\ \delta \geq \beta,\ \alpha+\beta \equiv 0(3)}} \frac{\tau(p^{\delta+\alpha-\beta})\, S(p^\delta, p^\beta)}{\|p\|^{(3s-3)\alpha/2+(3s-1)\beta/2}},$$

$$\mathcal{C} = \Big(1 - \frac{1}{\|p\|}\Big) \frac{S(p^\delta, p^{\delta+1})}{\|p\|^{(3s-1)(\delta+1)/2}} \sum_{\substack{\alpha \geq 1 \\ \alpha \equiv -1-\delta(3)}} \frac{\tau(p^{\alpha-1})}{\|p\|^{(3s-3)\alpha/2}}.$$

One can express $\mathcal{A}$ as

$$\mathcal{A} = \tau(p^\delta)\Big(1 - \frac{1}{\|p\|}\Big) \sum_{\substack{1 \leq \beta \leq \delta \\ \beta \equiv 0(3)}} \frac{1}{\|p\|^{(3s/2-4/3)\beta}}, \tag{1.5.10}$$

since $\tau(p^{\delta-\beta}) = \tau(p^\delta)\|p\|^{-\beta/6}$ if $\beta \equiv 0 \ (\mathrm{mod}\ 3)$. For the sum $\mathcal{B}$ we have

$$\mathcal{B} = \tau(p^\delta)\Big(1 - \frac{1}{\|p\|}\Big)\widetilde{\mathcal{B}} \quad \text{with}$$
$$\widetilde{\mathcal{B}} = \sum_{\substack{\alpha \geq 1 \\ \alpha \equiv 0(3)}} \frac{1}{\|p\|^{(3s/2-5/3)\alpha}} + \Big(1 - \frac{1}{\|p\|}\Big) \sum_{\substack{\alpha \geq 1 \\ \delta \geq \beta \geq 1,\ \alpha \equiv \beta \equiv 0(3)}} \frac{1}{\|p\|^{(3s/2-5/3)\alpha+(3s/2-4/3)\beta}}, \tag{1.5.11}$$

since $\tau(p^{\delta+\alpha-\beta}) = \tau(p^\delta)\|p\|^{(\alpha-\beta)/6}$ for $\alpha \equiv \beta \equiv 0 \ (\mathrm{mod}\ 3)$, and $S(p^\delta, p^\beta) = 0$ for $\delta \geq \beta$, $\beta \not\equiv 0 \ (\mathrm{mod}\ 3)$. To evaluate $\mathcal{C}$ notice that, if $\alpha \equiv -1 - \delta \ (\mathrm{mod}\ 3)$,

$$\tau(p^{\alpha-1}) = \begin{cases} \overline{S(1,p)}\|p\|^{(\alpha-5)/6} & \text{if } \delta \equiv 0(3), \\ \|p\|^{(\alpha-1)/6} & \text{if } \delta \equiv 1(3), \\ 0 & \text{if } \delta \equiv 2(3). \end{cases}$$

At the same time, $|S(1,p)|^2 = \|p\|$, $\tau(p^\delta) = 0$ if $\delta \equiv 2 \ (\mathrm{mod}\ 3)$, and

$$S(p^\delta, p^{\delta+1}) = \tau(p^\delta)\begin{cases} S(1,p)\|p\|^{5\delta/6} & \text{if } \delta \equiv 0(3), \\ \|p\|^{5\delta/6+2/3} & \text{if } \delta \equiv 1(3) \end{cases}$$

see 0.1.4 and 0.3.8. By means of these formulae one can express $\mathcal{C}$ as

$$\mathcal{C} = \frac{\tau(p^\delta)}{\|p\|^{(3s/2-4/3)\delta+3s/2}}\Big(1 - \frac{1}{\|p\|}\Big)\widetilde{\mathcal{C}} \quad \text{with}$$

$$\widetilde{\mathcal{C}} = \sum_{\substack{\alpha\geq 1 \\ \alpha\equiv -1-\delta(3)}} \frac{1}{\|p\|^{(3s/2-5/3)\alpha}}\begin{cases} \|p\|^{2/3} & \text{if } \delta \equiv 0(3), \\ \|p\| & \text{if } \delta \equiv 1(3), \\ 0 & \text{if } \delta \equiv 2(3). \end{cases} \tag{1.5.12}$$

Next, summing the geometric series in (1.5.10), (1.5.11), (1.5.12), and taking into account (1.5.9), we get, after elementary transformations, the final result

$$Q^{(p)}(\delta; s) = \tau(p^\delta)\Big(1 - \frac{1}{\|p\|^{9s/2-5}}\Big)^{-1}\Big(1 - \frac{1}{\|p\|^{9s/2-3}}\Big).$$

With this expression for $Q^{(p)}(\delta; s)$, we see that (1.5.7), Corollary 0.3.10 and (0.1.17) yield

$$\mathcal{D}_{\mu,0}(s) = k(\mu)\Big\{\prod_p \tau(p^{\vartheta_p})\Big\}\frac{\zeta_*(9s/2-5)}{\zeta_*(9s/2-3)} = \tau(\mu)\frac{\zeta_*(9s/2-5)}{\zeta_*(9s/2-3)},$$

that is (1.5.3), as required. ■

1.5.4 Fourier coefficients $c_{\mu,\nu}$ with $\mu\nu \neq 0$. Let us set $\vartheta_p = \operatorname{ord}_p \mu$ and $\zeta_p = \operatorname{ord}_p \nu$, for any prime $p \equiv 1 \pmod 3$, and let define $\mathcal{D}^{(p)}_{\mu,\nu}(s)$ by

$$\Big(1 - \frac{1}{\|p\|^{9s/2-3}}\Big)\mathcal{D}^{(p)}_{\mu,\nu}(s)$$

$$= \sum_{\alpha,\beta}\Big(\frac{\mu\nu^{-1}p^{\zeta_p-\vartheta_p}}{p}\Big)^{\alpha+\beta}\frac{\tau(p^{\vartheta_p+\alpha-\beta})S(p^{\vartheta_p}, p^\beta)S(p^{\zeta_p}, p^\alpha, p^\beta)}{\|p\|^{(3s-1)(\alpha+\beta)/2}},$$

where the summation is carried out over non-negative rational integers α and β subject to the condition $\vartheta_p + \alpha - \beta \geq 0$, τ is the Pattterson's function defined in 0.3.8, the Gauß sums $S(\ldots)$ are defined in 0.1.4. Only a finite number of the terms in the right-hand side are non-zero.

Theorem 1.5.4 *For the Eisenstein series* $\mathrm{E}(\sigma(\cdot), s; \Theta_{\text{K-P}})$ *Fourier coefficients* $c_{\mu,\nu}$, *with* $\mu\nu \neq 0$, *we have*

$$c_{\mu,\nu} = h\mathcal{D}_{\mu,\nu}(s)I_{\mu,\nu}(u, v, s),$$

$$\mathcal{D}_{\mu,\nu}(s) = k(\mu)\Big\{\prod_{p|\mu\nu}\mathcal{D}^{(p)}_{\mu,\nu}(s)\Big\}\frac{L_{\mu,\nu}(3s/2-1)}{\zeta_*(9s/2-3)},$$

where $h = 2^2/3^5$, $k(\mu)$ *means the product defined in Corollary* 0.3.10, $L_{\mu,\nu}$ *is the cubic Hecke series (see* 0.1.5), *the product is taken over prime* $p \equiv 1 \pmod 3$ *under the condition* $p \mid \mu\nu$ *(i.e.* $\mathrm{ord}_p(\mu\nu) \geq 1$), *and*

$$I_{\mu,\nu}(u,v,s) = \frac{2^{3s-2}\pi^{3s-1}|\mu|^{s-3}|\nu|^{2s-4}}{\Gamma(3s/2-2/3)\Gamma(3s/2-1/3)} R_{s-4/9,8/9}(|\mu|u,|\nu|v)$$

with $R_{s-4/9,8/9}$ *as in Theorem* 1.4.9 *and Theorem* 1.4.11. □

Proof. From Theorem 1.4.1, (1.5.2) and Theorem 1.4.9, it follows that

$$c_{\mu,\nu} = 2^2 3^{-5} \mathcal{D}_{\mu,\nu}(s) J_{\mu,\nu}(u,v,s,4/3) \quad \text{and} \quad J_{\mu,\nu}(u,v,s,4/3) = I_{\mu,\nu}(u,v,s).$$

Consequently we have

$$c_{\mu,\nu} = h\mathcal{D}_{\mu,\nu}(s) I_{\mu,\nu}(u,v,s) \quad \text{with} \quad \mathcal{D}_{\mu,\nu}(s) \quad \text{as in Theorem 1.4.1.} \tag{1.5.13}$$

To express $\mathcal{D}_{\mu,\nu}$ as the product we can apply Theorem 1.4.4 (with $\rho = \tau$). For this we take into account the factorization (1.5.5) for $\tau(-\mu n_1/n_2)$, the factorization (1.5.4) for $S(\mu, n_2)$, and the factorization

$$S(\nu, n_1, n_2) = \{\prod_{p \neq p'} \left(\frac{p'}{p}\right)^{\alpha_p \alpha_{p'} - \alpha_p \beta_{p'} - \alpha_{p'} \beta_p}\}\{\prod_p \left(\frac{\nu p^{-\zeta_p}}{p}\right)^{-\alpha_p - \beta_p} S(p^{\zeta_p}, p^{\alpha_p}, p^{\beta_p})\}$$

which follows from 0.1.4 (g), and multiplying all of them we find

$$\begin{aligned} &\tau(-\mu n_1/n_2) S(\mu, n_2) S(\nu, n_1, n_2) \\ &= k(\mu) \prod_p \left(\frac{\mu\nu^{-1} p^{\zeta_p - \vartheta_p}}{p}\right)^{\alpha_p + \beta_p} \tau(p^{\vartheta_p + \alpha_p - \beta_p}) S(p^{\vartheta_p}, p^{\beta_p}) S(p^{\zeta_p}, p^{\alpha_p}, p^{\beta_p}), \end{aligned} \tag{1.5.14}$$

where $\alpha_p = \mathrm{ord}_p n_1$, $\beta_p = \mathrm{ord}_p n_2$. From Theorem 1.4.4 and (1.5.14), we obtain the decomposition

$$\mathcal{D}_{\mu,\nu}(s) = k(\mu) \prod_p Q^{(p)}_{\mu,\nu}(s) \tag{1.5.15}$$

with

$$Q^{(p)}_{\mu,\nu}(s) = \sum_{\alpha,\beta} \left(\frac{\mu\nu^{-1} p^{\zeta_p - \vartheta_p}}{p}\right)^{\alpha+\beta} \frac{\tau(p^{\vartheta_p + \alpha - \beta}) S(p^{\vartheta_p}, p^{\beta}) S(p^{\zeta_p}, p^{\alpha}, p^{\beta})}{\|p^{\alpha+\beta}\|^{(3s-1)/2}}, \tag{1.5.16}$$

where the summation is carried out over non-negative rational integers α, β subject to the condition $\vartheta_p + \alpha - \beta \geq 0$. If $p \nmid \mu\nu$, then $\zeta_p = \vartheta_p = 0$ and

$$Q^{(p)}_{\mu,\nu}(s) = \sum_{\alpha \geq \beta \geq 0} \left(\frac{\mu\nu^{-1}}{p}\right)^{\alpha+\beta} \frac{\tau(p^{\alpha-\beta}) S(1, p^{\beta}) S(1, p^{\alpha}, p^{\beta})}{\|p^{\alpha+\beta}\|^{(3s-1)/2}}. \tag{1.5.17}$$

If $\beta \geq 2$, then $S(1, p^\beta) = 0$. Hence, one can omit the terms in (1.5.35) with $\beta \geq 2$. The formulae in 0.1.4 yield

$$S(1,1) = 1, \quad |S(1,p)|^2 = \|p\|, \quad \tau(1) = 1, \quad \tau(p) = \overline{S(1,p)}\|p\|^{-1/2},$$

$$S(1, p^\alpha, p^\beta) = \begin{cases} 1 & \text{if } \alpha = \beta = 0, \\ S(1,p) & \text{if } \alpha = 1,\ \beta = 0, \\ \overline{S(1,p)} & \text{if } \alpha = \beta = 1, \\ 0 & \text{if } \alpha \geq 2 \geq \beta. \end{cases}$$

This allows us to evaluate the sum in (1.5.17), and after some computation we find that

$$Q^{(p)}_{\mu,\nu}(s) = 1 + R + R^2 = (1 - R^3)(1 - R)^{-1},$$

where

$$R = \left(\frac{\mu\nu^{-1}}{p}\right)\frac{1}{\|p\|^{3s/2-1}} \quad \text{and so} \quad R^3 = \frac{1}{\|p\|^{9s/2-3}}.$$

Consequently,

$$Q^{(p)}_{\mu,\nu}(s) = \Big(1 - \frac{1}{\|p\|^{9s/2-3}}\Big)\Big(1 - \left(\frac{\mu\nu^{-1}}{p}\right)\frac{1}{\|p\|^{3s/2-1}}\Big)^{-1} \tag{1.5.18}$$

for all $p \nmid \mu\nu$. Also, comparing (1.5.16) with the definition of $\mathcal{D}^{(p)}_{\mu,\nu}(s)$, we see that

$$Q^{(p)}_{\mu,\nu}(s) = \Big(1 - \frac{1}{\|p\|^{9s/2-3}}\Big)\mathcal{D}^{(p)}_{\mu,\nu}(s) \tag{1.5.19}$$

for all p. From (1.5.19), (1.5.18) and (1.5.15), we obtain

$$\begin{aligned}
\mathcal{D}_{\mu,\nu}(s) &= k(\mu)\Big\{\prod_{p \mid \mu\nu} Q^{(p)}_{\mu,\nu}(s)\Big\}\Big\{\prod_{p \nmid \mu\nu} Q^{(p)}_{\mu,\nu}(s)\Big\} \\
&= k(\mu)\Big\{\prod_{p \mid \mu\nu} \mathcal{D}^{(p)}_{\mu,\nu}(s)\Big\}\Big\{\prod_{p}\Big(1 - \frac{1}{\|p\|^{9s/2-3}}\Big)\Big\}\Big\{\prod_{p \nmid \mu\nu}\Big(1 - \left(\frac{\mu\nu^{-1}}{p}\right)\frac{1}{\|p\|^{3s/2-1}}\Big)^{-1}\Big\} \\
&= k(\mu)\Big\{\prod_{p \mid \mu\nu} \mathcal{D}^{(p)}_{\mu,\nu}(s)\Big\}\frac{L_{\mu,\nu}(3s/2-1)}{\zeta_*(9s/2-3)}, \qquad \text{as required.} \qquad \blacksquare
\end{aligned}$$

1.5.5 Poles of $c_{0,0}$. Let us turn to Theorem 1.5.1 and look at the Eisenstein series $E(\sigma(\cdot), s; \Theta_{K\text{-}P})$ Fourier coefficient $c_{0,0}$. As a function on $\mathbf{C}$ (with fixed $u, v \in \mathbf{R}^*_+$) coefficient $c_{0,0}$ is meromorphic and we would like to look at its singularities. With 0.1.5 and Theorem 1.5.1 we find that

$c_{0,0}$ has a simple pole at the point $4/3$,

just because of

ζ_* has a simple pole at the point 1,
ζ_* is regular and non-zero at the point 3,
$I_{0,0}(u, v, \cdot)$ is regular and non-zero at the point $4/3$.

We find also that the point $4/3$ is the only pole of $c_{0,0}$ in $\{ s \in \mathbf{C} \mid \Re(s) \geq 8/9 \}$.

As an obviouse consequence, we have that the Eisenstein series $\mathrm{E}(w, \cdot; \Theta_{\text{K-P}})$ has a pole at the point $4/3$. Actually, one can show that this pole is simple and it is the largest one in the sense that all other poles lay in $\{ s \in \mathbf{C} \mid \Re(s) < 4/3 \}$. According to the general Eisenstein series theory, to prove these facts one need to look at the constant term of $\mathrm{E}(w, \cdot; \Theta_{\text{K-P}})$, that is the function

$$(\lambda; u, v; s) \mapsto \iiint_{\mathcal{F}_q} \mathrm{E}(\lambda w, s; \Theta_{\text{K-P}})\, dz_1 dz_2 dz_3,$$

where $s \in \mathbf{C}$, $u, v \in \mathbf{R}_+^*$ and $\lambda \in \mathrm{SL}(3, \mathcal{O})$ (compare with (1.3.3), (1.4.3), (1.3.23)). One has that any $s' \in \mathbf{C}$ is m^{th} order pole of the Eisenstein series $\mathrm{E}(w, \cdot; \Theta_{\text{K-P}})$ if and only if it is m^{th} order pole of the constant term for at least some λ. The $c_{0,0}$ we study corresponds to $\lambda = \sigma$. The computation one needs to evaluate the constant term for all $\lambda \in \mathrm{SL}(3, \mathcal{O})$ is rather massive, even if we take into account that (in view of Piatetski-Shapiro–Shalika expansion in 1.3.8, Proposition 1.3.7 and Theorem 0.3.8) it is sufficient to deal with

$$\lambda = \begin{pmatrix} \gamma & & 0 \\ & & 0 \\ 0 & 0 & 1 \end{pmatrix}, \quad \gamma \in \widetilde{\Gamma}^{(2)}_{\text{princ}}(3) \backslash \mathrm{SL}(2, \mathcal{O}),$$

only. We omit this computation.

1.5.6 Cubic theta function. Now we are in a position to define the main object of our research. General Eisenstein series theory and our considerations in 1.5.5 lead to the following theorem.

Theorem 1.5.5 *The Eisenstein series $\mathrm{E}(\sigma(\cdot), s; \Theta_{\text{K-P}})$ has a simple pole at the point $4/3$ and its residue at this pole is a cubic metaplectic form on $\mathbf{X}$ under the group $\Gamma(3)$.* ∎

We define the cubic theta function $\Theta\colon \mathbf{X} \to \mathbf{C}$ as

$$\Theta = 2^{-8/3} 3^{22/3} \pi^{-11/3} \zeta_*(3) \operatorname*{Res}_{s=4/3} \mathrm{E}(\sigma(\cdot), s; \Theta_{\text{K-P}}). \tag{1.5.20}$$

Up to the constant factor in (1.5.20), the Fourier coefficients $c_{\mu,\nu}(u, v; \Theta)$ of the cubic theta function Θ are residues of the Fourier coefficients $c_{\mu,\nu}$ of the Eisenstein series $\mathrm{E}(\sigma(\cdot), s; \Theta_{\text{K-P}})$.

Given $\lambda, \mu, \nu \in (\sqrt{-3})^{-3}\mathcal{O} \setminus \{0\}$, we set

$$\tilde{\tau}(\lambda) = \frac{\tau(\lambda)}{|\lambda|^{1/3}} \quad \text{and then}$$

$$\tilde{\tau}(\mu, \nu) = 3\frac{\tau(\mu)}{|\mu|^{1/3}} = 3\frac{\tau(\nu)}{|\nu|^{1/3}} \quad \text{if } \mu/\nu \text{ is a cube in } \mathbf{Q}(\sqrt{-3}), \tag{1.5.21}$$

$$\tilde{\tau}(\mu, \nu) = 0 \quad \text{otherwise,}$$

where τ means Patterson's function (0.3.15).

In the theorem below we give explicit expressions for $c_{\mu,\nu}(u, v; \Theta)$.

Theorem 1.5.6 *The cubic theta function* $\Theta\colon \mathbf{X} \to \mathbf{C}$ *is a cubic metaplectic form under* $\Gamma(3)$. *For its Fourier coefficients* $c_{\mu,\nu}(u, v; \Theta)$, $\mu, \nu \in (\sqrt{-3})^{-3}\mathcal{O}$, *one has the following formulae.*

(a) $c_{0,0}(u, v; \Theta) = (6\pi)^{-2/3}(uv)^{4/3}$,

(b) $c_{\mu,0}(u, v; \Theta) = \tilde{\tau}(\mu)(uv)^{4/3} Ai((6\pi|\mu|u)^{2/3})$ *if* $\mu \neq 0$,

(c) $c_{0,\nu}(u, v; \Theta) = \tilde{\tau}(\nu)(uv)^{4/3} Ai((6\pi|\nu|v)^{2/3})$ *if* $\nu \neq 0$,

(d) $c_{\mu,\nu}(u, v; \Theta) = \tilde{\tau}(\mu, \nu)(uv)^{4/3} Ai((6\pi|\mu|u)^{2/3} + (6\pi|\nu|v)^{2/3})$ *if* $\mu\nu \neq 0$.

Here by $\tilde{\tau}$ *we mean the function defined in* (1.5.21), *and* Ai *means the Airy function. In particular, we have* $c_{\mu,\nu}(u, v; \Theta) = 0$, *whenever* μ/ν *is not a cube in* $\mathbf{Q}(\sqrt{-3})$, $\mu\nu \neq 0$. □

Proof. (a), (b) and (c). Taking into account the factor involved in the definition (1.5.20), we find from Theorem 1.5.1, Theorem 1.5.2 and Theorem 1.5.3 that

$$\begin{aligned} c_{0,0}(u, v; \Theta) &= h(uv)^{4/3} \operatorname*{Res}_{s=4/3} \zeta_*(9s/2 - 5) I_{0,0}(u, v, 4/3), \\ c_{\mu,0}(u, v; \Theta) &= h'\tau(\mu) \operatorname*{Res}_{s=4/3} \zeta_*(9s/2 - 5) I_{\mu,0}(u, v, 4/3), \\ c_{0,\nu}(u, v; \Theta) &= h'' \operatorname*{Res}_{s=4/3} \Psi(3s/2 - 2/3, \nu) I_{0,\nu}(u, v, 4/3), \end{aligned} \tag{1.5.22}$$

where $h = 2^{1/3}3^{11/6}\pi^{-5/3}$, $h' = 2^{4/3}3^{7/3}\pi^{-5/3}$, $h'' = 2^{1/3}3^{11/6}\pi^{-5/3}\zeta_*(2)$ and

$$\begin{aligned} I_{0,0}(u, v, 4/3) &= 2^{-3}3^2u^{4/3}v^{4/3}, \\ I_{\mu,0}(u, v, 4/3) &= 2^{-3}3^2u^{5/3}v^{4/3}K_{1/3}(4\pi|\mu|u), \\ I_{0,\nu}(u, v, 4/3) &= 2^{-5/3}3\pi^{1/3}\Gamma(4/3)^{-1}|\nu|^{1/3}u^{4/3}v^{5/3}K_{1/3}(4\pi|\nu|v). \end{aligned} \tag{1.5.23}$$

We have also

$$\operatorname*{Res}_{s=4/3} \zeta_*(9s/2 - 5) = 2^2\pi/3^{9/2},$$

$$\operatorname*{Res}_{s=4/3} \Psi(3s/2 - 2/3, \nu) = 2^{5/3}3^{-3}\pi^{2/3}\zeta_*(2)^{-1}\Gamma(4/3)\tau(\nu)|\nu|^{-1/3},$$

see 0.1.5, Theorem 0.4.3 (b) and use the triplication formula in 0.1.6. Substituting these expressions for the residues and (1.5.23) into (1.5.22), and applying the Wirtinger formula (0.1.26) we get the formulae stated in (a), (b) and (c).

It is not so easy to prove the assertion (d). It follows from Theorem 1.5.4, Theorem 1.4.13 and (1.5.20) that

$$c_{\mu,\nu}(u,v;\Theta) = hk(\mu)\prod_{p|\mu\nu}\mathcal{D}^{(p)}_{\mu,\nu}(4/3)\operatorname*{Res}_{s=4/3} L_{\mu,\nu}(3s/2-1)\, I_{\mu,\nu}(u,v,4/3),$$
$$I_{\mu,\nu}(u,v,4/3) = h'|\mu|^{-1/3}(uv)^{4/3} Ai((6\pi|\mu|u)^{2/3} + (6\pi|\nu|v)^{2/3}), \qquad (1.5.24)$$

where $h = 2^{-2/3}3^{7/3}\pi^{-11/3}$ and $h' = 2^{-4/3}3^{13/6}\pi^{8/3}$.

If μ/ν is not a cube in $\mathbf{Q}(\sqrt{-3})$, then $L_{\mu,\nu}$ is an entire function, hence (1.5.24) yields $c_{\mu,\nu}(u,v;\Theta) = 0$, that is in agreement with assertion (d), in view of the definition (1.5.21). Then, we known from 0.1.5 that

$$\operatorname*{Res}_{s=4/3} L_{\mu,\nu}(3s/2-1) = 2^2 3^{-7/2}\pi \prod_{p|\mu\nu}\Big(1-\frac{1}{\|p\|}\Big), \qquad (1.5.25)$$

$$\text{if } \mu/\nu \text{ is a cube in } \mathbf{Q}(\sqrt{-3}). \qquad (1.5.26)$$

From now on we assume that μ and ν satisfy (1.5.26). We have to evaluate $\mathcal{D}^{(p)}_{\mu,\nu}(4/3)$. Immediately from the definition in 1.5.4 we get

$$\mathcal{D}^{(p)}_{\mu,\nu}(4/3) = \Big(1-\frac{1}{\|p\|^3}\Big)^{-1} Q^{(p)}(\vartheta_p,\zeta_p) \qquad (1.5.27)$$

where $\vartheta_p = \operatorname{ord}_p\mu$, $\zeta_p = \operatorname{ord}_p\nu$, and

$$Q^{(p)}(\vartheta,\zeta) = \sum_{\alpha,\beta}\frac{\tau(p^{\vartheta+\alpha-\beta})S(p^\vartheta,p^\beta)S(p^\zeta,p^\alpha,p^\beta)}{\|p\|^{3(\alpha+\beta)/2}}, \qquad (1.5.28)$$

under the assumption (1.5.26). The summation in (1.5.28) is carried out over non-negative rational integers α,β subject to the condition $\vartheta+\alpha-\beta\geq 0$ (just like in the definition of $\mathcal{D}^{(p)}_{\mu,\nu}(s)$). Let

$\mathcal{A}$ be the sum of the terms in (1.5.28) with $\alpha\leq\zeta$, $\beta\leq\vartheta$;

$\mathcal{B}$ be the sum of the terms in (1.5.28) with $\alpha=\zeta+1$, $\beta\leq\vartheta$;

$\mathcal{C}$ be the sum of the terms in (1.5.28) with $\alpha\leq\zeta$, $\beta=\vartheta+1$;

$\mathcal{D}$ be the term in (1.5.28) with $\alpha=\zeta+1$, $\beta=\vartheta+1$.

All other terms in (1.5.28) equals 0, since $S(p^\vartheta,p^\beta)=0$ for $\beta\geq\vartheta+2$ and $S(p^\zeta,p^\alpha,p^\beta)=0$ for $\alpha\geq\zeta+2$. Thus we have

$$Q^{(p)}(\vartheta,\zeta) = \mathcal{A}+\mathcal{B}+\mathcal{C}+\mathcal{D}. \qquad (1.5.29)$$

We shall compute (1.5.29) for $\zeta\equiv\vartheta \pmod 3$. Notice, see 0.1.4, that

$$\tilde{\varphi}(1) = 1, \quad \tilde{\varphi}(p^l) = \|p\|^l\Big(1-\frac{1}{\|p\|}\Big) \text{ if } l\geq 1,$$
$$\tau(p^{\vartheta+\gamma}) = \tau(p^\vartheta)\|p\|^{\gamma/6} \text{ if } \gamma\equiv 0(3),$$

$$S(p^\vartheta, p^\beta) = \begin{cases} \tilde\varphi(p^\beta) & \text{if } \beta \equiv 0(3),\ \beta \le \vartheta, \\ 0 & \text{if } \beta \not\equiv 0(3),\ \beta \le \vartheta, \end{cases}$$

$$S(p^\zeta, p^\alpha, p^\beta) = \begin{cases} \tilde\varphi(p^\alpha) & \text{if } \alpha+\beta \equiv 0(3),\ \alpha \le \zeta, \\ 0 & \text{if } \alpha+\beta \not\equiv 0(3),\ \alpha \le \zeta, \end{cases}$$

$$\tau(p^{\vartheta+\zeta+1})\, S(p^\zeta, p^{\zeta+1}) = \tau(p^\vartheta)\|p\|^{7\zeta/6} \begin{cases} 1 & \text{if } \vartheta \equiv \zeta \equiv 2(3), \\ \|p\|^{5/6} & \text{if } \vartheta \equiv \zeta \equiv 1(3), \\ \|p\|^{1/2} & \text{if } \vartheta \equiv \zeta \equiv 0(3), \end{cases}$$

$$\tau(p^{\alpha-1})\, S(p^\vartheta, p^{\vartheta+1}) = \tau(p^\vartheta)\|p\|^{\alpha/6+5\vartheta/6} \begin{cases} 1 & \text{if } \vartheta \equiv 2(3),\ \alpha \equiv 0(3), \\ \|p\|^{-1/6} & \text{if } \vartheta \equiv 1(3),\ \alpha \equiv 1(3), \\ \|p\|^{1/6} & \text{if } \vartheta \equiv 0(3),\ \alpha \equiv 2(3). \end{cases}$$

Applying these formulae we find

$$\mathcal{A} = \tau(p^\vartheta)\Big\{ \sum_{\substack{0\le\alpha\le\zeta \\ \alpha\equiv 0(3)}} \frac{\tilde\varphi(p^\alpha)}{\|p\|^{4\alpha/3}} \Big\}\mathcal{E} = \tau(p^\vartheta)\Big\{ 1 + \Big(1 - \frac{1}{\|p\|}\Big) \sum_{1\le l\le \zeta/3} \|p\|^{-l} \Big\}\mathcal{E},$$

$$\mathcal{B} = \mathcal{E}\,\frac{\tau(p^{\vartheta+\zeta+1})\, S(p^\zeta, p^{\zeta+1})}{\|p\|^{3(\zeta+1)/2}} = \mathcal{E}\,\frac{\tau(p^\vartheta)}{\|p\|^{\zeta/3+2/3}} \begin{cases} 1 & \text{if } \vartheta \equiv \zeta \equiv 2(3), \\ \|p\|^{5/6} & \text{if } \vartheta \equiv \zeta \equiv 1(3), \\ \|p\|^{1/2} & \text{if } \vartheta \equiv \zeta \equiv 0(3), \end{cases}$$

where

$$\mathcal{E} = \sum_{\substack{0\le\beta\le\vartheta \\ \beta\equiv 0(3)}} \frac{\tilde\varphi(p^\beta)}{\|p\|^{5\beta/3}} = 1 + \Big(1 - \frac{1}{\|p\|}\Big) \sum_{1\le l\le \vartheta/3} \|p\|^{-2l},$$

and, consequently,

$$\mathcal{A} + \mathcal{B} = \tau(p^\vartheta)\Big(1 + \frac{1}{\|p\|}\Big)\mathcal{E} = \tau(p^\vartheta)\Big(1 + \frac{1}{\|p\|}\Big)\Big\{ 1 + \Big(1 - \frac{1}{\|p\|}\Big) \sum_{1\le l\le \vartheta/3} \|p\|^{-2l} \Big\}.$$

Also we find that

$$\mathcal{D} = \frac{\tau(p^\vartheta)\, S(p^\vartheta, p^{\vartheta+1})\, S(p^\zeta, p^{\zeta+1}, p^{\vartheta+1})}{\|p\|^{4\zeta/3+5\vartheta/3+3}} = \frac{\tau(p^\vartheta)}{\|p\|^{\zeta/3+2\vartheta/3+2}},$$

$$\mathcal{C} = \frac{S(p^\vartheta, p^{\vartheta+1})}{\|p\|^{3(\vartheta+1)/2}} \sum_{\substack{1\le\alpha\le\zeta \\ \alpha+\vartheta+1\equiv 0(3)}} \frac{\tau(p^{\alpha-1})\tilde\varphi(p^\alpha)}{\|p\|^{3\alpha/2}} = \Big(1 - \frac{1}{\|p\|}\Big) \frac{S(p^\vartheta, p^{\vartheta+1})}{\|p\|^{3(\vartheta+1)/2}} \sum_{\substack{1\le\alpha\le\zeta \\ \alpha+\vartheta+1\equiv 0(3)}} \frac{\tau(p^{\alpha-1})}{\|p\|^{\alpha/2}}$$

$$= \Big(1 - \frac{1}{\|p\|}\Big) \frac{\tau(p^\vartheta)}{\|p\|^{2\vartheta/3}} \sum_{\substack{1\le\alpha\le\zeta \\ \alpha+\vartheta+1\equiv 0(3)}} \frac{1}{\|p\|^{\alpha/3}} \begin{cases} 1 & \text{if } \vartheta \equiv 2(3), \\ \|p\|^{-1} & \text{if } \vartheta \equiv 1(3), \\ \|p\|^{-4/3} & \text{if } \vartheta \equiv 0(3), \end{cases}$$

and thus

$$\mathcal{C}+\mathcal{D}=\frac{\tau(p^{\vartheta})}{\|p\|^{2\vartheta/3+2}}\begin{cases}1 & \text{if } \vartheta\not\equiv 1(3),\\ \|p\|^{2/3} & \text{if } \vartheta\equiv 1(3).\end{cases}$$

As a consequence, we have

$$\mathcal{A}+\mathcal{B}+\mathcal{C}+\mathcal{D}=\tau(p^{\vartheta})\Big(1+\frac{1}{\|p\|}+\frac{1}{\|p\|^2}\Big)$$

and, see (1.5.27), (1.5.29), the final expression

$$\mathcal{D}^{(p)}_{\mu,\nu}(4/3)=\tau(p^{\vartheta_p})\Big(1-\frac{1}{\|p\|}\Big)^{-1}. \tag{1.5.30}$$

Multiplying (1.5.30) over all prime $p\,|\,\mu\nu$, $p\equiv 1 \pmod 3$, and applying Corollary 0.3.10 we get

$$k(\mu)\prod_{p|\mu\nu}\mathcal{D}^{(p)}_{\mu,\nu}(4/3)=\tau(\mu)\prod_{p|\mu\nu}\Big(1-\frac{1}{\|p\|}\Big)^{-1}. \tag{1.5.31}$$

Now (d) follows from (1.5.24), (1.5.25) and (1.5.31). ■

1.5.7 Commentary. The cubic theta function Θ has been constructed and its Fourier coefficients have been evaluated in [78]. Our present exposition differs in certain respects from the original one, and we would like here to clarify the differences and the common points of them. First of all let us notice that the cubic Gauß sums, the Kubota and the Bass–Milnor–Serre homomorphisms and cubic metaplectic forms in [78] are complex conjugate to that in these notes. Also, we have used in [78] the so called Langlands coordinates on $\mathbf{X}\simeq \mathrm{SL}(3,\mathbf{C})/\mathrm{SU}(3)$, while the slightly changed coordinates introduced in 1.3.1 (being not so convenient for the definition of Eisenstein series in 1.3.7) give rise to more symmetric formulae in many cases. With these changes in mind, one can say the main object we dealt with in [78] is, in our present notations, the cubic metaplectic Eisenstein series

$$\mathrm{E}_{\min}(\cdot;s-t/3,2t/3), \tag{1.5.32}$$

defined for the group $\Gamma(3)$. Applying the method proposed originally by Vinogradov and Tahtajan in [100] (where the minimal parabolic Eisenstein series on $\mathrm{SL}(3,\mathbf{R})/\mathrm{SO}(3)$ automorphic under $\mathrm{SL}(3,\mathbf{Z})$ has been considered), we have evaluated its Fourier coefficients. It is clear now this method is not the best one to treate metaplectic forms, in particular because its application request explicit formulae (as that given in 0.2.6) for the Bass–Milnor–Serre homomorphism and, in this connection, causes a lot of unpleasant computations. Our present '$sl(2)$-triples technique' allows us to get Fourier coefficients of Eisenstein series $\mathrm{E}(\cdot,s;f)$, $\widetilde{\mathrm{E}}(\cdot,s;f)$, attached to any form f and to the group $\Gamma(q)$ with any $q\subset(3)$ with minimum computations. However, the Fourier coefficients of (1.5.32) were obtained, and it has became clear that, for fixed $s\in\mathbf{C}$, the series (1.5.32), as a function of t, has a simple pole at the point $t=4/3$, and that

$$\operatorname*{Res}_{t=4/3}\mathrm{E}_{\min}(\cdot;s-t/3,2t/3), \tag{1.5.33}$$

being considered as a function of s, has a simple pole at the point $s = 4/3$, that gives rise to the cubic theta function defined as

$$\operatorname*{Res}_{s=4/3} \operatorname*{Res}_{t=4/3} E_{\min}(\cdot; s - t/3, 2t/3). \tag{1.5.34}$$

Certainly, both (1.5.33) and (1.5.34) are cubic metaplectic forms under $\Gamma(3)$, and their Fourier coefficients are the residues of that of (1.5.32). In the meantime, one knows rearranging the terms of the series (1.5.32) one can treate it as the maximal parabolic Eisenstein series $E(\cdot, s; f)$ attached to $f = E_*(\cdot, t)$ (see Theorem 1.3.6 (c)). In view of (0.3.14), we have that (1.5.33) is, up to a constant factor, nothing but the Eisenstein series $E(\cdot, s; \Theta_{K\text{-}P})$. The formulae for the Fourier coefficients of (1.5.32) involve the series (0.3.11) and using Patterson's results (in the form in Theorem 0.4.3 (c)) we could able to find the Fourier coefficients of (1.5.33). The formulae in [78] for the Fourier coefficients of (1.5.33) differ in some respect from that stated in Theorem 1.5.4 because of the matrix σ involved in (1.5.1) and because of we did not evaluate in [78] the integrals defining the radial parts of Whittaker functions. Instead of the Euler product for μ, ν^{th} Fourier coefficients, $\mu\nu \neq 0$, given in Theorem 1.5.4, we had in [78] the Euler product

$$\frac{L_{\mu,\nu}(3s/2-1)}{L_{\nu,\mu}(3s-2)} \prod_{p \nmid \mu\nu} \left(1 + \frac{\overline{\chi}(p)}{\|p\|^{3s-2}}\left(1 + \frac{\chi(p)}{\|p\|^{3s/2-1}}\right)^{-1}\right) \nabla(\mu, \nu; s), \tag{1.5.35}$$

where χ is the cubic character defined by μ, ν as in (0.1.19), and we write $\nabla(\mu, \nu; s)$ for certain finite Dirichlet series. (Since $\chi(p)^3 = 1$ for prime $p \nmid \mu\nu$, and $\chi(p) = 0$ for other primes, one can rewrite (1.5.35) as

$$\frac{L_{\mu,\nu}(3s/2-1)}{\zeta_*(9s/2-3)} \prod_{p \mid \mu\nu} \left(1 - \frac{1}{\|p\|^{9s/2-3}}\right)^{-1} \nabla(\mu, \nu; s),$$

so essentially as in Theorem 1.5.4.) The significant point here is the occurence of the cubic Hecke series $L_{\mu,\nu}$ as a factor in the formulae for the Fourier coefficients. We have that the point $s = 4/3$ is or is not a pole of the first factor in (1.5.35) according to μ/ν is or is not a cube, while other factors in (1.5.35) are regular and non-zero at this point. Moreover, it occurs that $\nabla(\mu, \nu; s)$ reduces to very simple expression when $s = 4/3$ and μ/ν is a cube. In such a way we have got the formulae for the Fourier coefficients of the cubic theta function (1.5.34).

It is worth to say that the factor denoted by $\nabla(\mu, \nu; s)$ in (1.5.35) can be written as a product over prime $p \mid \mu\nu$ of the factors similar to $\mathcal{D}^{(p)}_{\mu,\nu}(s)$ in 1.5.4. Summarizing, we have the μ, ν^{th} Fourier coefficients, $\mu\nu \neq 0$, of the series $E(\cdot, s; \Theta_{K\text{-}P})$ 'about the cusp ∞' (as in [78]) as well as 'about the cusp 0' (as in Theorem 1.5.4) can be written as Euler products.

One can replace in the definition (1.5.20) the Eisenstein series $E(\cdot, s; \Theta_{K\text{-}P})$ attached to the maximal parabolic subgroup P by the series $\widetilde{E}(\cdot, s; \Theta_{K\text{-}P})$ attached to the maximal parabolic subgroup $\widetilde{P}$. However, this gives rise to just the same Θ, and not to one more cubic theta function, as it is clear in view of the relations in Theorem 1.3.6.

Now let us look on our subject from Kazhdan–Patterson's theory [46] point of view. Let A_F be adele ring of an algebraic number field F. Assuming F contains primitive m^{th} root of 1, one can define m-fold central extentions $\widetilde{\text{GL}}(n, A_F)$ of the nth order general linear groups $\text{GL}(n, A_F)$. These $\widetilde{\text{GL}}(n, A_F)$ are known as metaplectic groups, or as m-fold metaplectic covers of $\text{GL}(n, A_F)$. (When $n = 2$ one can describe the cocycle defining $\widetilde{\text{GL}}(n, A_F)$ explicitly; this was given by Kubota [52], [54]. The general notion of metaplectic groups was given by C. Moore [65], H. Matsumoto [63].) Then one can define automorphic forms on metaplectic groups, that are metaplectic forms in the sense of Kazhdan and Patterson. (The case $n = 2$ has been considered earlier by Kubota [54], Deligne [18], Flicker [27].) Among other metaplectic forms we find in [46] metaplectic Eisenstein series. It is shown in [46] the metaplectic Eisenstein series have some 'exceptional' poles, that gives rise to existence of theta functions of degree m on $\widetilde{\text{GL}}(n, A_F)$.

In the well known fashion, with any adelic automorphic form on $\text{GL}(n, A_F)$ one can relate classical automorphic form of Hilbert type. In entirely similar way, with adelic metaplectic forms on $\widetilde{\text{GL}}(n, A_F)$ one can relate classical metaplectic forms, that are nothing but automorphic forms with the Bass–Milnor–Serre homomorphisms as multiplier systems. In this sense, $\Theta_{\text{K-P}}$ and Θ are just classical counterparts of Kazhdan-Patterson's adelic theta functions attached to $F = \mathbf{Q}(\sqrt{-3}\,)$, $m = 3$ and, respectively, $n = 2$, $n = 3$.

Concerning the theta functions Fourier coefficients one of the most significant achivements is the general Kazhdan–Patterson periodicity theorem. We can illustrate this point by two examples. First, let us come back to 0.3.8, where we dealt with the Kubota–Patterson cubic theta function $\Theta_{\text{K-P}}$. Suppose for the moment that we do not know the formulae (0.3.15), and so we have the Fourier expansion in Theorem 0.3.8 (c) with some unknown coefficients $\tau(\nu)$. Let $\tilde{\tau}(\nu) = \tau(\nu)||\nu||^{-1/6}$, $\nu \in (\sqrt{-3}\,)^{-3}\mathcal{O} \setminus \{0\}$, as in 0.3.8 and 1.5.6. This case the periodicity theorem says that, given any prime $p \equiv 1 \pmod 3$ and $\nu \in (\sqrt{-3}\,)^{-3}\mathcal{O} \setminus \{0\}$, $l \mapsto \tilde{\tau}(\nu p^l)$ is a periodic function with period 3 in l. So, we have

$$\tau(\nu p^3) = ||p||^{1/2}\tau(\nu) \tag{1.5.36}$$

for all prime $p \equiv 1 \pmod 3$ and $\nu \in (\sqrt{-3}\,)^{-3}\mathcal{O}\setminus\{0\}$. As one knows, $\Theta_{\text{K-P}}$ is an eigenfunction of the Hecke operators T_{p^3}, that yields certain relations satisfied by τ-function, see 0.3.12. These relations alone are not sufficient to determine τ-function, but being combined with (1.5.36) they give rise to $\tau(p) = \overline{S(1,p)}$ for prime $p \equiv 1 \pmod 3$ and $\tau(\nu) = \overline{S(1,c)}|d/c|$, where $\nu = cd^3$, c is cube-free and $c \equiv d \equiv 1 \pmod 3$. So, with the periodicity theorem one can determine τ-function almost explicitly, and one will need some complementary considerations only to find $\tau(\xi(\sqrt{-3}\,)^m)$ for $\xi \in \mathcal{O}^*$, $m \in \mathbf{Z}$. In a similar way, let us assume for the moment we have the formula in Theorem 1.5.6 (d) for the Fourier coefficients of the cubic theta function Θ with some unknown coefficients $\tilde{\tau}(\mu, \nu)$. This case the periodicity theorem says

$$\tilde{\tau}(\mu p^3, \nu) = \tilde{\tau}(\mu, \nu p^3) = \tilde{\tau}(\mu, \nu) \tag{1.5.37}$$

for all prime $p \equiv 1 \pmod 3$ and $\mu, \nu \in (\sqrt{-3})^{-3}\mathcal{O} \setminus \{0\}$. Also, we have Hecke operators acting on cubic metaplectic forms on $\mathbf{X}$, and we have that the Eisenstein series $\mathrm{E}(\cdot, s; \Theta_{\text{K-P}})$, and Θ also, are eigenfunctions of the Hecke operators. As in the first example, this yields some relations satisfyed by $\tilde{\tau}(\mu, \nu)$, which, being combined with (1.5.37), determine the coefficients $\tilde{\tau}(\mu, \nu)$, except, as in the first example, the remaining problem with units and 'bad' prime $\sqrt{-3}$.

One important fact which is contained implicitly in [46] but was not found by the author in [78], is that Θ is invariant under $\mathrm{SL}(3, \mathbf{Z})$. This fact one can get also combining Theorem 1.3.6 with Proposition 1.3.7 and taking into account that $\Theta_{\text{K-P}}$ is invariant under $\mathrm{SL}(2, \mathbf{Z})$. So one can replace the group $\Gamma(3)$ in Theorem 1.5.6 by the group $\widetilde{\Gamma}^{(3)}_{\text{princ}}(3)$.

It was shown by Patterson and Piatetski-Shapiro [74] that, besides Θ, there exist some other cubic metaplectic forms on $\mathbf{X}$ which are cuspidal and joint eigenfunctions of $\mathbf{D}(\mathbf{X})$ attached to the same character as Θ does. It seems these functions maybe very interesting objects for further research.

Part 2

2.1 Group Sp(4, C)

2.1.1 Group Sp(4, C). The symplectic group $\mathrm{Sp}(4, \mathbf{C})$ is a simple complex Lie group of dimension 10 and rank 2. It is the group of type C_2 in Cartan–Killing classification. One has

$$\mathrm{Sp}(4, \mathbf{C}) = \{\, \gamma \in \mathrm{SL}(4, \mathbf{C}) \mid \gamma\sigma{}^t\gamma = \sigma \,\} \tag{2.1.1}$$

where

$$\sigma = \begin{pmatrix} 0 & 0 & -1 & 0 \\ 0 & 0 & 0 & -1 \\ 1 & 0 & 0 & 0 \\ 0 & 1 & 0 & 0 \end{pmatrix}. \tag{2.1.2}$$

Notice that $\sigma^2 = e_4$. If $\gamma \in \mathrm{SL}(4, \mathbf{C})$ is expressed as

$$\gamma = \begin{pmatrix} \mathcal{A} & \mathcal{B} \\ \mathcal{C} & \mathcal{D} \end{pmatrix}, \tag{2.1.3}$$

where $\mathcal{A}$, $\mathcal{B}$, $\mathcal{C}$, $\mathcal{D}$ are 2×2 matrices, then the condition $\gamma\sigma{}^t\gamma = \sigma$ in (2.1.1) is equivalent to

$$\mathcal{A}{}^t\mathcal{B} = \mathcal{B}{}^t\mathcal{A}, \quad \mathcal{C}{}^t\mathcal{D} = \mathcal{D}{}^t\mathcal{C}, \quad \mathcal{A}{}^t\mathcal{D} - \mathcal{B}{}^t\mathcal{C} = \mathcal{D}{}^t\mathcal{A} - \mathcal{C}{}^t\mathcal{B} = e_2. \tag{2.1.4}$$

The centre of $\mathrm{Sp}(4, \mathbf{C})$ is finite, and $\mathrm{Sp}(4, \mathbf{C})$ is invariant under transposition: if $\gamma \in \mathrm{Sp}(4, \mathbf{C})$, then ${}^t\gamma \in \mathrm{Sp}(4, \mathbf{C})$. Some more data on $\mathrm{Sp}(4, \mathbf{C})$ are collected below.

2.1.2 Iwasawa decomposition. One has $\mathrm{Sp}(4,\mathbf{C}) = NAK$, where

$$N \text{ is the group of } n(l) = \begin{pmatrix} 1 & l_1 & l_2-l_1l_4 & l_4 \\ 0 & 1 & l_4-l_1l_3 & l_3 \\ 0 & 0 & 1 & 0 \\ 0 & 0 & -l_1 & 1 \end{pmatrix}, \quad l = (l_1, l_2, l_3, l_4) \in \mathbf{C}^4, \tag{2.1.5}$$

$$A \text{ is the group of } \operatorname{diag}(t_1, t_2, t_1^{-1}, t_2^{-1}), \quad t_1, t_2 \in \mathbf{R}_+^*, \tag{2.1.6}$$

$K = \mathrm{Sp}(4) = \mathrm{Sp}(4,\mathbf{C}) \cap \mathrm{SU}(4)$ is a maximal compact subgroup of $\mathrm{Sp}(4,\mathbf{C})$.

More precisely, each matrix in $\mathrm{Sp}(4,\mathbf{C})$ can be factored uniquely as the product of the matrix in N, of the matrix in A, and of the matrix in K.

One has also that the product mapping $N \times A \times K \to NAK$ is a real analytic manifolds isomorphism (= an analytic diffeomorphism). With this supplement, the decomposition $\mathrm{Sp}(4,\mathbf{C}) = NAK$ is a special case of the Iwasawa decomposition.

2.1.3 Weyl group. Let T be the group of all diagonal matrices in $\mathrm{Sp}(4,\mathbf{C})$, thus of the matrices (2.1.6) with $t_1, t_2 \in \mathbf{C}^*$. Let us write $Norm(T)$ and $Centr(T)$ for normalizer and centralizer of T in $\mathrm{Sp}(4,\mathbf{C})$. Weyl group W of $\mathrm{Sp}(4,\mathbf{C})$, being defined as $W = Norm(T)/Centr(T)$ is dihedral group of order 8. Representatives of cosets in $Norm(T)/Centr(T)$ can be choosen in $Norm(T) \cap K$.

The Weyl group acts on T by conjugation. More precisely, given $\delta \in T$ and $\varepsilon \in W$, the image of δ under ε is defined to be $\delta^\varepsilon = \varepsilon'^{-1}\delta\varepsilon'$, where ε' mean any representative of ε in $Norm(T)$. The group $A \subset T$ is invariant under W, and by restriction we get the action of W on A. This action induces the action of the Weyl group on the group of characters of A. More precisely, given character (= homomorphism) $\Phi\colon A \to \mathbf{C}^*$ and $\varepsilon \in W$, the image ${}^\varepsilon\Phi\colon A \to \mathbf{C}^*$ of Φ under ε is defined to be ${}^\varepsilon\Phi(a) = \Phi(\varepsilon'^{-1}a\varepsilon')$ for any $a \in A$.

2.1.4 Bruhat decomposition. With T and N as above, semidirect product $B = TN = NT$ is a minimal parabolic (or Borel) subgroup of $\mathrm{Sp}(4,\mathbf{C})$. As a special case of Bruhat decomposition we have

$$\mathrm{Sp}(4,\mathbf{C}) = \bigcup_{r \in S} BrB \tag{2.1.7}$$

where S is any representative system in K of the Weyl group. One can take

$$S = \{\, e_4,\ p,\ \tilde{p},\ p\tilde{p},\ \tilde{p}p,\ p\tilde{p}p,\ \tilde{p}p\tilde{p},\ \sigma \,\}. \tag{2.1.8}$$

Here e_4 is 4×4 identity matrix,

$$p = \begin{pmatrix} 0 & 1 & 0 & 0 \\ 1 & 0 & 0 & 0 \\ 0 & 0 & 0 & 1 \\ 0 & 0 & 1 & 0 \end{pmatrix}, \qquad \tilde{p} = \begin{pmatrix} 1 & 0 & 0 & 0 \\ 0 & 0 & 0 & -1 \\ 0 & 0 & 1 & 0 \\ 0 & 1 & 0 & 0 \end{pmatrix}$$

and σ is the matrix in (2.1.2), which we can express also as

$$\sigma = (p\tilde{p})^2. \tag{2.1.9}$$

2.1.5 Involutions. Sp(4, C) possess involutions

$$\gamma \mapsto \begin{pmatrix} \mathcal{D} & -\mathcal{C} \\ -\mathcal{B} & \mathcal{A} \end{pmatrix} \quad \text{and} \quad \gamma \mapsto \begin{pmatrix} \mathcal{A} & -\mathcal{B} \\ -\mathcal{C} & \mathcal{D} \end{pmatrix}, \tag{2.1.10}$$

where we mean $\gamma \in \mathrm{Sp}(4, \mathbf{C})$ is given as in (2.1.3). The first one is the Cartan involution, and it may be defined equivalently as $\gamma \mapsto {}^t\gamma^{-1}$, and also as $\gamma \mapsto \sigma^{-1}\gamma\sigma$. The second one is just $\gamma \mapsto \tilde{\sigma}\gamma\tilde{\sigma}^{-1}$ with $\tilde{\sigma} = \mathrm{diag}(-1, -1, 1, 1)$. The groups A and K in the Iwasawa decomposition are invariant under both of them, but N is invariant under the second one only.

2.1.6 On the group N. If $l = (l_1, l_2, l_3, l_4) \in \mathbf{C}^4$ and $l' = (l'_1, l'_2, l'_3, l'_4) \in \mathbf{C}^4$, then

$$n(l)n(l') = n(t_1, t_2, t_3, t_4), \quad n(l)^{-1} = n(r_1, r_2, r_3, r_4)$$

with

$$t_1 = l_1 + l'_1, \quad t_2 = l_2 + l'_2 + l_1^2 l'_3 + 2l_1 l'_4, \quad t_3 = l_3 + l'_3, \quad t_4 = l_4 + l'_4 + l_1 l'_3,$$

$$r_1 = -l_1, \quad r_2 = -l_2 + 2l_1 l_4 - l_1^2 l_3, \quad r_3 = -l_3, \quad r_4 = -l_4 + l_1 l_3,$$

(see (2.1.5)). Given any $\mu, \nu \in \mathbf{C}$, the mapping

$$n(l_1, l_2, l_3, l_4) \mapsto e(\mu l_1 + \nu l_3) \tag{2.1.11}$$

is an unitary character of N, i.e., the homomorphism $N \to \mathbf{C}^*$ with the values in the unit circle. These are the only unitary characters of N. The character (2.1.11) is said to be degenerate if $\mu\nu = 0$, and non-degenerate otherwise.

2.1.7 Embeddings SL(2, C) → Sp(4, C). For

$$\gamma = \begin{pmatrix} a & b \\ c & d \end{pmatrix} \in \mathrm{SL}(2, \mathbf{C})$$

we set

$$h_1(\gamma) = \begin{pmatrix} 1 & 0 & 0 & 0 \\ 0 & a & 0 & b \\ 0 & 0 & 1 & 0 \\ 0 & c & 0 & d \end{pmatrix}, \quad h_2(\gamma) = \begin{pmatrix} a & 0 & 0 & b \\ 0 & a & b & 0 \\ 0 & c & d & 0 \\ c & 0 & 0 & d \end{pmatrix},$$

$$h_3(\gamma) = \begin{pmatrix} a & 0 & b & 0 \\ 0 & 1 & 0 & 0 \\ c & 0 & d & 0 \\ 0 & 0 & 0 & 1 \end{pmatrix}, \quad h_4(\gamma) = \begin{pmatrix} a & b & 0 & 0 \\ c & d & 0 & 0 \\ 0 & 0 & d & -c \\ 0 & 0 & -b & a \end{pmatrix},$$

Proposition 2.1.1 *h_j are group monomorphisms* $\mathrm{SL}(2,\mathbf{C}) \to \mathrm{Sp}(4,\mathbf{C})$. □

Proof. It is obvious that h_j are group monomorphisms $\mathrm{SL}(2,\mathbf{C}) \to \mathrm{SL}(4,\mathbf{C})$. So, it only remains to check that $h_j(\gamma) \in \mathrm{Sp}(4,\mathbf{C})$ for all $\gamma \in \mathrm{SL}(2,\mathbf{C})$. This can be done by using (2.1.4). ■

2.1.8 Maximal parabolic subgroups. Let P and $\widetilde{P}$ be the groups of all matrices in $\mathrm{Sp}(4,\mathbf{C})$ of the form

$$\begin{pmatrix} * & * & * & * \\ * & * & * & * \\ 0 & 0 & * & * \\ 0 & 0 & * & * \end{pmatrix} \quad \text{and} \quad \begin{pmatrix} * & * & * & * \\ 0 & * & * & * \\ 0 & 0 & * & 0 \\ 0 & * & * & * \end{pmatrix},$$

respectively. Let us write

N_P for the group $\{n(0,l_2,l_3,l_4) \mid l_2,l_3,l_4 \in \mathbf{C}\} \subset N \subset P$,

$N_{\widetilde{P}}$ for the group $\{n(l_1,l_2,0,l_4) \mid l_1,l_2,l_4 \in \mathbf{C}\} \subset N \subset \widetilde{P}$.

The so defined P and $\widetilde{P}$ are maximal parabolic subgroups (associated with two different positive simle roots) of $\mathrm{Sp}(4,\mathbf{C})$. The P and $\widetilde{P}$ are known also as the Siegel and, respectively, as the Jacobi subgroups. The groups N_P and $N_{\widetilde{P}}$ are their unipotent radicals. In view of the formulae in 2.1.6 it is clear that N_P is abelian group, while $N_{\widetilde{P}}$ is a 3-dimentional complex Heisenberg group. In the notations of 2.1.4 we have: P is generated by B and p; $\widetilde{P}$ is generated by B and $\tilde{p}$.

Proposition 2.1.2 *Each $\gamma \in P$ and $\tilde{\gamma} \in \widetilde{P}$ can be factored in* $\mathrm{Sp}(4,\mathbf{C})$ *uniquely as*

$$\gamma = \delta n(l) h_4(\lambda), \qquad \tilde{\gamma} = \delta n(\tilde{l}) h_1(\lambda),$$

where $n(l) \in N_P$, $n(\tilde{l}) \in N_{\widetilde{P}}$, $\lambda \in \mathrm{SL}(2,\mathbf{C})$, $\delta = \mathrm{diag}(t,1,t^{-1},1)$, $t \in \mathbf{C}^*$. □

Proof. It follows immediately from (2.1.3), (2.1.4) that each matrix $\gamma \in P$ can be written as

$$\gamma = \begin{pmatrix} {}^t\mathcal{D}^{-1} & \mathcal{C}\mathcal{D} \\ 0 & \mathcal{D} \end{pmatrix} \quad \text{with} \quad {}^t\mathcal{C} = \mathcal{C}.$$

Let $t = \det \mathcal{D}^{-1}$. Then $\gamma = \delta\vartheta$ with

$$\vartheta = \begin{pmatrix} {}^t\widetilde{\mathcal{D}}^{-1} & \widetilde{\mathcal{C}}\widetilde{\mathcal{D}} \\ 0 & \widetilde{\mathcal{D}} \end{pmatrix} = \begin{pmatrix} 1 & 0 & \multicolumn{2}{c}{\widetilde{\mathcal{C}}} \\ 0 & 1 & & \\ 0 & 0 & 1 & 0 \\ 0 & 0 & 0 & 1 \end{pmatrix} \begin{pmatrix} \multicolumn{2}{c}{{}^t\widetilde{\mathcal{D}}^{-1}} & 0 & 0 \\ & & 0 & 0 \\ 0 & 0 & \multicolumn{2}{c}{\widetilde{\mathcal{D}}} \\ 0 & 0 & & \end{pmatrix}.$$

where $\det \widetilde{\mathcal{D}} = 1$ and $\widetilde{\mathcal{C}} = {}^t\,\widetilde{\mathcal{C}}$. Thus we have the desired expression for γ with $\lambda = {}^t\,\widetilde{\mathcal{D}}^{-1}$ and with $n(l) \in N_P$, $l = (0,l_2,l_3,l_4)$, defined by

$$\begin{pmatrix} l_2 & l_4 \\ l_4 & l_3 \end{pmatrix} = \widetilde{\mathcal{C}}.$$

Uniqueness is obvious.

To get the desired expression for $\tilde{\gamma}$, choose $t \in \mathbf{C}^*$ in such a way that the upper left entry of the matrix $\tilde{\vartheta} = \delta^{-1}\tilde{\gamma}$ is equal to 1. Then, from (2.1.4), it follows that

$$\tilde{\vartheta} = \begin{pmatrix} 1 & * & * & * \\ 0 & a & * & b \\ 0 & 0 & 1 & 0 \\ 0 & c & * & d \end{pmatrix}, \quad \text{where} \quad \begin{pmatrix} a & b \\ c & d \end{pmatrix} \in \mathrm{SL}(2, \mathbf{C}).$$

Setting $\lambda = \begin{pmatrix} a & b \\ c & d \end{pmatrix}$ we get $\tilde{\gamma} = \delta\vartheta h_1(\lambda)$ with

$$\vartheta = \begin{pmatrix} 1 & * & * & * \\ 0 & 1 & * & 0 \\ 0 & 0 & 1 & 0 \\ 0 & 0 & * & 1 \end{pmatrix} = n(l_1, l_2, 0, l_4) \in N_{\widetilde{P}}$$

(see (2.1.4) and the definition of N), as desired. Uniqueness is obvious. ■

2.1.9 An auxiliary decomposition.

Proposition 2.1.3 *Let*

$$\gamma = \begin{pmatrix} a_1 & a_2 & b_1 & b_2 \\ a_3 & a_4 & b_3 & b_4 \\ c_1 & c_2 & d_1 & d_2 \\ c_3 & c_4 & d_3 & d_4 \end{pmatrix} \in \mathrm{Sp}(4, \mathbf{C}). \tag{2.1.12}$$

If

$$d_1 \neq 0, \qquad d_1 d_4 - d_2 d_3 \neq 0, \tag{2.1.13}$$

and σ is as in (2.1.2), *then one has the decomposition*

$$\gamma\sigma = n(l)\,\mathrm{diag}(t_1, t_2, t_1^{-1}, t_2^{-1})\sigma n(l'), \tag{2.1.14}$$

in which $l = (l_1, l_2, l_3, l_4) \in \mathbf{C}^4$, $l' = (l'_1, l'_2, l'_3, l'_4) \in \mathbf{C}^4$ and

$$\begin{aligned} l_1 &= -\frac{d_3}{d_1}, \quad t_1 = \frac{1}{d_1}, \quad t_2 = \frac{d_1}{d_1 d_4 - d_2 d_3}, \quad l'_1 = \frac{d_2}{d_1}, \\ l'_1 l'_4 - l'_2 &= \frac{c_1}{d_1}, \qquad l'_3 = \frac{c_2 d_3 - c_4 d_1}{d_1 d_4 - d_2 d_3}, \qquad l'_4 = -\frac{c_2}{d_1}. \end{aligned} \tag{2.1.15}$$

□

Proof. Multiplying the matrices in (2.1.14), we find (2.1.14) is equivalent to

$$\begin{pmatrix} * & * & * & * \\ * & * & * & * \\ d_1 & d_2 & -c_1 & -c_2 \\ d_3 & d_4 & -c_3 & -c_4 \end{pmatrix} = t_1^{-1} \begin{pmatrix} * & * & * & * \\ * & * & * & * \\ 1 & l'_1 & l'_2 - l'_1 l'_4 & l'_4 \\ -l_1 & t_1 t_2^{-1} - l_1 l'_1 & * & t_1 t_2^{-1} l'_3 - l_1 l'_4 \end{pmatrix}.$$

Comparing the entries we get the equations which lead to (2.1.15). ■

2.2 Discrete subgroups

2.2.1 Preliminaries. As usual, let $q \subset (3)$ be an ideal of the integers ring $\mathcal{O}$ of the field $\mathbf{Q}(\sqrt{-3})$ and let $\Gamma_4(q)$ be the congruence subgroup mod q of $\mathrm{SL}(n,\mathcal{O})$ defined in 0.2.1. Let $\Gamma(q) = \Gamma_4(q) \cap \mathrm{Sp}(4,\mathbf{C})$. The so defined group $\Gamma(q)$ is a congruence subgroup mod q of $\mathrm{Sp}(4,\mathcal{O})$.

We shall prove in this section several propositions that describe the group $\Gamma(q)$ itself and some related groups and cosets. Our aim is to prepare the technique we need to evaluate Eisenstein series Fourie coefficients. Our approach is based on consideration of the monomorphisms $\Gamma_2(q) \to \Gamma(q)$ given by restriction of h_j in 2.1.7.

2.2.2 Groups Δ, $\Delta_\times$, $\widetilde{\Delta}_\times$. Let B and P, $\widetilde{P}$ are the minimal and maximal parabolic subgroups of $\mathrm{Sp}(4,\mathbf{C})$, as defined in 2.1.4 and 2.1.8. We set $\Delta = \Gamma(q) \cap B$, $\Delta_\times = \Gamma(q) \cap P$, $\widetilde{\Delta}_\times = \Gamma(q) \cap \widetilde{P}$.

Proposition 2.2.1 (a) $\Delta = \{\, n(l) \mid l = (l_1, l_2, l_3, l_4) \in q^4 \,\}$;

(b) *Each* $\gamma \in \Delta_\times$ *can be factored uniquely as* $\gamma'\gamma''$ *with* $\gamma' = n(0,*,*,*) \in \Delta$, $\gamma'' = h_4(\lambda)$, $\lambda \in \Gamma_2(q)$;

(c) $\Delta_\times = \Delta h_4(\Gamma_2(q))$;

(d) *Each* $\tilde{\gamma} \in \widetilde{\Delta}_\times$ *can be factored uniquely as* $\gamma'\gamma''$ *with* $\gamma' = n(*,*,0,*) \in \Delta$, $\gamma'' = h_1(\lambda)$, $\lambda \in \Gamma_2(q)$;

(e) $\widetilde{\Delta}_\times = \Delta h_1(\Gamma_2(q))$. □

Proof. Part (a) is obvious. Following the proof of Proposition 2.1.2 one can find that for $\gamma \in \Delta_\times$ and $\tilde{\gamma} \in \widetilde{\Delta}_\times$ the factorizations given by Proposition 2.1.2 have place with $\lambda \in \Gamma_2(q)$ and $\delta = e_4$. Thus we get (b) and (d). Part (c) follows from (b), since Δ and $h_4(\Gamma_2(q))$ are subgroups of $\Delta_\times$. Part (e) follows from (d), since Δ and $h_1(\Gamma_2(q))$ are subgroups of $\widetilde{\Delta}_\times$. ■

2.2.3 On the group $\Gamma(q)$. For each $n \in \mathcal{O}$, $n \equiv 1 \pmod 3$, and each $m \in q$, satisfying $\gcd(m,n) = 1$, choose the matrix

$$\lambda_{m,n} = \begin{pmatrix} * & * \\ m & n \end{pmatrix} \in \Gamma_2(q).$$

Denote by Λ the set of all vectors

$$c = (m_1, n_1, m_2, n_2, m_3, n_3) \tag{2.2.1}$$

that satisfy the conditions

$$m_j \in q, \quad n_j \in \mathcal{O}, \quad n_j \equiv 1 \pmod 3, \quad \gcd(m_j, n_j) = 1, \tag{2.2.2}$$

$j = 1, 2, 3$. For $c \in \Lambda$ define $\gamma_c \in \Gamma(q)$ as the product $\gamma_c = \varepsilon_3\varepsilon_2\varepsilon_1$ with

$$\varepsilon_j = h_j(\lambda_{m_j,n_j}). \tag{2.2.3}$$

Proposition 2.2.2 *Each coset in $\Delta_\times\backslash\Gamma(q)$ contains some γ_c, $c \in \Lambda$.* □

Proof. We only have to show that, given $\gamma \in \Gamma(q)$, there exist $\delta \in \Delta_\times$ and $c \in \Lambda$ such that $\gamma = \delta\gamma_c$. For this, let

$$\gamma = \begin{pmatrix} * & * & * & * \\ * & * & * & * \\ * & * & d_1 & * \\ * & * & d_3 & * \end{pmatrix} \in \Gamma(q),$$

and choose the matrix $\lambda' \in \Gamma_2(q)$ in such a way that $\lambda'^{-1}\begin{pmatrix} d_1 \\ d_3 \end{pmatrix} = \begin{pmatrix} * \\ 0 \end{pmatrix}$.

Then define the matrix $\gamma' \in \Gamma(q)$ by $\gamma = h_4(\lambda')\gamma'$. For the bottom row $c(\gamma')$ of γ' we have $c(\gamma') = (*, *, 0, *)$. Denote by m_2 the first entry of $c(\gamma')$, and by n_2 the greatest common divisor of the second and forth entries of $c(\gamma')$. So, $c(\gamma') = (m_2, m_1n_2, 0, n_1n_2)$ where $n_1 \equiv n_2 \equiv 1 \pmod 3$, $m_1, m_2 \in q$, $\gcd(m_j, n_j) = 1$, $j = 1, 2$. Next define ε_j for n_j, m_j by (2.2.3). We see that the bottom row of the matrix $\varepsilon_2\varepsilon_1$ is the same as the bottom row of γ'. Consequently, $\gamma' = \gamma''\varepsilon_2\varepsilon_1$ where $\gamma'' \in \Gamma(q)$ has the bottom row $(0, 0, 0, 1)$. Using (2.1.4) we find that the third row of γ'' is $(m_3, 0, n_3, *)$ where (since $\gamma'' \in \Gamma(q)$) $m_3 \in q$ and $n_3 \in \mathcal{O}$, $n_3 \equiv 1 \pmod 3$. It is clear that each common divisor of m_3 and n_3 divides $\det\gamma'' = 1$. Thus, $\gcd(m_3, n_3) = 1$ and one can define ε_3 for m_3, n_3 as in (2.2.3). We see that $\gamma'' = \tilde{\gamma}\varepsilon_3$ with

$$\tilde{\gamma} = \begin{pmatrix} * & * & * & * \\ * & * & * & * \\ 0 & 0 & 1 & * \\ 0 & 0 & 0 & 1 \end{pmatrix} \in \Gamma(q). \tag{2.2.4}$$

After all we have the expression $\gamma = h_4(\lambda')\tilde{\gamma}\varepsilon_3\varepsilon_2\varepsilon_1$ where $\tilde{\gamma}$ is as in (2.2.4), $\lambda' \in \Gamma_2(q)$ and ε_j are as indicated in (2.2.3). So, $\gamma_c = \varepsilon_3\varepsilon_2\varepsilon_1$ with c in (2.2.1) and it only remains to notice that $\tilde{\gamma} \in \Delta_\times$ and $h_4(\lambda') \in \Delta_\times$. ■

Proposition 2.2.3 *For each matrix $\gamma \in \Gamma(q)$*

(a) *There exist matrices $\delta \in \Delta$ and $\lambda_j \in \Gamma_2(q)$ such that*

$$\gamma = \delta\varepsilon_4\varepsilon_3\varepsilon_2\varepsilon_1, \quad \textit{where} \quad \varepsilon_j = h_j(\lambda_j); \tag{2.2.5}$$

(b) *There exist matrices $\delta \in {}^t\Delta$ and $\lambda_j \in \Gamma_2(q)$ such that*

$$\gamma = \varepsilon_1\varepsilon_2\varepsilon_3\varepsilon_4\delta, \quad \textit{where} \quad \varepsilon_j = h_j(\lambda_j). \quad \square \tag{2.2.6}$$

Proof. (a) By Proposition 2.2.2, γ can be factored as $\alpha\varepsilon_3\varepsilon_2\varepsilon_1$, where $\alpha \in \Delta_\times$. By Proposition 2.2.1 (c), α can be factored as $\delta\varepsilon_4$. Thus we get (2.2.5).

(b) This follows by transposition from (a) applied to the matrix ${}^t\gamma$. ■

Analyzing the proof of Proposition 2.2.2 we find the following complement.

Proposition 2.2.4 *If a vector like $(*, 0, 0, *)$ is the bottom row of $\gamma \in \Gamma(q)$, then γ can be factored as in (2.2.5) with identity matrix ε_1.* ■

2.2.4 A special remark.

Proposition 2.2.5 *Let*

$$\gamma = \begin{pmatrix} a_1 & a_2 & b_1 & b_2 \\ a_3 & a_4 & b_3 & b_4 \\ c_1 & c_2 & d_1 & d_2 \\ c_3 & c_4 & d_3 & d_4 \end{pmatrix} \in \Gamma(q). \tag{2.2.7}$$

If there exist matrices

$$\varepsilon_j = h_j(\lambda_j), \quad \lambda_j = \begin{pmatrix} k_j & l_j \\ m_j & n_j \end{pmatrix} \in \Gamma_2(q), \quad j = 1, 2, 3, \tag{2.2.8}$$

such that $\gamma = \varepsilon_3\varepsilon_2\varepsilon_1$, then

(a) $\begin{pmatrix} d_1 & * \\ d_3 & d_4 \end{pmatrix} = \begin{pmatrix} n_2 n_3 & * \\ 0 & n_1 n_2 \end{pmatrix}$;

(b) $c_3 = m_2, \quad c_4 = m_1 n_2$;

(c) $\gcd(c_4, d_4) = n_2$. □

Proof. It is sufficient to multiply the matrices ε_j in (2.2.8) and then to compare the product with (2.2.7). ■

2.2.5 Cosets $\Delta_\times \backslash \Gamma(q)$. We will need a convenient representative system for $\Delta_\times \backslash \Gamma(q)$ and we can receive such a system by using Proposition 2.2.2 and Proposition 2.2.5. Let us retain the notations introduced in the beginning of 2.2.3.

Proposition 2.2.6 *The set $\{\gamma_c \mid c \in \Lambda\}$ contains a unique representative of each coset in $\Delta_\times \backslash \Gamma(q)$.* □

Proof. Proposition 2.2.2 provides the presence of the matrix γ_c in each coset in $\Delta_\times \backslash \Gamma(q)$. Therefore it is sufficient to prove uniqueness. Let $c, c' \in \Lambda$,

$$c = (m_1, n_1, m_2, n_2, m_3, n_3), \quad c' = (m_1', n_1', m_2', n_2', m_3', n_3'). \tag{2.2.9}$$

If γ_c and $\gamma_{c'}$ lie in one coset in $\Delta_\times \backslash \Gamma(q)$, then $\gamma_{c'} = \delta\gamma_c$ with some $\delta \in \Delta_\times$. Next notice that (see Proposition 2.2.5 (a) and Proposition 2.2.1)

the bottom row of $\gamma_{c'}$ is $(*, *, 0, \star)$,
the third column of γ_c is ${}^t(*, *, *, 0)$,
the bottom row of δ is $(0, 0, *, \star)$,

where $\star$ mean the entries $\equiv 1 \pmod 3$, so, $\neq 0$. With this remark, we find that $\gamma_{c'} = \delta\gamma_c$ yields the bottom row of δ is $(0,0,0,1)$ and, so, the bottom row of $\gamma_{c'}$ is the same as of γ_c one. Thus, if $\gamma_{c'}$ and γ_c lie in one coset in $\Delta_\times\backslash\Gamma(q)$, then the bottom row of $\gamma_{c'}$ coincides with the bottom row of γ_c. Expressing such matrices as $\gamma_c = \varepsilon_3\varepsilon_2\varepsilon_1$, $\gamma_{c'} = \varepsilon'_3\varepsilon'_2\varepsilon'_1$ with

$$\varepsilon_j = h_j(\lambda_{m_j,n_j}), \qquad \varepsilon'_j = h_j(\lambda_{m'_j,n'_j})$$

and applying Proposition 2.2.5 we find that

$$n'_1 n'_2 = n_1 n_2, \quad n'_2 = n_2, \quad m'_1 n'_2 = m_1 n_2, \quad m'_2 = m_2, \tag{2.2.10}$$

notations are as in (2.2.9). From (2.2.10), it follows that $\varepsilon'_1 = \varepsilon_1$ and $\varepsilon'_2 = \varepsilon_2$. Consequently, $\gamma_{c'} = \delta\gamma_c$ implies $\delta\varepsilon'_3 = \varepsilon_3$, and thus we have

$$\varepsilon_3\varepsilon_3'^{-1} \in \Delta_\times \cap h_3(\Gamma_2(q)) = \left\{ h_3\left(\begin{pmatrix} 1 & l \\ 0 & 1 \end{pmatrix}\right) \mid l \in q \right\}.$$

This yields further $n'_3 = n_3$, $m'_3 = m_3$. Thus we get $m'_j = m_j$, $n'_j = n_j$ for $j = 1,2,3$ and, so, $c' = c$, as claimed. ■

For $\gamma \in \Gamma(q)$ define the group Δ^γ by ${}^t\Delta^\gamma = \gamma^{-1}\Delta_\times\gamma \cap {}^t\Delta$. This group can be characterized also as the group of all $\delta \in \Delta$ which satisfies $\Delta_\times\gamma{}^t\delta = \Delta_\times\gamma$. By means of simple but somewhat tedious computation we find that

$$\Delta^\gamma = \{ n(l,0,0,0) \mid l \equiv 0 \ (\mathrm{mod}\ \vartheta q) \}, \tag{2.2.11}$$

where $\vartheta = d_4/\gcd(2, c_4, d_4)$ if $(*, c_4, *, d_4)$ is the bottom row of γ. With the same notations let

$$\mathcal{F}_\gamma = \mathbf{C}/\vartheta q \times \mathbf{C}^3, \tag{2.2.12}$$

$\mathbf{C}/\vartheta q$ being a fundamental domain of the lattice ϑq. Also, for $\gamma \in \Gamma(q)$ we denote by Λ_γ some set that contains a unique representative in ${}^t\Delta$ of each coset in ${}^t\Delta^\gamma\backslash{}^t\Delta$, and we write σ for the matrix defined in (2.1.2).

Proposition 2.2.7 *Let a set $\Omega \subset \Gamma(q)$ contains a unique representative of each double coset in $\Delta_\times\backslash\Gamma(q)/{}^t\Delta$, and let $\mathcal{P} = \{ (\gamma, h) \mid \gamma \in \Omega,\ h \in \Lambda_\gamma \}$. One has*

(a) *The mapping $\mathcal{P} \to \Delta_\times\backslash\Gamma(q)$ defined by $(\gamma, h) \mapsto \Delta_\times\gamma h$ is bijective;*

(b) *$\sigma^{-1}\Lambda_\gamma\sigma$ contains a unique representative of each coset in $\Delta^\gamma\backslash\Delta$;*

(c) *$n(\mathcal{F}_\gamma) = \{ n(l_1, l_2, l_3, l_4) \mid l_1 \in \mathbf{C}/\vartheta q;\ l_2, l_3, l_4 \in \mathbf{C} \}$ contains a unique representative of each coset in $\Delta^\gamma\backslash N$.* □

Proof. (a) Surjectivity is obvious. To prove injectivity, let $\gamma, \gamma' \in \Omega$, $h \in \Lambda_\gamma$, $h' \in \Lambda_{\gamma'}$. If $\Delta_\times\gamma h = \Delta_\times\gamma' h'$, then there exists $\delta \in \Delta_\times$ such that

$$\delta\gamma h = \gamma' h', \tag{2.2.13}$$

and so $\gamma' = \delta\gamma h h'^{-1}$. Since $\delta \in \Delta_\times$ and $hh'^{-1} \in {}^t\Delta$, this yields that γ' and γ lie in one double coset in $\Delta_\times\backslash\Gamma(q)/{}^t\Delta$, hence $\gamma' = \gamma$. Next, (2.2.13) yields $hh'^{-1} = \gamma^{-1}\delta^{-1}\gamma \in \gamma^{-1}\Delta_\times\gamma$. Combining this with $hh'^{-1} \in {}^t\Delta$, we get $hh'^{-1} \in {}^t\Delta^\gamma$ and furthermore, since $h, h' \in \Lambda_\gamma$, we receive $h' = h$, as required.

(b) Immediately from the definition of Λ_γ, we find that the set $\sigma^{-1}\Lambda_\gamma\sigma$ contains a unique representative of each coset in $\sigma^{-1}{}^t\Delta^\gamma\sigma\backslash\sigma^{-1}{}^t\Delta\sigma$. It remains to notice that $\sigma^{-1}{}^t\Delta\sigma = \Delta$ and $\sigma^{-1}{}^t\Delta^\gamma\sigma = \Delta^\gamma$.

(c) Follows from (2.2.11) and the formulae in 2.1.6. ∎

2.2.6 Homomorphism ψ: $\widetilde{\Gamma}^{(4)}_{\text{princ}}(3) \to \mathbf{C}^*$. Let κ be the Kubota homomorphism, and let κ_n, $n \geq 2$, be the Bass–Milnor–Serre homomorphisms (see 0.2.2, 0.2.4, 0.2.5). We put $\psi = \kappa_4$. We shall prove two propositions to describe the restrictions of the homomorphism ψ on some subgroups of $\Gamma(q) \subset \widetilde{\Gamma}^{(4)}_{\text{princ}}(3)$.

Proposition 2.2.8 *One has*

(a) $\psi(\gamma) = 1$ *for* $\gamma \in \Delta$ *and for* $\gamma \in {}^t\Delta$;

(b) *If* $\gamma = \gamma'\gamma'' \in \Delta_\times$ *with* $\gamma' \in \Delta$ *and* $\gamma'' = h_4(\lambda) \in \Delta_\times$, $\lambda \in \Gamma_2(q)$, *then* $\psi(\gamma) = \kappa(\lambda)$;

(c) *If* $\gamma = \gamma'\gamma'' \in \Delta_\times$ *with* $\gamma' \in \Delta$ *and* $\gamma'' = h_1(\lambda) \in \Delta_\times$, $\lambda \in \Gamma_2(q)$, *then* $\psi(\gamma) = \kappa(\lambda)^{-1}$. □

Proof. (a) In view of the formulae in 2.1.6 and Proposition 2.2.1 (a), each $\gamma \in \Delta$ can be factored as $\gamma_1\gamma_2\gamma_3\gamma_4$, where $\gamma_j = n(l_1, l_2, l_3, l_4)$, $l_j \in q$ and $l_k = 0$ for $k \neq j$. The matrices $\gamma_1, \gamma_2, \gamma_3$ are elementary, while γ_4 is the product of two elementary matrices. By Theorem 0.2.2 (c), $\psi(\gamma_j) = \psi({}^t\gamma_j) = 1$. Hence, $\psi(\gamma) = \psi({}^t\gamma) = 1$, as required.

To prove (b) and (c) notice that (a) yields $\psi(\gamma) = \psi(\gamma'')$ for both cases. The needed formulae for $\psi(\gamma'')$ are given by the next proposition. ∎

Proposition 2.2.9 *Let* $\lambda \in \Gamma_2(q)$ *and* $\varepsilon_j = h_j(\lambda)$. *Then*

$$\psi(\varepsilon_j) = \kappa(\lambda)^{-1} \quad \text{for } j = 1, 3; \quad \psi(\varepsilon_j) = \kappa(\lambda) \quad \text{for } j = 2, 4. \qquad \square$$

Proof. Applying Theorem 0.2.2 (a), (b), (f) we deduce

$$\psi(\varepsilon_1) = \psi(\varepsilon_3) = \kappa_3\left(\begin{pmatrix} a & 0 & b \\ 0 & 1 & 0 \\ c & 0 & d \end{pmatrix}\right) = \kappa_3\left(\gamma\begin{pmatrix} a & 0 & b \\ 0 & 1 & 0 \\ c & 0 & d \end{pmatrix}\gamma^{-1}\right) = \kappa_3\left(\begin{pmatrix} a & b & 0 \\ c & d & 0 \\ 0 & 0 & 1 \end{pmatrix}\right)$$

$$= \kappa_2(\lambda) = \kappa(\lambda)^{-1}, \quad \text{if} \quad \lambda = \begin{pmatrix} a & b \\ c & d \end{pmatrix}, \quad \gamma = \begin{pmatrix} 1 & 0 & 0 \\ 0 & 0 & 1 \\ 0 & -1 & 0 \end{pmatrix}. \tag{2.2.14}$$

For the case $j = 4$ applying Theorem 0.2.2 (a), (b), (d) we obtain

$$\psi(\varepsilon_4) = \kappa_4\left(\begin{pmatrix} \lambda & \begin{matrix} 0 & 0 \\ 0 & 0 \end{matrix} \\ \begin{matrix} 0 & 0 \\ 0 & 0 \end{matrix} & {}^t\lambda^{-1} \end{pmatrix}\right) = \kappa_4\left(\begin{pmatrix} \lambda & \begin{matrix} 0 & 0 \\ 0 & 0 \end{matrix} \\ \begin{matrix} 0 & 0 \\ 0 & 0 \end{matrix} & \begin{matrix} 1 & 0 \\ 0 & 1 \end{matrix} \end{pmatrix}\right)\kappa_4\left(\begin{pmatrix} \begin{matrix} 1 & 0 \\ 0 & 1 \end{matrix} & \begin{matrix} 0 & 0 \\ 0 & 0 \end{matrix} \\ \begin{matrix} 0 & 0 \\ 0 & 0 \end{matrix} & {}^t\lambda^{-1} \end{pmatrix}\right)$$

$$= \kappa_2^2(\lambda) = \kappa(\lambda).$$

As to $j = 2$, we have $\varepsilon_2 = \delta\delta'$ and thus $\psi(\varepsilon_2) = \psi(\delta)\psi(\delta')$ with

$$\delta = \begin{pmatrix} a & 0 & 0 & b \\ 0 & 1 & 0 & 0 \\ 0 & 0 & 1 & 0 \\ c & 0 & 0 & d \end{pmatrix}, \qquad \delta' = \begin{pmatrix} 1 & 0 & 0 & 0 \\ 0 & a & b & 0 \\ 0 & c & d & 0 \\ 0 & 0 & 0 & 1 \end{pmatrix},$$

if λ is written as in (2.2.14). Applying, as above, Theorem 0.2.2 we obtain $\psi(\delta) = \psi(\delta') = \kappa(\lambda)^{-1}$ and, consequently, $\psi(\varepsilon_2) = \kappa(\lambda)$. ■

2.2.7 Representative system for $\Delta_{\times}\backslash\Gamma(q)/{}^t\Delta$. Let n be a triple (n_1, n_2, n_3) with $n_j \equiv 1 \pmod 3$ and $\gcd(n_j, q) = 1$, $j = 1, 2, 3$. Given n, we choose subsets $\mathfrak{m}_1(n)$ and $\mathfrak{l}_1(n)$ of q each of which contains a unique representative of each class mod n_1 coprime with n_1. Given n and $m_1, l_1 \in q$, we choose subsets $\mathfrak{m}_2(n; m_1, l_1)$ and $\mathfrak{l}_2(n; m_1, l_1)$ of q each of which contains a unique representative of each class $\bmod n_2$ coprime with n_2. Given n and $m_1, l_1, m_2, l_2 \in q$, we choose a subset $\mathfrak{m}_3(n; m_1, m_2, l_1, l_2)$ of q which contains a unique representative of each coprime with n_3 class $\bmod(n_2 n_3 \gcd(2, n_2))$ and a subset $\mathfrak{l}_3(n; m_1, m_2, l_1, l_2)$ of q which contains a unique representative of each class $\bmod n_3$ coprime with n_3.

The next proposition gives us very convenient representative system of the double cosets $\Delta_{\times}\backslash\Gamma(q)/{}^t\Delta$ which we shall use in subsection 2.4.4 to evaluate the Eisenstein series Fourier coefficients.

Theorem 2.2.10 *Each double coset in $\Delta_{\times}\backslash\Gamma(q)/{}^t\Delta$ contains a unique matrix γ which can be factored as $\gamma = \varepsilon_3\varepsilon_2\varepsilon_1$ where*

$$\varepsilon_j = h_j(\lambda_j), \qquad \lambda_j = \begin{pmatrix} k_j & l_j \\ m_j & n_j \end{pmatrix} \in \Gamma_2(q),$$

$$m_3 \in \mathfrak{m}_3(n; m_1, m_2, l_1, l_2), \qquad l_3 \in \mathfrak{l}_3(n; m_1, m_2, l_1, l_2),$$

$$m_2 \in \mathfrak{m}_2(n; m_1, l_1), \qquad l_2 \in \mathfrak{l}_2(n; m_1, l_1),$$

$$m_1 \in \mathfrak{m}_1(n), \qquad l_1 \in \mathfrak{l}_1(n), \qquad n = (n_1, n_2, n_3). \qquad \square$$

We shall prove this statement in 2.2.9, 2.2.10.

2.2.8 Some auxiliary propositions.

Proposition 2.2.11 *If the matrices*

$$\gamma = \begin{pmatrix} * & * & * & * \\ * & * & * & * \\ * & * & d_1 & * \\ c_3 & c_4 & 0 & d_4 \end{pmatrix}, \qquad \gamma' = \begin{pmatrix} * & * & * & * \\ * & * & * & * \\ * & * & d'_1 & * \\ c'_3 & c'_4 & 0 & d'_4 \end{pmatrix}, \tag{2.2.15}$$

lie in one double coset in $\Delta_{\times}\backslash\Gamma(q)/{}^t\Delta$, then

$$\begin{aligned} &d'_1 = d_1, \qquad d'_4 = d_4, \qquad \gcd(c'_4, d'_4) = \gcd(c_4, d_4), \\ &c'_4 \equiv c_4 \pmod{d_4 q}, \qquad c'_3 \equiv c_3 \pmod{\gcd(c_4, d_4) q}. \qquad \square \end{aligned} \tag{2.2.16}$$

Proof. If γ and γ' lie in one double coset in $\Delta_\times\backslash\Gamma(q)/{}^t\Delta$, then

$$\gamma' = \lambda\gamma\delta \tag{2.2.17}$$

with some

$$\lambda = \begin{pmatrix} * & * & * & * \\ * & * & * & * \\ 0 & 0 & * & * \\ 0 & 0 & * & * \end{pmatrix} \in \Delta_\times, \qquad \delta = \begin{pmatrix} 1 & 0 & 0 & 0 \\ * & 1 & 0 & 0 \\ * & * & 1 & * \\ * & * & 0 & 1 \end{pmatrix} \in {}^t\Delta. \tag{2.2.18}$$

The third entry of the bottom row of the matrix γ' is 0 and for (2.2.17) to take place the third entry of the bottom row of the matrix $\lambda\gamma\delta$ has to be 0 also, but this is possible only if the bottom row of the matrix λ is $(0,0,0,*)$. Then Proposition 2.2.1 yields

$$\lambda = \begin{pmatrix} 1 & 0 & * & * \\ * & 1 & * & * \\ 0 & 0 & 1 & * \\ 0 & 0 & 0 & 1 \end{pmatrix} \quad \text{and} \quad \lambda\gamma = \begin{pmatrix} * & * & * & * \\ * & * & * & * \\ * & * & d_1 & * \\ c_3 & c_4 & 0 & d_4 \end{pmatrix}. \tag{2.2.19}$$

Since (see (2.2.18) and Proposition 2.2.1)

$$\delta = \begin{pmatrix} 1 & 0 & 0 & 0 \\ l & 1 & 0 & 0 \\ * & * & 1 & -l \\ t & v & 0 & 1 \end{pmatrix} \quad \text{with} \quad l, t, v \in q,$$

from (2.2.17), (2.2.15), (2.2.19) follows $d_1' = d_1$, $d_4' = d_4$, $c_4' = c_4 + d_4 v$ and $c_3' = c_3 + c_4 l + d_4 t$, that gives (2.2.16). ∎

Proposition 2.2.12 *If the matrices γ and γ' lie in one double coset in $\Delta_\times\backslash\Gamma(q)/{}^t\Delta$ and can be factored as*

$$\gamma = \varepsilon_3\varepsilon_2\varepsilon_1, \qquad \gamma' = \varepsilon_3'\varepsilon_2'\varepsilon_1'$$

with $\varepsilon_j = h_j(\lambda_j)$, $\varepsilon_j' = h_j(\lambda_j')$,

$$\lambda_j = \begin{pmatrix} k_j & l_j \\ m_j & n_j \end{pmatrix} \in \Gamma_2(q), \qquad \lambda_j' = \begin{pmatrix} k_j' & l_j' \\ m_j' & n_j' \end{pmatrix} \in \Gamma_2(q),$$

then $n_j = n_j'$ for all j and $m_j \equiv m_j' \pmod{n_j q}$ for $j = 1, 2$. □

Proof. From Proposition 2.2.5, it follows that γ and γ' have the form (2.2.15). So we can apply Proposition 2.2.11. Combining the formulae of Proposition 2.2.5 and Proposition 2.2.11, we find

$$n_2' n_3' = n_2 n_3, \qquad n_1' n_2' = n_1 n_2, \qquad n_2' = n_2,$$
$$m_1' n_2' \equiv m_1 n_2 \pmod{n_1 n_2 q}, \qquad m_2' \equiv m_2 \pmod{n_2 q},$$

as required. ∎

Proposition 2.2.13 *Let* $\varepsilon_j = h_j(\lambda_j)$ *with*

$$\lambda_j = \begin{pmatrix} k_j & l_j \\ m_j & n_j \end{pmatrix} \in \Gamma_2(q), \quad j = 1, 2.$$

We have

(a) $\varepsilon_1^{-1} h_2(\begin{pmatrix} 1 & 0 \\ r & 1 \end{pmatrix})\varepsilon_1 \in {}^t\Delta$ *for all* $r \in q$;

(b) *If* $v = (-n_1 r, -m_1 n_1 r^2 + s, 0, m_1 r)$ *with* $r, s \in q$, *then*

$$\varepsilon_2 \varepsilon_1 {}^t n(v) \varepsilon_1^{-1} \varepsilon_2^{-1} = \begin{pmatrix} 1 & 0 & 0 & * \\ * & 1 & 0 & * \\ z & 0 & 1 & * \\ 0 & 0 & 0 & 1 \end{pmatrix} \quad \text{with } z = -2m_2 n_2 r + n_2^2,$$

where $n(v)$ *as defined in* 2.1.2.

Proof. One can check this by simple computation. ∎

2.2.9 Proof of existence in Theorem 2.2.10. Let Z be a double coset in $\Delta_x \backslash \Gamma(q)/{}^t\Delta$. According to Proposition 2.2.2, Z contains the matrix γ which can be factored as the product $\gamma = \nu_3 \nu_2 \nu_1$ with some

$$\nu_j = h_j(\mu_j), \quad \mu_j = \begin{pmatrix} * & * \\ * & n_j \end{pmatrix} \in \Gamma_2(q). \tag{2.2.20}$$

Let $n = (n_1, n_2, n_3)$. Put $\varepsilon_1 = \alpha \nu_1 \beta$ with

$$\alpha = h_1(\begin{pmatrix} 1 & \xi' \\ 0 & 1 \end{pmatrix}), \quad \beta = h_1(\begin{pmatrix} 1 & 0 \\ \xi & 1 \end{pmatrix}), \quad \xi, \xi' \in q.$$

With suitably chosen ξ, ξ', the matrix ε_1 satisfies the conditions of Theorem 2.2.10, so

$$\varepsilon_1 = h_1(\lambda_1), \quad \lambda_1 = \begin{pmatrix} k_1 & l_1 \\ m_1 & n_1 \end{pmatrix} \in \Gamma_2(q),$$

with $m_1 \in \mathfrak{m}_1(n)$, $l_1 \in \mathfrak{l}_1(n)$. The bottom row of the matrix $\gamma' = \nu_3 \nu_2 \alpha^{-1}$ is as indicated in Proposition 2.2.4. Consequently, γ' can be factored as the product $\delta \nu_3' \nu_2'$ with $\delta \in \Delta_x$, $\nu_j' = h_j(\mu_j')$, $\mu_j' \in \Gamma_2(q)$. Since $\gamma = \gamma' \varepsilon_1 \beta^{-1}$ and $\beta \in {}^t\Delta$, the matrix $\nu_3' \nu_2' \varepsilon_1$ lies in the same double coset Z as γ does. From Proposition 2.2.11, it follows that

$$\mu_j' = \begin{pmatrix} k_j' & l_j' \\ m_j' & n_j \end{pmatrix}$$

with n_j as in (2.2.20). Let $\varepsilon_2 = \alpha' \nu_2' \beta'$ with

$$\alpha' = h_2(\begin{pmatrix} 1 & \vartheta \\ 0 & 1 \end{pmatrix}), \quad \beta' = h_2(\begin{pmatrix} 1 & 0 \\ \vartheta' & 1 \end{pmatrix}), \quad \vartheta, \vartheta' \in q.$$

With suitably chosen ϑ, ϑ', the matrix ε_2 satisfies the conditions of Theorem 2.2.10, so

$$\varepsilon_2 = h_2(\lambda_2), \quad \lambda_2 = \begin{pmatrix} k_2 & l_2 \\ m_2 & n_2 \end{pmatrix} \in \Gamma_2(q),$$

with $m_2 \in \mathfrak{m}_2(n; m_1, l_1)$, $l_2 \in \mathfrak{l}_2(n; m_1, l_1)$. Let us set

$$\tilde{\gamma} = \nu_3' \alpha'^{-1} = \begin{pmatrix} k_3' & 0 & l_3' & -k_3'\vartheta \\ 0 & 1 & -\vartheta & 0 \\ m_3' & 0 & n_3 & -m_3'\vartheta \\ 0 & 0 & 0 & 1 \end{pmatrix}. \tag{2.2.21}$$

From Proposition 2.2.13 (a), it follows that $\beta'^{-1}\varepsilon_1 = \varepsilon_1 \omega$ with some $\omega \in {}^t\Delta$. Consequently, $\nu_3'\nu_2'\varepsilon_1 = \tilde{\gamma}\varepsilon_2\beta'^{-1}\varepsilon_1 \in \tilde{\gamma}\varepsilon_2\varepsilon_1{}^t\Delta$ and each matrix $\tilde{\gamma}\varepsilon_2\varepsilon_1{}^t\delta$ with $\delta \in \Delta$ lie in the same double coset Z as γ does. Let us take

$$\delta = n(-n_1 r, -m_1 n_1 r^2 + s, 0, m_1 r)$$

with $r, s \in q$. From Proposition 2.2.13 (b), it follows that $\tilde{\gamma}\varepsilon_2\varepsilon_1{}^t\delta = \tilde{\gamma}\tilde{\delta}\varepsilon_2\varepsilon_1$ with

$$\tilde{\delta} = \begin{pmatrix} 1 & 0 & 0 & * \\ * & 1 & 0 & * \\ z & 0 & 1 & * \\ 0 & 0 & 0 & 1 \end{pmatrix}, \qquad z = -2m_2 n_2 r + n_2^2 s.$$

From this and (2.2.21), it follows that

$$\tilde{\gamma}\tilde{\delta} = \begin{pmatrix} * & * & * & * \\ * & * & * & * \\ m_3' + n_3 z & 0 & n_3 & * \\ 0 & 0 & 0 & 1 \end{pmatrix}.$$

Since $\gcd(m_2, n_2) = 1$, one can find $r, s \in q$ such that

$$m_3' + n_3 z \in \mathfrak{m}_3(n; m_1, m_2, l_1, l_2).$$

For such r, s, we put $m_3 = m_3' + n_3 z$ and $\varepsilon_3 = h_3(\lambda_3)$ with

$$\lambda_3 = \begin{pmatrix} k_3 & l_3 \\ m_3 & n_3 \end{pmatrix} \in \Gamma_2(q), \quad l_3 \in \mathfrak{l}_3(n; m_1, m_2, l_1, l_2).$$

The so defined ε_3 satisfies the conditions of Theorem 2.2.10 and $\tilde{\gamma}\tilde{\delta}\varepsilon_3^{-1}$ looks like

$$\begin{pmatrix} * & * & * & * \\ * & * & * & * \\ 0 & 0 & 1 & * \\ 0 & 0 & 0 & 1 \end{pmatrix}$$

and, consequently, lie in $\Delta_\times$. Hence, we have $\tilde{\gamma}\tilde{\delta} \in \Delta_\times \varepsilon_3$. Being combined with $\tilde{\gamma}\tilde{\delta}\varepsilon_2\varepsilon_1 \in Z$, this yields $\varepsilon_3\varepsilon_2\varepsilon_1 \in Z$, as required.

2.2.10 Proof of uniqueness in Theorem 2.2.10. Let Z mean any double coset in $\Delta_\times\backslash\Gamma(q)/{}^t\Delta$. Let γ be a matrix in Z and let $\gamma = \varepsilon_3\varepsilon_2\varepsilon_1$ be the decomposition of γ indicated in Theorem 2.2.10. Suppose that the matrix γ' in Z is expressed in a similar way, so $\gamma' = \varepsilon_3'\varepsilon_2'\varepsilon_1'$ with $\varepsilon_j' = h_j(\lambda_j')$ and

$$\lambda_j' = \begin{pmatrix} k_j' & l_j' \\ m_j' & n_j' \end{pmatrix} \in \Gamma_2(q), \qquad n' = (n_1', n_2', n_3'),$$

$$m_1' \in \mathfrak{m}_1(n'), \qquad l_1' \in \mathfrak{l}_1(n'), \qquad m_2' \in \mathfrak{m}_2(n'; m_1', l_1'), \qquad l_2' \in \mathfrak{l}_2(n'; m_1', l_1'),$$

$$m_3' \in \mathfrak{m}_3(n'; m_1', m_2', l_1', l_2'), \qquad l_3' \in \mathfrak{l}_3(n'; m_1', m_2', l_1', l_2').$$

To prove $\gamma' = \gamma$ we shall prove $\varepsilon_j' = \varepsilon_j$ successively for $j = 1, 2, 3$.

By Proposition 2.2.12 we have $n' = n$, $n_j' = n_j$ for $j = 1, 2, 3$, and also $m_1' \equiv m_1 \pmod{n_1 q}$. Then our assumptions yield $m_1, m_1' \in \mathfrak{m}_1(n)$, while $l_1, l_1' \in \mathfrak{l}_1(n)$ are determined by m_1', m_1. This yields further that $m_1' = m_1$, $l_1' = l_1$, which, being combined with $n_1' = n_1$, gives rise to $\lambda_1' = \lambda_1$. Hence $\varepsilon_1' = \varepsilon_1$.

We have now $m_2, m_2' \in \mathfrak{m}_2(n; m_1, l_1)$, $l_2, l_2' \in \mathfrak{l}_2(n; m_1, l_1)$, and we know that l_2, l_2' are determined by m_2', m_2. From Proposition 2.2.12, we have also $m_2' \equiv m_2 \pmod{n_2 q}$. Therefore $m_2' = m_2$, $l_2' = l_2$, which, being combined with $n_2' = n_2$, gives rise to $\lambda_2' = \lambda_2$. Hence $\varepsilon_2' = \varepsilon_2$.

Thus, $\gamma' = \varepsilon_3'\varepsilon_2\varepsilon_1$ and $\gamma = \varepsilon_3\varepsilon_2\varepsilon_1$ with

$$\varepsilon_3 = h_3\left(\begin{pmatrix} k_3 & l_3 \\ m_3 & n_3 \end{pmatrix}\right), \quad \varepsilon_3' = h_3\left(\begin{pmatrix} k_3' & l_3' \\ m_3' & n_3 \end{pmatrix}\right)$$

$$m_3, m_3' \in \mathfrak{m}_3(n; m_1, m_2, l_1, l_2), \quad l_3, l_3' \in \mathfrak{l}_3(n; m_1, m_2, l_1, l_2).$$

Now it remains only to prove that $\varepsilon_3' = \varepsilon_3$, and this will be done if we shall prove that $m_3' \equiv m_3 \pmod{n_2 n_3 \gcd(2, n_2) q}$. Since γ' and γ lay in one double coset in $\Delta_\times\backslash\Gamma(q)/{}^t\Delta$, we have $\gamma' = \lambda\gamma\delta$ with $\lambda \in \Delta_\times$, $\delta \in {}^t\Delta$. From Proposition 2.2.5, it follows that the bottom row of the matrix γ' as well as the bottom row of the matrix γ is equal to $(m_2, m_1 n_2, 0, n_1 n_2)$. This makes some restrictions on λ and δ. To be precise, λ has to be like

$$\begin{pmatrix} 1 & 0 & * & * \\ * & 1 & * & * \\ 0 & 0 & 1 & * \\ 0 & 0 & 0 & 1 \end{pmatrix} \tag{2.2.22}$$

(if not, the third entry of the bottom row of γ or γ' will be $\neq 0$) and δ has to be like

$$\begin{pmatrix} 1 & 0 & 0 & 0 \\ -n_1\vartheta & 1 & 0 & 0 \\ \eta & m_1\vartheta & 1 & n_1\vartheta \\ m_1\vartheta & 0 & 0 & 1 \end{pmatrix} \tag{2.2.23}$$

with some $\eta, \vartheta \in q$. Now one can rewrite $\gamma' = \lambda\gamma\delta$ as

$$\varepsilon_3'\varepsilon_2\varepsilon_1 = \lambda\varepsilon_3\varepsilon_2\varepsilon_1\delta; \tag{2.2.24}$$

with $\lambda \in \Delta_\times$, $\delta \in {}^t\Delta$ as in (2.2.22), (2.2.23). Let us set

$$\alpha = \begin{pmatrix} 1 & 0 & 0 & 0 \\ -\vartheta & 1 & 0 & 0 \\ \eta & 0 & 1 & \vartheta \\ \vartheta & 0 & 0 & 1 \end{pmatrix}, \quad \beta = \begin{pmatrix} 1 & 0 & 0 & 0 \\ \omega_1 & 1 & 0 & \omega_2 \\ \omega_3 & 0 & 1 & \omega_4 \\ \vartheta & 0 & 0 & 1 \end{pmatrix}$$

with η, ϑ as in (2.2.23) and with

$$\omega_4 = \vartheta(k_2n_2 + l_2m_2) - l_2n_2\eta, \quad \omega_3 = -2\vartheta n_2m_2 + \eta n_2^2,$$
$$\omega_2 = 2k_2l_2, \quad \omega_1 = -\vartheta(k_2n_2 + l_2m_2).$$

It is easy to check that $\varepsilon_1\delta = \alpha\varepsilon_1$, $\varepsilon_2\alpha = \beta\varepsilon_2$ and, consequently,

$$\varepsilon_2\varepsilon_1\delta = \beta\varepsilon_2\varepsilon_1. \tag{2.2.25}$$

Substituting (2.2.25) into (2.2.24) we receive $\varepsilon_3' = \lambda\varepsilon_3\beta$. Evaluating and comparing the first entries of the third rows of $\lambda\varepsilon_3\beta$ and ε_3' we obtain $m_3' = m_3 + n_2^2n_3\eta - 2m_2n_2n_3\vartheta$ and, consequently, $m_3' \equiv m_3 \pmod{n_2n_3\gcd(2, n_2)q}$, as required.

2.3 Cubic metaplectic forms on $\mathbf{X} \simeq \mathrm{Sp}(4, \mathbf{C})/\mathrm{Sp}(4)$

2.3.1 Space X. We set $\mathbf{X} = \mathbf{C}^4 \times \mathbf{R}_+^{*2}$, considering the right-hand side as a product of real analytic manifolds. For

$$w = (z_1, z_2, z_3, z_4, u, v) \in \mathbf{X} \tag{2.3.1}$$

we set (see 2.1.2)

$$n(w) = n(z_1, z_2, z_3, z_4), \quad a(w) = \mathrm{diag}(uv^{1/2}, v^{1/2}, u^{-1}v^{-1/2}, v^{-1/2}),$$
$$U(w) = u, \quad V(w) = v, \quad Z_j(w) = z_j, \quad Q(w) = (z_1, u), \quad \widetilde{Q}(w) = (z_3, v).$$

We consider $Q(w)$ and $\widetilde{Q}(w)$ as points in $\mathbf{H}$. According to Iwasawa we have the real analytic manifolds isomorphism

$$\mathbf{X} \to \mathrm{Sp}(4, \mathbf{C})/\mathrm{Sp}(4), \quad w \mapsto n(w)a(w)\mathrm{Sp}(4),$$

which transfers on $\mathbf{X}$ the action by left multiplication of the group $\mathrm{Sp}(4,\mathbf{C})$ on $\mathrm{Sp}(4,\mathbf{C})/\mathrm{Sp}(4)$. We shall write γw for the image of the point $w \in \mathbf{X}$ under the action of $\gamma \in \mathrm{Sp}(4,\mathbf{C})$. This action will be described in subsections 2.3.3, 2.3.4 and 2.3.5.

The space $\mathbf{X}$, being equipped with a Riemannian $\mathrm{Sp}(4,\mathbf{C})$-invariant metric (which is determined uniquely up to a constant factor), is a Riemannian globally symmetric space of type 4 in Cartan classification.

We shall write $\mathbf{D}(\mathbf{X})$ for the algebra of invariant differential operators on $\mathbf{X}$. One knows $\mathbf{D}(\mathbf{X})$ is a commutative $\mathbf{C}$-algebra isomorphic to the algebra of polynomials of 2 variables over $\mathbf{C}$. As generators of $\mathbf{D}(\mathbf{X})$ one can take the Laplace–Beltrami operator (attached to the Riemannian metric on $\mathbf{X}$) and one operator of order 4.

From now on we save the notations (2.3.1).

2.3.2 Cubic metaplectic forms. Let Γ be a discrete subgroup of $\mathrm{Sp}(4,\mathbf{C})$. Let $\tilde{\psi}: \Gamma \to \mathbf{C}^*$ be an unitary character. We say $F: \mathbf{X} \to \mathbf{C}$ is an automorphic form under Γ with multiplier system $\tilde{\psi}$ if F is a C^∞-function satisfying the conditions

(a) $F(\gamma w) = \tilde{\psi}(\gamma)F(w)$ for all $\gamma \in \Gamma$, $w \in \mathbf{X}$;

(b) For any $D \in \mathbf{D}(\mathbf{X})$ one has $DF = \lambda_F(D)F$ with some $\lambda_F(D) \in \mathbf{C}$;

(c) There is $c \in \mathbf{R}$ such that $|F(w)| < \mathit{Norm}\big(n(w)a(w)\big)^c$ for all $w \in \mathbf{X}$.

In part (c), by $\mathit{Norm}(\delta)$ for $\delta \in \mathrm{Sp}(4,\mathbf{C})$ we mean $\mathit{trace}(\delta^t\bar{\delta})$. One can consider λ_F in (b) as a function $\mathbf{D}(\mathbf{X}) \to \mathbf{C}$, and as that λ_F is a $\mathbf{C}$-algebras homomorphism, i.e., λ_F is a character of $\mathbf{D}(\mathbf{X})$. One say F satisfying (b) is a joint eigenfunction of $\mathbf{D}(\mathbf{X})$ attached to the character λ_F. In the meantime, one gets more general notion of automorphic forms by replacing (b) in the definition above by

(b′) One has $DF = 0$ for all $D \in I_F$, where I_F is an ideal of finite codimention of $\mathbf{D}(\mathbf{X})$.

We shall need this in subsection 2.5.6 only in connection with cubic symplectic theta functions, while all other automorphic forms we shall deal with satisfy (b).

Let $\psi = \kappa_4: \widetilde{\Gamma}^{(4)}_{\mathrm{princ}}(3) \to \mathbf{C}^*$ be the Bass–Milnor–Serre homomorphism, see 2.2.6, and let Γ be a subgroup of finite index in $\widetilde{\Gamma}^{(4)}_{\mathrm{princ}}(3) \cap \mathrm{Sp}(4,\mathcal{O})$. We say $F: \mathbf{X} \to \mathbf{C}$ is a cubic metaplectic form under Γ if F is an automorphic form under Γ with multiplier system ψ.

In that follows we restrict ourselves mainly by consideration of metaplectic forms under the groups $\Gamma(q)$, $q \subset (3)$.

Notice that the isomorphism of real analytic manifolds

$$(z_1, z_2, z_3, z_4) \mapsto n(z_1, z_2, z_3, z_4) \tag{2.3.2}$$

carries Lebesgue measure on $\mathbf{C}^4$ to Haar measure on N. Let Lebesgue measurable set $\mathcal{F}_q \subset \mathbf{C}^4$ be such that its image under (2.3.2) contains a unique representative of each coset in $\Delta\backslash N$; $\Delta = \Gamma(q) \cap N$. One can take, for example, $\mathcal{F}_q = (\mathbf{C}/q)^4$ where $\mathbf{C}/q$ is a fundamental domain of the lattice q. We write $\mathrm{vol}(\mathcal{F}_q)$ for the volume of $\mathcal{F}_q$ and set $c = \mathrm{vol}(\mathcal{F}_q)^{-1}$. It is obvious that $\mathrm{vol}(\mathcal{F}_q)$ does not depend on the particular choice of $\mathcal{F}_q$. Now let $F \colon \mathbf{X} \to \mathbf{C}$ be an automorphic form under $\Gamma(q)$ with multilier system $\tilde{\psi}$. Assume that Δ is contained in the kernel of the homomorphims $\tilde{\psi}$. Given $\mu, \nu \in q^*$, let

$$c_{\mu,\nu}(u, v; F) = c \iiiint\limits_{\mathcal{F}_q} F(w)\overline{e(\mu z_1 + \nu z_3)}\, dz_1 dz_2 dz_3 dz_4, \tag{2.3.3}$$

with w as in (2.3.1). The integral (2.3.3) does not depend on the particular choice of $\mathcal{F}_q$ since Δ is contained in both the kernel of $\tilde{\psi}$ and the kernel of the homomorphism (2.1.11). We call $c_{\mu,\nu}(u, v; F)$ the Fourier coefficients of the form F. In view of Proposition 2.2.8 (a), this definition is valid for cubic metaplectic forms F.

If F is an automorphic form under $\Gamma(q)$ and $\lambda \in \mathrm{Sp}(4, \mathbf{C})$ is any matrix satisfying the conditions

$$\lambda\Gamma(q)\lambda^{-1} = \Gamma(q) \quad \text{and} \quad \tilde{\psi}(\lambda\gamma\lambda^{-1}) = \tilde{\psi}(\gamma) \quad \text{for all} \quad \gamma \in \Gamma,$$

then the function $w \mapsto F(\lambda w)$, $w \in \mathbf{X}$, also is an automorphic form under the group $\Gamma(q)$, with just the same multiplier system $\tilde{\psi}$ as F is. This can be easily checked. In particular, if F is a cubic metaplectic form under $\Gamma(q)$ and $\lambda \in \mathrm{Sp}(4, \mathbf{Z})$, then $w \mapsto F(\lambda w)$, $w \in \mathbf{X}$, is a cubic metaplectic form under $\Gamma(q)$ too, see Theorem 0.2.2 (f).

2.3.3 The action of unipotent and diagonal matrices.

Proposition 2.3.1 (a) *If $\gamma = n(l)$, $l = (l_1, l_2, l_3, l_4) \in \mathbf{C}^3$, then*

$$Z_j(\gamma w) = z_j + l_j + \begin{cases} 0 & \text{for } j = 1, 3, \\ l_1^2 z_3 + 2 l_1 z_4 & \text{for } j = 2, \\ l_1 z_3 & \text{for } j = 4, \end{cases} \qquad \begin{aligned} U(\gamma w) &= u, \\ V(\gamma w) &= v, \\ Q(\gamma w) &= (z_1 + l_1, u), \\ \tilde{Q}(\gamma w) &= (z_3 + l_3, v). \end{aligned}$$

(b) *If $\gamma = \mathrm{diag}(t_1, t_2, t_1^{-1}, t_2^{-1})$, $t_1, t_2 \in \mathbf{C}^*$, then*

$$Z_j(\gamma w) = z_j \begin{cases} t_1 t_2^{-1} & \text{if } j = 1, \\ t_1^2 & \text{if } j = 2, \\ t_2^2 & \text{if } j = 3, \\ t_1 t_2 & \text{if } j = 4, \end{cases} \qquad \begin{aligned} U(\gamma w) &= |t_1 t_2^{-1}| u, \\ V(\gamma w) &= |t_2^2| v, \\ Q(\gamma w) &= (t_1 t_2^{-1} z_1, |t_1 t_2^{-1}| u), \\ \tilde{Q}(\gamma w) &= (t_2^2 z_3, |t_2^2| v). \end{aligned} \qquad \square$$

Proof. (a) By definition of the $\mathrm{Sp}(4, \mathbf{C})$ action on $\mathbf{X}$ one has $n(\gamma w)a(\gamma w) = \gamma n(w) a(w) k$ with some $k \in K = \mathrm{Sp}(4)$. Since $\gamma \in N$, this implies $n(\gamma w) =$

$\gamma n(w)$, $a(\gamma w) = a(w)$ and we receive the desired results applying the formulae in 2.1.5.

(b) Let $w_\gamma = (t_1 t_2^{-1} z_1,\ t_1^2 z_2,\ t_2^2 z_3,\ t_1 t_2 z_4,\ |t_1 t_2^{-1}| u, |t_2^2| v)$, $\varepsilon_1 = t_1/|t_1|$, $\varepsilon_2 = t_2/|t_2|$. We only have to prove $w_\gamma = \gamma w$. One can easily check that $\gamma n(w) = n(w_\gamma)\gamma$ and $\gamma a(w) = a(w_\gamma)k$, with some $k = \mathrm{diag}(\varepsilon_1, \varepsilon_2, \varepsilon_1^{-1}, \varepsilon_2^{-1}) \in K = \mathrm{Sp}(4)$. This yields $\gamma n(w) a(w) = n(w_\gamma) a(w_\gamma) k$ with the same k, and thus $w_\gamma = \gamma w$. ■

2.3.4 The action of p and $\tilde{p}$. The next proposition describes the action on the space $\mathbf{X}$ of the matrices p and $\tilde{p}$ defined in 2.1.3.

Proposition 2.3.2 *For $w = (z_1, z_2, z_3, z_4, u, v) \in \mathbf{X}$ we have*

(a)
$$\begin{aligned} &Z_1(pw) = \bar{z}_1 L^{-1}, && Z_2(pw) = z_3, \\ &Z_3(pw) = z_2, && Z_4(pw) = z_4, \\ &U(pw) = uL^{-1}, && V(pw) = vL, \quad \textit{where } L = u^2 + |z_1|^2; \end{aligned}$$

(b)
$$\begin{aligned} &Z_1(\tilde{p}w) = z_1 z_3 - z_4, && (Z_2 - Z_1 Z_4)(\tilde{p}w) = z_2 - z_1 z_4, \\ &Z_3(\tilde{p}w) = -\bar{z}_3 L^{-1}, && (Z_4 - Z_1 Z_3)(\tilde{p}w) = z_1, \\ &U(\tilde{p}w) = uL^{1/2}, && V(\tilde{p}w) = vL^{-1}, \quad \textit{where } L = v^2 + |z_3|^2. \end{aligned}$$ □

Proof. To prove (a) we set

$$u_p = U(pw), \quad v_p = V(pw), \quad s_j = Z_j(pw), \quad w_p = pw = (s_1, s_2, s_3, s_4, u_p, v_p).$$

Immediately from the definition of the $\mathrm{Sp}(4, \mathbf{C})$ action on $\mathbf{X}$ follows

$$n(w_p) a(w_p) = p n(w) a(w) k \quad \text{with some } k \in K = \mathrm{Sp}(4). \tag{2.3.4}$$

Since $k{}^t\bar{k}$ is the identity matrix and $a(w_p)$, $a(w)$ are real symmetric matrices, (2.3.4) implies

$$n(w_p) a(w_p)^2\, \overline{{}^t n(w_p)} = (pn(w))\, a(w)^2\, \overline{{}^t(pn(w))}. \tag{2.3.5}$$

By means of routine computation we find that

$$n(w_p) a(w_p)^2\, \overline{{}^t n(w_p)} = (m_{ij}), \quad (pn(w))\, a(w)^2\, \overline{{}^t(pn(w))} = (l_{ij}) \tag{2.3.6}$$

with

$$\begin{aligned} &m_{13} = (s_2 - s_1 s_4) u_p^{-2} v_p^{-1}, && l_{13} = (z_1 z_3 - z_4)\bar{z}_1 u^{-2} v^{-1} + z_3 v^{-1}, \\ &m_{23} = (s_4 - s_1 s_3) u_p^{-2} v_p^{-1}, && l_{23} = (z_1 z_4 - z_2)\bar{z}_1 u^{-2} v^{-1} + z_4 v^{-1}, \\ &m_{33} = u_p^{-2} v_p^{-1}, && l_{33} = v^{-1} + |z_1|^2 u^{-2} v^{-1}, \\ &m_{24} = -\bar{s}_1 (s_4 - s_1 s_3) u_p^{-2} v_p^{-1} + s_3 v_p^{-1}, && l_{24} = (z_2 - z_1 z_4) u^{-2} v^{-1}, \\ &m_{34} = -\bar{s}_1 u_p^{-2} v_p^{-1}, && l_{34} = -z_1 u^{-2} v^{-1}, \\ &m_{44} = |s_1|^2 u_p^{-2} v_p^{-1} + v_p^{-1}, && l_{44} = u^{-2} v^{-1}. \end{aligned}$$

It follows from (2.3.5) and (2.3.6) that $m_{ij} = l_{ij}$ for all i, j. Therefore we have 6 equations which give rise to the expressions for $s_1, \ldots, u_p, v_p$ in terms of $z_1, \ldots, u, v$. These expressions are just the ones stated in part (a). Part (b) can be proved by similar arguments, and we omit the details. ∎

2.3.5 The mappings Q and $\tilde{Q}$. The mappings $Q\colon \mathbf{X} \to \mathbf{H}$ and $\tilde{Q}\colon \mathbf{X} \to \mathbf{H}$ are defined by the formulae

$$Q(w) = (z_1, u), \qquad \tilde{Q}(w) = (z_3, v).$$

We have that $\mathrm{SL}(2, \mathbf{C})$ acts on $\mathbf{H}$, and $\mathrm{Sp}(4, \mathbf{C})$ acts on $\mathbf{X}$. These actions are related one with another as it is described in the following propositions.

Proposition 2.3.3 *Let* $\gamma = h_4(\alpha)$, $\alpha = \begin{pmatrix} a & b \\ c & d \end{pmatrix} \in \mathrm{SL}(2, \mathbf{C})$. *Then we have* $(UV)(\gamma w) = uv$ *and*

$$Q(\gamma w) = \begin{pmatrix} a & b \\ c & d \end{pmatrix} Q(w), \qquad \begin{pmatrix} Z_2(\gamma w) \\ Z_4(\gamma w) \\ Z_3(\gamma w) \end{pmatrix} = \begin{pmatrix} a^2 & 2ab & b^2 \\ ac & ad+bc & bd \\ c^2 & 2cd & d^2 \end{pmatrix} \begin{pmatrix} z_2 \\ z_4 \\ z_3 \end{pmatrix}. \quad \square$$

Proposition 2.3.4 *Let* $\gamma = h_1(\alpha)$, $\alpha = \begin{pmatrix} a & b \\ c & d \end{pmatrix} \in \mathrm{SL}(2, \mathbf{C})$. *Then we have* $(U^2V)(\gamma w) = u^2 v$ *and*

$$\tilde{Q}(\gamma w) = \begin{pmatrix} a & b \\ c & d \end{pmatrix} \tilde{Q}(w), \qquad \begin{pmatrix} (Z_4 - Z_1 Z_3)(\gamma w) \\ Z_1(\gamma w) \end{pmatrix} = \begin{pmatrix} a & -b \\ -c & d \end{pmatrix} \begin{pmatrix} z_4 - z_1 z_3 \\ z_1 \end{pmatrix},$$

$$(Z_2 - Z_1 Z_4)(\gamma w) = z_2 - z_1 z_4. \quad \square$$

Proof (of Propositions 2.3.3, 2.3.4). If $c \neq 0$, let us express α as the product $\delta_1 \delta_2 \delta_3 \delta_4$ with

$$\delta_1 = \begin{pmatrix} 1 & a/c \\ 0 & 1 \end{pmatrix}, \quad \delta_2 = \begin{pmatrix} 1/c & 0 \\ 0 & c \end{pmatrix}, \quad \delta_3 = \begin{pmatrix} 0 & -1 \\ 1 & 0 \end{pmatrix}, \quad \delta_4 = \begin{pmatrix} 1 & d/c \\ 0 & 1 \end{pmatrix}.$$

Correspondingly, one can write $\gamma = \varepsilon_1 \varepsilon_2 \varepsilon_3 \varepsilon_4$, where $\varepsilon_i = h_j(\delta_i)$, $j = 4$ for the case of Proposition 2.3.3, and $j = 1$ for the case of Proposition 2.3.4. The action of the matrices ε_1, ε_2 and ε_4 on $\mathbf{X}$ is described by Proposition 2.3.1. The action of the matrix ε_3 is described by Proposition 2.3.2. Applying successively the formulae of these propositions we obtain the desired formulae. If $c = 0$, express α as the product

$$\alpha = \begin{pmatrix} 1 & b/d \\ 0 & 1 \end{pmatrix} \begin{pmatrix} a & 0 \\ 0 & d \end{pmatrix}$$

and do as in the previous case. ∎

2.3.6 Eisenstein series. Let $\tilde{\psi}: \Gamma(q) \to \mathbf{C}^*$ be an unitary character. Let us assume that its kernel is a subgroup of finite index in $\Gamma(q)$ which contains $\Delta = \Gamma(q) \cap N$. Then, let us define two unitary characters $\tilde{\kappa}': \Gamma_2(q) \to \mathbf{C}^*$ and $\tilde{\kappa}'': \Gamma_2(q) \to \mathbf{C}^*$ setting

$$\tilde{\kappa}'(\gamma) = \tilde{\psi}(h_4(\gamma)), \quad \tilde{\kappa}''(\gamma) = \overline{\tilde{\psi}}(h_1(\gamma)) \quad \text{for any} \quad \gamma \in \Gamma_2(q).$$

Given an automorphic form $f: \mathbf{H} \to \mathbf{C}$ under $\Gamma_2(q)$ with multiplier system $\tilde{\kappa}'$ or $\tilde{\kappa}''$, one can attach to f the series

$$\mathrm{E}(w, s; f) = \sum_{\gamma} \overline{\tilde{\psi}}(\gamma) U(\gamma w)^s V(\gamma w)^s f(Q(\gamma w)), \tag{2.3.7}$$

or, respectively,

$$\widetilde{\mathrm{E}}(w, s; f) = \sum_{\gamma} \overline{\tilde{\psi}}(\gamma) U(\gamma w)^{2s} V(\gamma w)^s \overline{f}(\widetilde{Q}(\gamma w)), \tag{2.3.8}$$

$w \in \mathbf{X}$, $s \in \mathbf{C}$. The summation is carried out over a set of matrices $\gamma \in \Gamma(q)$ that contains a unique representative of each coset in $\Delta_\times \backslash \Gamma(q)$ or $\widetilde{\Delta}_\times \backslash \Gamma(q)$, according to the case (2.3.7) or (2.3.8). In both cases, the terms do not depend on the choice of representatives. To prove, notice first that $\Delta_\times = \Delta h_4(\Gamma_2(q))$ and $\widetilde{\Delta}_\times = \Delta h_1(\Gamma_2(q))$, see Proposition 2.2.1. Recall also that Δ is a subgroup of N. So we have:

$(UV)(\delta w) = (UV)(w)$ for $\delta \in \Delta$ (by Proposition 2.3.1), for $\delta \in h_4(\Gamma_2(q))$ (by Proposition 2.3.3) and thus for all $\delta \in \Delta_\times$;

$(U^2V)(\delta w) = (U^2V)(w)$ for $\delta \in \Delta$ (by Proposition 2.3.1), for $\delta \in h_1(\Gamma_2(q))$ (by Proposition 2.3.4), and thus for all $\delta \in \widetilde{\Delta}_\times$.

At the same time:

For (2.3.7), expressing $\delta \in \Delta_\times$ as $\delta' h_4(\lambda)$ with $\delta' \in \Delta$, $\lambda \in \Gamma_2(q)$ (see 2.2.2), we find $\tilde{\psi}(\delta) = \tilde{\kappa}'(\lambda)$, and then $f(Q(\delta w)) = f(\lambda Q(w)) = \tilde{\kappa}'(\lambda) f(Q(w))$ (see Proposition 2.3.1, Proposition 2.3.3), so we have $\overline{\tilde{\psi}}(\delta) f(Q(\delta w)) = f(Q(w))$ for all $\delta \in \Delta_\times$;

For (2.3.8), expressing $\delta \in \widetilde{\Delta}_\times$ as $\delta' h_1(\lambda)$ with $\delta' \in \Delta$, $\lambda \in \Gamma_2(q)$ (see 2.2.2), we find $\tilde{\psi}(\delta) = \overline{\tilde{\kappa}}''(\lambda)$, and then $f(\widetilde{Q}(\delta w)) = f(\lambda \widetilde{Q}(w)) = \tilde{\kappa}''(\lambda) f(\widetilde{Q}(w))$ (see Proposition 2.3.1, Proposition 2.3.4), so we have $\overline{\tilde{\psi}}(\delta) \overline{f}(\widetilde{Q}(\delta w)) = \overline{f}(\widetilde{Q}(w))$ for all $\delta \in \widetilde{\Delta}_\times$.

Thus it is proved that the series (2.3.7), (2.3.8) are defined correctly. These series are analoque of Eisenstein-Klingen series in the Siegel modular forms theory [49]. In framework of the Selberg–Langlands theory these are the maximal parabolic Eisenstein series attached to the form f and the maximal parabolic subgroups P and $\widetilde{P}$.

Let us now define one more series

$$\mathrm{E}_{\min}(w;s,t) = \sum_{\gamma} \overline{\tilde{\psi}}(\gamma) U(\gamma w)^{2t+s} V(\gamma w)^{s+t}, \qquad w \in \mathbf{X}, \quad s,t \in \mathbf{C}, \quad (2.3.9)$$

known as a minimal parabolic Eisenstein series. Here the summation is carried out over a set of matrices $\gamma \in \Gamma(q)$ that contains a unique representative of each coset in $\Delta\backslash\Gamma(q)$. Just like as above one can prove the independence of the terms on the choice of representatives.

The next theorem is known from the general Eisenstein series theory [59].

Theorem 2.3.5 (a) *The minimal parabolic Eisenstein series* (2.3.9) *converges absolutely and locally uniformly in* $\mathbf{X} \times \{(s,t) \in \mathbf{C}^2 \mid \Re(s) > \tau,\ \Re(t) > \tau\}$ *for some* $\tau \in \mathbf{R}$. *For each* $w \in \mathbf{X}$ *the function* $\mathrm{E}_{\min}(w;\cdot,\cdot)$ *is regular on* $\{(s,t) \in \mathbf{C}^2 \mid \Re(s) > \tau,\ \Re(t) > \tau\}$ *and can be extended meromorphically to* $\mathbf{C}^2$.

(b) *The maximal parabolic Eisenstein series* (2.3.7), (2.3.8) *converge absolutely and locally uniformly in* $\mathbf{X} \times \{s \in \mathbf{C} \mid \Re(s) > \tau\}$ *for some* $\tau \in \mathbf{R}$. *Given* $w \in \mathbf{X}$, *the functions* $\mathrm{E}(w,\cdot;f)$, $\widetilde{\mathrm{E}}(w,\cdot;f)$ *are regular on* $\{s \in \mathbf{C} \mid \Re(s) > \tau\}$ *and can be extended meromorphically to* $\mathbf{C}$.

(c) *All the functions* $\mathrm{E}(\cdot,s;f)$, $\widetilde{\mathrm{E}}(\cdot,s;f)$ *and* $\mathrm{E}_{\min}(\cdot;s,t)$ *are automorphic forms on* $\mathbf{X}$ *under* $\Gamma(q)$ *with multiplier system* $\tilde{\psi}$, *whenever* s *and* (s,t) *are regular points.* ■

It is seen from Proposition 2.2.8 (a) that one can take $\tilde{\psi}$ to be Bass–Milnor–Serre's ψ, and thus to get well defined cubic metaplectic minimal parabolic Eisenstein series (2.3.9). With such a choice of $\tilde{\psi}$, we get that both $\tilde{\kappa}'$ and $\tilde{\kappa}''$ are nothing but Kubota's κ, see Proposition 2.2.9. So, given a cubic metaplectic form $f\colon \mathbf{H} \to \mathbf{C}$ under $\Gamma_2(q)$, we have two well defined maximal parabolic cubic metaplectic Eisenstein series (1.3.8), (1.3.9) attached to f.

Let us now restrict our attention by the cubic metaplectic Eisenstein series. By the general theory, the minimal parabolic Eisenstein series can be described as the maximal parabolic Eisenstein series attached to $f = \mathrm{E}_*(\cdot,t)$. The next theorem gives us precies relations.

Theorem 2.3.6 *Let* $\mathrm{E}_*(\cdot,t)\colon \mathbf{H} \to \mathbf{C}$ *be the cubic metaplectic Eisenstein series defined in* 0.3.5. *Then for the cubic metaplectic Eisenstein series under* $\Gamma(q)$ *(defined by* (2.3.7), (2.3.8), (2.3.9) *with* $\tilde{\psi} = \psi$*) one has the relations*

(a) $\mathrm{E}(w,s;\mathrm{E}_*(\cdot,t)) = \mathrm{E}_{\min}(w;s-t,t)$,

(b) $\widetilde{\mathrm{E}}(w,s;\mathrm{E}_*(\cdot,\bar{t})) = \mathrm{E}_{\min}(w;2t,s-t)$

are valid for all $s,t \in \mathbf{C}$ *and* $w \in \mathbf{X}$ *under the only condition that the series at the right-hand sides converge absolutely.* □

Proof. (a) Let Λ_* be a subset of $\Gamma_2(q)$ which contains a unique representative of each coset in $\Delta_*\backslash\Gamma_2(q)$, where Δ_* is a subgroup of upper triangular matrices in $\Gamma_2(q)$. It is easy to find, see Proposition 2.2.1, that the set $\Lambda = h_4(\Lambda_*)$ contains a unique representative of each coset in $\Delta\backslash\Delta_\times$. So, we have

$$\sum_{\lambda\in\Lambda}\overline{\psi}(\lambda)U(\lambda w)^t = \sum_{\delta\in\Lambda_*}\overline{\kappa}(\delta)U(h_4(\delta)w)^t = \mathrm{E}_*(Q(w),t),$$

by Proposition 2.2.8 (b), Proposition 2.3.3 and the definitions given in 0.3.1. Let Λ' be a subset of $\Gamma(q)$ which contains a unique representative of each coset in $\Delta_\times\backslash\Gamma(q)$. We set $\nabla = \{\lambda\lambda' \mid \lambda\in\Lambda,\ \lambda'\in\Lambda'\}$ and notice that the mapping $\Lambda\times\Lambda'\to\nabla$ defined by $(\lambda,\lambda')\mapsto\lambda\lambda'$ is bijective and that ∇ contains a unique representative of each coset in $\Delta\backslash\Gamma(q)$. Also notice that, by Proposition 2.3.3, $(UV)(\lambda\lambda' w) = (UV)(\lambda' w)$. These notices allow us to rearrange the terms in (2.3.9), and we find

$$\begin{aligned}\mathrm{E}_{\min}(w;s-t,t) &= \sum_{\lambda'\in\Lambda',\ \lambda\in\Lambda}\overline{\psi}(\lambda\lambda')U(\lambda\lambda' w)^{s+t}V(\lambda\lambda' w)^s\\ &= \sum_{\lambda'\in\Lambda'}\overline{\psi}(\lambda')(UV)(\lambda' w)^s\sum_{\lambda\in\Lambda}\overline{\psi}(\lambda)U(\lambda\lambda' w)^t\\ &= \sum_{\lambda'\in\Lambda'}\overline{\psi}(\lambda')(UV)(\lambda' w)^s\mathrm{E}_*(Q(\lambda' w),t) = \mathrm{E}(w,s;\mathrm{E}_*(\cdot,t)).\end{aligned}$$

(b) We can do just like in the part (a). So, with Λ_* as above, we have that the set $\Lambda = h_1(\Lambda_*)$ contains a unique representative of each coset in $\Delta\backslash\widetilde{\Delta}_\times$ and we have

$$\sum_{\lambda\in\Lambda}\overline{\psi}(\lambda)V(\lambda w)^t = \sum_{\delta\in\Lambda_*}\kappa(\delta)V(h_1(\delta)w)^t = \overline{\mathrm{E}}_*(\widetilde{Q}(w),\overline{t}),$$

by Proposition 2.2.8 (c), Proposition 2.3.3 and the definitions given in 0.3.1. Let Λ be as above and let Λ' be a subset of $\Gamma(q)$ which contains a unique representative of each coset in $\widetilde{\Delta}_\times\backslash\Gamma(q)$. Set $\nabla = \{\lambda\lambda' \mid \lambda\in\Lambda,\ \lambda'\in\Lambda'\}$ and notice that the mapping $\Lambda\times\Lambda'\to\nabla$ defined by $(\lambda,\lambda')\mapsto\lambda\lambda'$ is bijective and that ∇ contains a unique representative of each coset in $\Delta\backslash\Gamma(q)$. Also notice that, by Proposition 2.3.4, $(U^2V)(\lambda\lambda' w) = (U^2V)(\lambda' w)$. These notices allow us to rearrange the terms in (2.3.9), and we find

$$\begin{aligned}\mathrm{E}_{\min}(w;2t,s-t) &= \sum_{\lambda'\in\Lambda',\ \lambda\in\Lambda}\overline{\psi}(\lambda\lambda')U(\lambda\lambda' w)^{2s}V(\lambda\lambda' w)^{s+t}\\ &= \sum_{\lambda'\in\Lambda'}\overline{\psi}(\lambda')(U^2V)(\lambda' w)^s\sum_{\lambda\in\Lambda}\overline{\psi}(\lambda)V(\lambda\lambda' w)^t\\ &= \sum_{\lambda'\in\Lambda'}\overline{\psi}(\lambda')(U^2V)(\lambda' w)^s\overline{\mathrm{E}}_*(\widetilde{Q}(\lambda' w),\overline{t}) = \widetilde{\mathrm{E}}(w,s;\mathrm{E}_*(\cdot,\overline{t})),\end{aligned}$$

as required. ∎

Let us conclude this subsection by one more elementary observation.

Proposition 2.3.7 *Let $f\colon \mathbf{H} \to \mathbf{C}$ be a cubic metaplectic form under $\Gamma_2(3)$. Suppose further that f is invariant under some $\lambda \in \mathrm{SL}(2,\mathbf{Z})$, that is $f(\lambda(z,v)) = f(z,v)$ for all $(z,v) \in \mathbf{H}$. Then for the cubic metaplectic Eisenstein series attached to f and $\Gamma(3)$ we have*

(a) $\mathrm{E}(\delta w, s; f) = \mathrm{E}(w, s; f)$ *for* $\delta = h_4(\lambda)$;

(b) $\widetilde{\mathrm{E}}(\delta w, s; f) = \widetilde{\mathrm{E}}(w, s; f)$ *for* $\delta = h_1(\lambda)$. □

Proof. (a) By analytic continuation principle it is sufficient to prove the desired formula assuming $\Re(s)$ so large that the Eisenstein series converge absolutely. We have $\Gamma(3) = \Gamma^{(4)}_{\mathrm{princ}}(3) \cap \mathrm{Sp}(4,\mathcal{O})$ and $\Gamma_2(3) = \Gamma^{(2)}_{\mathrm{princ}}(3)$, that are normal subgroups of $\mathrm{Sp}(4,\mathcal{O})$ and $\mathrm{SL}(2,\mathcal{O})$, respectively. This yields further $\delta\Gamma(3)\delta^{-1} = \Gamma(3)$ and $\delta\Delta_{\mathsf{x}}\delta^{-1} = \Delta_{\mathsf{x}}$. Now, by definition (2.3.7) we have

$$\mathrm{E}(\delta w, s; f) = \sum_{\gamma\in\Omega} \overline{\psi}(\gamma)(UV)(\gamma\delta w)^s f(Q(\gamma\delta w)),$$

where $\Omega \subset \Gamma(3)$ contains a unique representative of each coset in $\Delta_{\mathsf{x}}\backslash\Gamma(3)$. For such Ω, the set $\delta\Omega\delta^{-1}$ also contains a unique representative of each coset in $\Delta_{\mathsf{x}}\backslash\Gamma(3)$, just because of $\delta\Gamma(3)\delta^{-1} = \Gamma(3)$ and $\delta\Delta_{\mathsf{x}}\delta^{-1} = \Delta_{\mathsf{x}}$. So, one can replace γ by $\delta\gamma\delta^{-1}$ in the right-hand side to obtain

$$\mathrm{E}(\delta w, s; f) = \sum_{\gamma\in\Omega} \overline{\psi}(\delta\gamma\delta^{-1})(UV)(\delta\gamma w)^s f(Q(\delta\gamma w)), \qquad \text{where}$$

$$\begin{aligned}
&\psi(\delta\gamma\delta^{-1}) = \psi(\gamma), \quad \text{by Theorem 0.2.2 (f)};\\
&(UV)(\delta\gamma w) = (UV)(\gamma w), \quad \text{by Proposition 2.3.3};\\
&f(Q(\delta\gamma w)) = f(Q(\gamma w)), \quad \text{by Proposition 2.3.3 and assumptions.}
\end{aligned}$$

So, the right-hand side is equal to

$$\sum_{\gamma\in\Omega} \overline{\psi}(\gamma)(UV)(\gamma w)^s f(Q(\gamma w)) = \mathrm{E}(w, s; f), \qquad \text{as required.}$$

(b) This can be proved by quite simular arguments, let us omit details. ■

2.3.7 Whittaker functions. Let $s, t \in \mathbf{C}$ and let $f\colon \mathbf{H} \to \mathbf{C}$ be an eigenfunction of the Laplace–Beltrami operator, say $\mathrm{D}_{\text{L-B}} f = r(2-r)$ with some $r \in \mathbf{C}$. For $w = (z_1, z_2, z_3, u, v) \in \mathbf{X}$ we set

$$\begin{aligned}
\nabla_{s,t}(w) &= u^{s+2t} v^{s+t},\\
\nabla_s(w; f) &= u^s v^s f(z_1, u), \qquad\qquad (2.3.10)\\
\widetilde{\nabla}_s(w; f) &= u^{2s} v^s \overline{f}(z_3, v).
\end{aligned}$$

The so defined functions are known to be joint eigenfunctions of $\mathbf{D}(\mathbf{X})$. Let us define a character $\chi_{s,t}\colon \mathbf{D}(\mathbf{X}) \to \mathbf{C}$ setting $D\nabla_{s,t} = \chi_{s,t}(D)\nabla_{s,t}$ for $D \in \mathbf{D}(\mathbf{X})$.

Then we have $\nabla_s(\cdot; f)$ and $\widetilde{\nabla}_s(\cdot; f)$ are joint eigenfunctions of $\mathbf{D}(\mathbf{X})$ attached to the characters $\chi_{s-r,r}$ and $\chi_{2\bar{r},s-\bar{r}}$, respectively. The translations of the functions in (2.3.10) occur as the terms in the Eisenstein series in 2.3.6, and the Eisenstein series are the eigenfunctions of all the invariant differential operators just because the functions (2.3.10) are. More precisely, we have that the Eisenstein series $\mathrm{E}_{\min}(\cdot; s, t)$, $\mathrm{E}(\cdot, s; f)$ and $\widetilde{\mathrm{E}}(\cdot, s; f)$ are the joint eigenfunctions of $\mathbf{D}(\mathbf{X})$ attached, respectively, to the characters $\chi_{s,t}$, $\chi_{s-r,r}$ and $\chi_{2\bar{r},s-\bar{r}}$.

Let $\mu, \nu \in \mathbf{C}$, and let $\phi_{\mu,\nu}: N \to \mathbf{C}^*$ be the unitary character of the group N defined as in 2.1.6. So we have $\phi_{\mu,\nu}(n(z_1, z_2, z_3, z_4)) = e(\mu z_1 + \nu z_3)$ for all $n(z_1, z_2, z_3, z_4) \in N$. We say a C^∞-function $W: \mathbf{X} \to \mathbf{C}$ is an Whittaker function attached to $\chi_{s,t}$ and $\phi_{\mu,\nu}$ if

(a) $W(\delta w) = \phi_{\mu,\nu}(\delta) W(w)$ for all $\delta \in N$, $w \in \mathbf{X}$;

(b) $DW = \chi_{s,t}(D) W$ for all $D \in \mathbf{D}(\mathbf{X})$;

(c) $W(z_1, z_2, z_3, z_4, u, v)$ is of at most polynomial growth as $u, v \to \infty$.

It follows immediately from (a) that Whittaker function W can be written uniquely as the product

$$W(w) = R(u, v) e(\mu z_1 + \nu z_3),$$

where $R: \mathbf{R}_+^* \times \mathbf{R}_+^* \to \mathbf{C}$ is a C^∞-function called the radial part of W, and, as usually, $w = (z_1, z_2, z_3, z_4, u, v) \in \mathbf{X}$.

Let $c_{\mu,\nu}(u, v; F)$ be the μ, ν^{th} Fourier coefficient of the form F attached to the character $\chi_{s,t}$, as it was defined in (2.3.3). Then one can show

$$w \mapsto c_{\mu,\nu}(u, v; F) e(\mu z_1 + \nu z_3)$$

is an Whittaker function attached to the characters $\chi_{s,t}$ and $\phi_{\mu,\nu}$. In particular, one can take F to be the Eisenstein series and thus to get some integral representation for the Whittaker function. This will be treated in Section 2.4.

The simplest Whittaker function is the function $\nabla_{s,t}(w)$ defined in (2.3.10). It is attached to the characters $\chi_{s,t}$ and $\phi_{0,0}$.

The Whittaker function W is said to be non-degenerate if it is attached to non-degenerate character $\phi_{\mu,\nu}$, i.e., if $\mu\nu \neq 0$. Otherwise, W is said to be degenerate. One knows the radial parts of degenerate Whittaker functions attached to the characters $\phi_{\mu,0}$ and $\phi_{0,\nu}$ with $\mu\nu \neq 0$ are, in essence, the Bessel–MacDonald functions.

Let us look more closely on non-degenerate Whittaker functions. Let $R_{s,t}$ be the radial part of an Whittaker function $W_{s,t}^{1,1}$ attached to the characters $\chi_{s,t}$ and $\phi_{1,1}$. We then have

$$W_{s,t}^{\mu,\nu}(w) = R_{s,t}(|\mu| u, |\nu| v) e(\mu z_1 + \nu z_3),$$

is an Whittaker function attached to the characters $\chi_{s,t}$ and $\phi_{\mu,\nu}$. Thus, the study of non-degenerate Whittaker functions can be reduced to study the functions $R_{s,t}$ only. It is clear, $W_{s,t}^{1,1}$ times any factor depending on s and t only is an Whittaker function attached to the same characters as $W^{1,1}_{s,t}$ does. One knows a factor above can be choosen in such a way, that, given any $u, v \in \mathbf{R}^*_+$, the function $(s,t) \mapsto R_{2+s,1+t}(u,v)$ is regular on $\mathbf{C} \times \mathbf{C}$ and is invariant under the action of the Weyl group on $\mathbf{C} \times \mathbf{C}$. To describe the action we mean let us turn to 2.1.3, and let $\Phi_{s,t}\colon A \to \mathbf{C}^*$ be defined by

$$\Phi_{s,t}(\operatorname{diag}(uv^{1/2}, v^{1/2}, u^{-1}v^{-1/2}, v^{-1/2}) = u^{s+2t}v^{s+t}.$$

Given ε in the Weyl group, we have ${}^{\varepsilon}\Phi_{s,t} = \Phi_{s',t'}$ with some $(s',t') \in \mathbf{C} \times \mathbf{C}$ uniquely determined by ε and (s,t). Setting (s',t') to be the image of (s,t) under ε we get the action of the Weyl group on $\mathbf{C} \times \mathbf{C}$. By simple computation we find

$$\begin{aligned} &(s,t) \overset{p}{\longmapsto} (s+2t, -t), && (s,t) \overset{\tilde{p}}{\longmapsto} (-s, s+t), \\ &(s,t) \overset{p\tilde{p}}{\longmapsto} (s+2t, -s-t), && (s,t) \overset{\tilde{p}p}{\longmapsto} (-s-2t, s+t), \\ &(s,t) \overset{p\tilde{p}p}{\longmapsto} (s, -s-t), && (s,t) \overset{\tilde{p}p\tilde{p}}{\longmapsto} (-s-2t, t), \\ &\qquad\qquad (s,t) \overset{\sigma}{\longmapsto} (-s,-t), \end{aligned} \tag{2.3.11}$$

with the notations in 2.1.4.

For the Whittaker functions theory see [43], [33], [90]. We shall continue the discussion of Whittaker functions in 2.4.7.

2.4 Eisenstein series Fourier coefficients

2.4.1 Basic theorem. 2.4.2 An auxiliary proposition. 2.4.3 Proof of Theorem 2.4.1. 2.4.4 On the series $\mathcal{D}_{\mu,\nu}$. 2.4.5 Evaluation of integrals. 2.4.6 Mellin transform. 2.4.7 More on Whittaker functions.

2.4.1 Basic theorem. Let $\tilde{\kappa}\colon \Gamma_2(q) \to \mathbf{C}^*$ and $\tilde{\psi}\colon \Gamma(q) \to \mathbf{C}^*$ $(q \subset (3))$ be unitary characters satisfying the conditions:

the kernel of $\tilde{\kappa}$ is a subgroup of finite index in $\Gamma_2(q)$;

the kernel of $\tilde{\psi}$ is a subgroup of finite index in $\Gamma(q)$;

$\tilde{\kappa}(\gamma) = \tilde{\psi}(h_4(\gamma))$ for $\gamma \in \Gamma_2(q)$;

$\tilde{\kappa}(\gamma) = 1$ for $\gamma \in \Delta_*$ and for $\gamma \in {}^t\Delta_*$;

$\tilde{\psi}(\gamma) = 1$ for $\gamma \in \Delta$ and for $\gamma \in {}^t\Delta$;

$\tilde{\psi}(\sigma\gamma\sigma^{-1}) = \tilde{\psi}(\gamma)$ for all $\gamma \in \Gamma(q)$.

Recall that Δ_* and Δ are the subgroups of upper triangular unipotent matrices in $\Gamma_2(q)$ and $\Gamma(q)$ respectively (see 2.2.2, 0.3.4), and that

$$\sigma = \begin{pmatrix} 0 & 0 & -1 & 0 \\ 0 & 0 & 0 & -1 \\ 1 & 0 & 0 & 0 \\ 0 & 1 & 0 & 0 \end{pmatrix} \quad \text{(as in 2.1.1, 2.1.4).}$$

Let $f\colon \mathbf{H} \to \mathbf{C}$ be an automorphic form under the group $\Gamma_2(q)$ with multiplier system $\tilde{\kappa}$, and let $\mathrm{E}(\cdot, s; f)$ be the maximal parabolic Eisenstein series attached to f and $\tilde{\psi}$, as it was defined in 2.3.6. More precisely, we mean that $\mathrm{E}(\cdot, s; f)$ is the series (2.3.7), so it is an automorphic form under $\Gamma(q)$ with multiplier system $\tilde{\psi}$. Then we have that

$$w \mapsto \mathrm{E}(\sigma w, s; f), \qquad w \in \mathbf{X}, \tag{2.4.1}$$

is an automorphic form under the group $\Gamma(q)$ with multiplier system $\tilde{\psi}$, as well as the Eisenstein series $\mathrm{E}(\cdot, s; f)$ itself. This is just because of we have $\sigma\Gamma(q)\sigma^{-1} = \Gamma(q)$ and $\tilde{\psi}(\sigma\gamma\sigma^{-1}) = \tilde{\psi}(\gamma)$ for all $\gamma \in \Gamma(q)$, see remark in 2.3.2. We shall evaluate the Fourier coefficients

$$c_{\mu,\nu}(u, v; \mathrm{E}(\sigma(\cdot), s; f)) \tag{2.4.2}$$

of the form (2.4.1). The result is stated below in Theorem 2.4.1. We assume that $\mu, \nu \in q^*$. Only for such μ and ν are the coefficients (2.4.2) defined. Also, in our computations we assume $\Re(s)$ to be sufficiently large for the integrals and the series involved were absolutely convergent. However, we have that $s \mapsto c_{\mu,\nu}(u, v; \mathrm{E}(\sigma w, s; f))$ is a meromorphic function on $\mathbf{C}$, because the Eisenstein series is. Also, the functions $(s,t) \mapsto J_{\mu,\nu}(u, v, s, t)$ in the formulae (2.4.11), (2.4.12) in Theorem 2.4.1 are meromorphic functions on $\mathbf{C}^2$. This is known from the Whittaker functions theory, and this will be seen from our formulae in 2.4.5. So, we have the formulae (2.4.6), (2.4.7) in Theorem 2.4.1 still valid without any restrictions on s, and we have the Dirihlet series $\mathcal{D}_{\mu,\nu}$ and $\mathcal{D}_{0,\nu}(\cdot, t)$ involved are meromorphic functions on $\mathbf{C}$.

It follows from Theorem 0.2.2 (c), (f) and Proposition 2.2.8 (a), (b) that the listed above conditions on $\tilde{\kappa}$ and $\tilde{\psi}$ are satisfied by the Kubota's κ and the Bass–Milnor–Serre's ψ. Therefore Theorem 2.4.1 is valid for cubic metaplectic Eisenstein series, and this is just the case we are interested in.

For the sake of brevity, let us write in that follows $c_{\mu,\nu}$ instead of (2.4.2). Immediately from the definitions we get

$$\begin{aligned} c_{\mu,\nu} &= c \iiiint\limits_{\mathcal{F}_q} \mathrm{E}(\sigma w, s; f)\overline{e(\mu z_1 + \nu z_3)}\, dz_1 dz_2 dz_3 dz_4 \\ &= c \sum_{\gamma} \bar{\tilde{\psi}}(\gamma) \iiiint\limits_{\mathcal{F}_q} \nabla(\gamma\sigma w)\overline{e(\mu z_1 + \nu z_3)}\, dz_1 dz_2 dz_3 dz_4. \end{aligned} \tag{2.4.3}$$

Here the summation is carried out over some set of matrices γ that contains a unique representative of each coset in $\Delta_\times\backslash\Gamma(q)$; $c = \mathrm{vol}(\mathcal{F}_q)$, $\mathcal{F}_q$ denotes $(\mathbf{C}/q)^4$ or any other set indicated in 2.3.2, and

$$\nabla(w) = (UV)(w)^s f(Q(w)), \qquad w = (z_1, z_2, z_3, z_4, u, v) \in \mathbf{X}. \tag{2.4.4}$$

We shall use the notations of Corollary 0.3.3 for the form $f\colon \mathbf{H} \to \mathbf{C}$. So, $\mathrm{D}_{\text{L-B}} f = r(2 - r)$ with $r \in \mathbf{C}$, and $\rho(\nu)$ means the νth Fourier coefficient of f about the cusp 0. In addition, let

$$m_+ = m\rho_+, \qquad m_- = m\rho_-, \qquad m = \mathrm{vol}(\mathbf{C}/q)^{-3}. \tag{2.4.5}$$

The notations introduced here are remained till the end of this section.

Theorem 2.4.1 *For the Eisenstein series* (2.4.1) *Fourier coefficients* $c_{\mu,\nu}$ *the following formulae hold. First, we have either*

$$c_{0,\nu} = m_+ \mathcal{D}_{0,\nu}(s,r) J_{0,\nu}(u,v,s,r) + m_- \mathcal{D}_{0,\nu}(s,2-r) J_{0,\nu}(u,v,s,2-r)$$

or

$$c_{0,\nu} = m_+ \frac{\partial}{\partial t} \mathcal{D}_{0,\nu}(s,t) J_{0,\nu}(u,v,s,t)\Big|_{t=1} + m_- \mathcal{D}_{0,\nu}(s,1) J_{0,\nu}(u,v,s,1), \tag{2.4.6}$$

according to $r \neq 1$ *or* $r = 1$. *Then, for any* r,

$$c_{\mu,\nu} = m \mathcal{D}_{\mu,\nu}(s) J_{\mu,\nu}(u,v,s,r). \tag{2.4.7}$$

Here $\mu, \nu \in q^*$, $\mu \neq 0$, $s \in \mathbf{C}$ *and*

$$\mathcal{D}_{0,\nu}(s,t) = \sum_{\gamma} \overline{\overline{\psi}}(\gamma) e(-\nu c_4/d_4) |d_1|^{t-s} |d_4|^{2-t-s} |l_\gamma|^{-2}, \tag{2.4.8}$$

$$\mathcal{D}_{\mu,\nu}(s) = \sum_{\gamma} \overline{\psi}(\gamma) e(\mu d_2/d_1 - \nu c_4/d_4) \rho(\mu d_4/d_1) |d_1 d_4|^{1-s} |l_\gamma|^{-2}. \tag{2.4.9}$$

The summation in (2.4.8), (2.4.9) *is over some set of matrices* $\gamma \in \Gamma(q)$ *that contains a unique representative of each double coset in* $\Delta_\times \backslash \Gamma(q) / {}^t\Delta$. *This set of representatives is chosen so that*

$$d_3 = 0, \tag{2.4.10}$$

and arbitrarily otherwise. We mean by c_j, d_j *the entries of the matrix* γ, *as in Proposition* 2.1.2 *and Proposition* 2.2.5; $l_\gamma = \gcd(2, c_4, d_4)$. *The coefficients* $\rho(\lambda)$, $\lambda \in q^*$, *and* r *are defined with respect to the form* f *as in Corollary* 0.3.3. *In addition, it is assumed that* $\rho(\lambda) = 0$ *if* $\lambda \notin q^*$.

$$J_{\mu,\nu}(u,v,s,t) = \iiint_{\mathbf{CCC}} (V^s U^{s+t})(\widehat{\sigma}\widehat{w}) \overline{e(\nu z_3)}\, dz_2 dz_3 dz_4 \quad \text{if } \mu = 0 \text{ and} \tag{2.4.11}$$

$$= \iiint_{\mathbf{CCC}} (V^s U^{s+1})(\widehat{\sigma}\widehat{w}) K_{t-1}(4\pi|\mu| U(\widehat{\sigma}\widehat{w})) \overline{e(\mu Z_1(\widehat{\sigma}\widehat{w}) + \nu z_3)}\, dz_2 dz_3 dz_4 \tag{2.4.12}$$

otherwise. In both formulae

$$\widehat{\sigma} = \begin{pmatrix} 0 & 0 & 0 & -1 \\ 0 & 0 & -1 & 0 \\ 0 & 1 & 0 & 0 \\ 1 & 0 & 0 & 0 \end{pmatrix}, \qquad \widehat{w} = (0, z_2, z_3, z_4, u, v). \tag{2.4.13}$$

The series (2.4.8), (2.4.9) *and the integrals* (2.4.11), (2.4.12) *are absolutely convergent whenever* $\Re(s)$ *is sufficiently large.* □

The existence of a representative system satisfying (2.4.10) is clear from Propositions 2.2.4, 2.2.3, 2.2.1.

We shall prove Theorem 2.4.1 in 2.4.3, and then, in 2.4.4 and 2.4.5, we shall find more manageable formulae for the series $\mathcal{D}_{\mu,\nu}$ and the integrals $J_{\mu,\nu}$.

2.4.2 An auxiliary proposition. Let σ and $\widehat{\sigma}$ be the matrices defined by (2.1.2), (2.4.13). It is easy to check that

$$\sigma = p\widehat{\sigma}, \tag{2.4.14}$$

where p is the matrix defined in 2.1.3.

Proposition 2.4.2 *Let* $a, b \in \mathbf{C}^*$ *and*

$$\tilde{w} = (0, \tilde{z}_2, z_3, \tilde{z}_4, u, y), \quad \tilde{z}_2 = z_2 - 2z_1z_4 + z_1^2 z_3, \quad \tilde{z}_4 = z_4 - z_1 z_3. \tag{2.4.15}$$

Then

(a) $Q(\sigma w) = \begin{pmatrix} 0 & -1 \\ 1 & 0 \end{pmatrix} (z_1 - Z_1(\widehat{\sigma}\tilde{w}), U(\widehat{\sigma}\tilde{w}))$;

(b) $\left(\frac{a}{b} Z_1(\sigma w), \left|\frac{a}{b}\right| U(\sigma w)\right) = \begin{pmatrix} 0 & -1 \\ 1 & 0 \end{pmatrix} \left(\frac{b}{a}(z_1 - Z_1(\widehat{\sigma}\tilde{w})), \left|\frac{b}{a}\right| U(\widehat{\sigma}\tilde{w})\right)$;

(c) $(UV)(\sigma w) = (UV)(\widehat{\sigma}\tilde{w})$. □

Proof. (a) By Proposition 2.3.3, Proposition 2.3.1 (b) and (2.4.14),

$$Q(\sigma w) = Q(p\widehat{\sigma} w) = \begin{pmatrix} 0 & -1 \\ 1 & 0 \end{pmatrix} (-Z_1(\widehat{\sigma} w), U(\widehat{\sigma} w)), \tag{2.4.16}$$

since

$$p = h_4\left(\begin{pmatrix} 0 & -1 \\ 1 & 0 \end{pmatrix}\right) \operatorname{diag}(1, -1, 1, -1).$$

One has $a(w) = a(\tilde{w})$ and $n(w) = \delta n(\tilde{w})$ with $\delta = n(z_1, 0, 0, 0)$. Obviously, $\widehat{\sigma}\delta = \delta'\widehat{\sigma}$ with $\delta' = n(-z_1, 0, 0, 0)$. Consequently $\widehat{\sigma} n(w) a(w) = \widehat{\sigma}\delta n(\tilde{w}) a(\tilde{w}) = \delta'\widehat{\sigma} n(\tilde{w}) a(\tilde{w})$ and, so,

$$\widehat{\sigma} w = \delta'\widehat{\sigma}\tilde{w}, \quad \text{where} \quad \delta' = n(-z_1, 0, 0, 0) = h_4\left(\begin{pmatrix} 1 & -z_1 \\ 0 & 1 \end{pmatrix}\right). \tag{2.4.17}$$

From (2.4.17) and Proposition 2.3.3, we obtain

$$\mathbf{Z}_1(\widehat{\sigma} w) = -z_1 + Z_1(\widehat{\sigma}\tilde{w}), \quad U(\widehat{\sigma} w) = U(\widehat{\sigma}\tilde{w}),$$

that, being composed with (2.4.16), gives (a).

(b) Let us take $t \in \mathbf{C}$ such that $t^2 = a/b$. Then, by means of (0.3.2), we find that

$$\text{the left-hand side of (b) equals} \quad \begin{pmatrix} t & 0 \\ 0 & 1/t \end{pmatrix} (Z_1(\sigma w), U(\sigma w)),$$

$$\text{the right-hand side of (b) equals} \quad \begin{pmatrix} 0 & -1 \\ 1 & 0 \end{pmatrix} \begin{pmatrix} 1/t & 0 \\ 0 & t \end{pmatrix} (z_1 - Z_1(\widehat{\sigma}\tilde{w}), U(\widehat{\sigma}\tilde{w})).$$

Now to prove (b) we notice that

$$\begin{pmatrix} t & 0 \\ 0 & 1/t \end{pmatrix} = \begin{pmatrix} 0 & -1 \\ 1 & 0 \end{pmatrix} \begin{pmatrix} 1/t & 0 \\ 0 & t \end{pmatrix} \begin{pmatrix} 0 & -1 \\ 1 & 0 \end{pmatrix}^{-1} \quad \text{and we apply (a).}$$

(c) Proposition 2.3.2 (a), Proposition 2.3.3, (2.4.17) and (2.4.14) yield

$$(UV)(\sigma w) = (UV)(p\widehat{\sigma} w) = (UV)(\widehat{\sigma} w) = (UV)(\delta' \widehat{\sigma} \tilde{w}) = (UV)(\widehat{\sigma} \tilde{w}). \quad \blacksquare$$

2.4.3 Proof of Theorem 2.4.1. From formula (2.4.3), Proposition 2.2.7 (a) and Proposition 2.2.8 (a), we get

$$c_{\mu,\nu} = c \sum_{\gamma} \bar{\bar{\psi}}(\gamma) \sum_{h} \iiiint_{\mathcal{F}_q} \nabla(\gamma\sigma h w) \overline{e(\mu z_1 + \nu z_3)} \, dz_1 dz_2 dz_3 dz_4. \tag{2.4.18}$$

Here w is as in (2.4.4) and the summation is over $\gamma \in \Omega$, $h \in \sigma^{-1}\Lambda_\gamma \sigma$, with the notations of Proposition 2.2.7). Then, by means of Proposition 2.2.7 (b), (c) we reduce (2.4.18) to the following form:

$$c_{\mu,\nu} = c \sum_{\gamma} \bar{\bar{\psi}}(\gamma) \iiiint_{\mathcal{F}_\gamma} \nabla(\gamma\sigma w) \overline{e(\mu z_1 + \nu z_3)} \, dz_1 dz_2 dz_3 dz_4; \tag{2.4.19}$$

here the summation is over $\gamma \in \Omega$ and

$$\mathcal{F}_\gamma = \{ (z_1, z_2, z_3, z_4) \mid z_1 \in \mathbf{C}/\vartheta q, \ z_2, z_3, z_4 \in \mathbf{C} \}, \qquad \vartheta = d_4/l_\gamma. \tag{2.4.20}$$

Now express $\gamma\sigma$ in (2.4.19) by formulae (2.1.14), (2.1.15) of Proposition 2.1.3. (The conditions of this proposition are hold, since $d_4 \equiv d_1 d_4 - d_2 d_3 \equiv 1 \pmod 3$ for $\gamma \in \Gamma(q)$.) Choose the representative system Ω in such a way that (2.4.10) take place. Then for $\gamma \in \Omega$ one has $l_1 = 0$ in the decomposition (2.1.14). By using Proposition 2.3.1 we find

$$\begin{aligned} (UV)(\gamma\sigma w) &= (UV)(\tilde{a}\sigma\tilde{n}w) = |t_1 t_2| (UV)(\sigma\tilde{n}w), \\ Q(\gamma\sigma w) &= Q(\tilde{a}\sigma\tilde{n}w) = \Big(\frac{t_1}{t_2} Z_1(\sigma\tilde{n}w), \Big|\frac{t_1}{t_2}\Big| U(\sigma\tilde{n}w) \Big); \end{aligned} \tag{2.4.21}$$

here $\tilde{a}$ and $\tilde{n}$ are respectively the second and the last matrices in decomposition (2.1.14), and t_1, t_2 are as in (2.1.15), so that

$$t_1 t_2 = (d_1 d_4)^{-1}, \qquad t_1/t_2 = d_4/d_1. \tag{2.4.22}$$

From (2.4.21), (2.4.22), we obtain

$$\nabla(\gamma\sigma w) = \frac{1}{|d_1 d_4|^s} (UV)(\sigma\tilde{n}w)^s f\Big(\frac{d_4}{d_1} Z_1(\sigma\tilde{n}w), \Big|\frac{d_4}{d_1}\Big| U(\sigma\tilde{n}w) \Big). \tag{2.4.23}$$

We now substitute the right-hand side of (2.4.23) into (2.4.19) and, using Proposition 2.3.1 (a), we make a change of variable in the integral (2.4.19) so that the integrand becomes not depending on $\tilde{n}$. As a result, we obtain

$$c_{\mu,\nu} = c\sum_{\gamma} \bar{\tilde{\psi}}(\gamma)e(\mu d_2/d_1 - \nu c_4/d_4)|d_1 d_4|^{-s} I, \tag{2.4.24}$$

$$I = \iiiint_{\mathcal{F}_\gamma} (UV)(\sigma w)^s f\Big(\frac{d_4}{d_1} Z_1(\sigma w), \Big|\frac{d_4}{d_1}\Big| U(\sigma w)\Big)\overline{e(\mu z_1 + \nu z_3)}\, dz_1 dz_2 dz_3 dz_4.$$

By means of Proposition 2.4.2 we find for I the expression

$$I = \int_{\mathbf{C}/\vartheta q}\int_{\mathbf{C}} \tilde{I}\, dz_1 dz_3, \quad \text{where} \quad \tilde{I} =$$

$$\iint_{\mathbf{C}\,\mathbf{C}} (UV)(\tilde{\sigma}\tilde{w})^s f\Big(\begin{pmatrix} 0 & -1 \\ 1 & 0 \end{pmatrix}\Big(\frac{d_1}{d_4}(z_1 - Z_1(\tilde{\sigma}\tilde{w})), \Big|\frac{d_1}{d_4}\Big| U(\tilde{\sigma}\tilde{w})\Big)\Big)\overline{e(\mu z_1 + \nu z_3)}\, dz_2 dz_4,$$

$\tilde{\sigma}$ as in (2.4.13) and $\tilde{w}$ as in (2.4.15). Changing variables

$$z_4 \mapsto z_4 + z_1 z_3, \qquad z_2 \mapsto z_2 + 2z_1 z_4 - z_1^2 z_3$$

one can replace $\tilde{w}$ by $\widehat{w} = (0, z_2, z_3, z_4, u, v)$ in the integral $\tilde{I}$, and thus to get

$$I = \iiint_{\mathbf{C}\mathbf{C}\mathbf{C}} (UV)(\widehat{\sigma}\widehat{w})^s\, \overline{e(\nu z_3)}\, \widehat{I}\, dz_2 dz_3 dz_4, \tag{2.4.25}$$

$$\widehat{I} = \int_{\mathbf{C}/\vartheta q} f\Big(\begin{pmatrix} 0 & -1 \\ 1 & 0 \end{pmatrix}\Big(\frac{d_1}{d_4}(z_1 - Z_1(\widehat{\sigma}\widehat{w})), \Big|\frac{d_1}{d_4}\Big| U(\widehat{\sigma}\widehat{w})\Big)\Big)\overline{e(\mu z_1)}\, dz_1 \tag{2.4.26}$$

with $\widehat{\sigma}$ as in (2.4.13). To evaluate the integral $\widehat{I}$ we should notice first that, in accordance with our stipulation (2.4.10) about Ω, we have† $\vartheta d_1/d_4 \in \mathcal{O}$. And, second,

$$\int_{\mathbf{C}/\vartheta q} e(\lambda z_1)\, dz_1 = \begin{cases} \mathrm{vol}(\mathbf{C}/q)|\vartheta|^2 & \text{if } \lambda = 0, \\ 0 \text{ if } \lambda\vartheta \in q^*, & \lambda \neq 0. \end{cases}$$

Now let us substitute the Fourier expansion of the form f from Corrolary 0.3.3 into the right-hand side of (2.4.26). Then integrating the Fourier series term-

---

† We have $\mathcal{C}\,{}^t\mathcal{D} = \mathcal{D}\,{}^t\mathcal{C}$, see (2.1.4). Combining this with (2.4.10) we find $c_3 d_1 = c_2 d_4 - c_4 d_2$. Thus, if $2\,|\,c_4$, $2\,|\,d_4$, then $2\,|\,c_3 d_1$. On the other hand, if $2\,|\,c_4$, $2\,|\,d_4$, then $2 \nmid c_3$ (in opposite case $2\,|\,\det\gamma = 1$). Thus we get $d_1/\gcd(2, c_4, d_4) \in \mathcal{O}$, as claimed.

by-term we obtain

$$\widehat{I} = \mathrm{vol}(\mathbf{C}/q)|\vartheta|^2 \cdot \tag{2.4.27}$$

$$\cdot \begin{cases} \rho_+|\frac{d_1}{d_4}|^r U(\widehat{\sigma}\widehat{w})^r + \rho_-|\frac{d_1}{d_4}|^{2-r} U(\widehat{\sigma}\widehat{w})^{2-r} & \text{if } \mu = 0,\ r \neq 1, \\ \rho_+|\frac{d_1}{d_4}|U(\widehat{\sigma}\widehat{w}) \log\left(|\frac{d_1}{d_4}|U(\widehat{\sigma}\widehat{w})\right) + \rho_-|\frac{d_1}{d_4}|U(\widehat{\sigma}\widehat{w}) & \text{if } \mu = 0,\ r = 1, \\ \rho(\mu\frac{d_4}{d_1})|\frac{d_4}{d_1}|U(\widehat{\sigma}\widehat{w})K_{r-1}(4\pi|\mu|U(\widehat{\sigma}\widehat{w}))\overline{e(\mu \widehat{Z_1}(\widehat{\sigma}\widehat{w}))} & \text{if } \mu \neq 0. \end{cases}$$

Formulae (2.4.6)†, ..., (2.4.9) and (2.4.11), (2.4.12) follow from (2.4.24), ..., (2.4.27) and (2.4.22), (2.4.5). ■

2.4.4 On the series $\mathcal{D}_{\mu,\nu}$. We can apply Theorem 2.2.10 to give more manageable form for the series $\mathcal{D}_{\mu,\nu}$ that arose in the formulae for the Eisenstein series Fourier coefficients given by Theorem 2.4.1. Indeed, by means of Theorem 2.2.10 we can rewrite (2.4.8), (2.4.9) as the sums over

$$\begin{gathered} n_1, n_2, n_3 \equiv 1 \ (\mathrm{mod}\ 3), \quad \gcd(n_1 n_2 n_3, q) = 1, \\ m_1 \in \mathfrak{m}_1(n), \quad m_2 \in \mathfrak{m}_2(n, m_1, l_1), \quad m_3 \in \mathfrak{m}_3(n, m_1, m_2, l_1, l_2), \end{gathered} \tag{2.4.28}$$

where $n = (n_1, n_2, n_3)$. Then, by means of Proposition 2.2.5 we can rewrite terms in (2.4.8), (2.4.9) in terms of $n_1, n_2, n_3, m_1, m_2, m_3, l_1, l_2$. This leads to the following formulae:

$$\mathcal{D}_{0,\nu}(s,t) = \sum_{\substack{n_1,n_2,n_3 \equiv 1(3) \\ \gcd(n_1 n_2 n_3, q)=1}} \frac{\mathcal{E}(0,\nu;n)}{|n_1|^{s+t-2}|n_2|^{2s-2}|n_3|^{s-t}}, \tag{2.4.29}$$

$$\mathcal{D}_{\mu,\nu}(s) = \sum_{\substack{n_1,n_2,n_3 \equiv 1(3) \\ \gcd(n_1 n_2 n_3, q)=1,\ \mu n_1/n_3 \in q^*}} \frac{\rho(\mu n_1/n_3)\mathcal{E}(\mu,\nu,n)}{|n_1 n_2^2 n_3|^{s-1}}, \quad \nu \neq 0, \tag{2.4.30}$$

where $\mathcal{E}(\mu,\nu,n) = \gcd(2,n_2)^{-2}\widetilde{\mathcal{E}}(\mu,\nu,n)$ with $\widetilde{\mathcal{E}}(\mu,\nu,n)$

$$= \sum_{\substack{m_1,m_2,m_3 \\ \text{as in (2.4.28)}}} \bar{\widetilde{\psi}}\Big(\prod_{j=1,2,3} h_j\Big(\begin{pmatrix} * & l_j \\ m_j & n_j \end{pmatrix}\Big)\Big) e\Big(\mu\frac{l_1 m_2}{n_2} + \mu\frac{n_1 l_2 m_3}{n_2 n_3} - \nu\frac{m_1}{n_1}\Big). \tag{2.4.31}$$

Let us now restrict our attention by the case of the cubic metaplectic Eisenstein series only, so by the case $\widetilde{\psi}$ is Bass–Milnor–Serre's ψ.

† For the exceptional case $\mu = 0$, $r = 1$ notice also that $v \log v = \frac{\partial}{\partial t} v^t \big|_{t=1}$.

Theorem 2.4.3 *For the cubic metaplectic Eisenstein series* $\mathrm{E}(\sigma(\cdot), s; f)$, *the series* $\mathcal{D}_{0,\nu}$ *in* (2.4.8) *can be expressed as*

$$\mathcal{D}_{0,\nu}(s,t) = \sum_{n_1,n_2,n_3} \frac{\overline{S(\nu,n_1)}\, S(0,n_2)\, \overline{S(0,n_3)}}{|n_1|^{s+t-2}|n_2|^{2s-4}|n_3|^{s-t}}.$$

Here: $\nu \in q^*$, $s,t \in \mathbf{C}$; *the summation is carried out over* $n_1, n_2, n_3 \in \mathcal{O}$, $n_1, n_2, n_3 \equiv 1 \pmod 3$, $\gcd(n_1 n_2 n_3, q) = 1$; $\Re(s)$ *is assumed to be sufficiently large for the absolute convergence of the series in the right-hand side.* □

Theorem 2.4.4 *For the cubic metaplectic Eisenstein series* $\mathrm{E}(\sigma(\cdot), s; f)$, *the series* $\mathcal{D}_{\mu,\nu}$ *in* (2.4.9) *can be expressed as*

$$\mathcal{D}_{\mu,\nu}(s) = \sum_{n_1,n_2,n_3} \frac{\rho(\mu n_1/n_3)\, \overline{S(\mu n_1/n_2, n_3)}\, S(\mu, n_2, n_3^2)\, \overline{S(\nu, n_1, n_2^2 n_3)}}{|n_1 n_3|^{s-1}|n_2|^{2s-4}}.$$

Here: $\mu, \nu \in q^*$, $\mu \neq 0$, $s \in \mathbf{C}$; *the summation is carried out over* $n_1, n_2 \in \mathcal{O}$, $n_1, n_2, n_3 \equiv 1 \pmod 3$, $\gcd(n_1 n_2 n_3, q) = 1$, *subject to the conditions*

$$\mu n_1/n_2 \in q^*, \qquad \mu n_1/n_3 \in q^*, \tag{2.4.32}$$

$$S(\mu n_1/n_2, n_3) \neq 0, \qquad S(\mu, n_2, n_3^2) \neq 0; \tag{2.4.33}$$

$\Re(s)$ *is assumed to be sufficiently large for the absolute convergence of the series in the right-hand side.* □

One remark has to be done in connection with Theorem 2.4.4 to ensure that the Gauß sums with supplementary modules are defined (see 0.1.3). Under the condition (2.4.32), if $S(\mu n_1/n_2, n_3) \neq 0$, then n_3^2 can be factored as ld^3 with $l \mid n_2^\infty$, $l, d \in \mathcal{O}$. So, the Gauß sum $S(\mu, n_2, n_3^2)$ is defined. It follows from (2.4.32), (2.4.33) that $n_2^2 n_3$ can be factored as ld^3 with $l \mid n_1^\infty$, $l, d \in \mathcal{O}$. So, the Gauß sum $S(\nu, n_1, n_2^2 n_3)$ is defined.

Proof (of Theorem 2.4.3 and Theorem 2.4.4). By means of Proposition 2.2.9 we can rewrite (2.4.31) as

$$\begin{aligned} \mathcal{E}(\mu,\nu,n) &= \gcd(2,n_2)^{-2}\widetilde{\mathcal{E}}(\mu,\nu,n) \quad \text{with} \quad \widetilde{\mathcal{E}}(\mu,\nu,n) \\ &= \sum_{\substack{m_1,m_2,m_3 \\ \text{as in (2.4.28)}}} \overline{\left(\frac{m_1}{n_1}\right)}\left(\frac{m_2}{n_2}\right)\overline{\left(\frac{m_3}{n_3}\right)} e\Big(\mu\frac{l_1 m_2}{n_2} + \mu\frac{n_1 l_2 m_3}{n_2 n_3} - \nu\frac{m_1}{n_1}\Big). \end{aligned} \tag{2.4.34}$$

For $\mu = 0$ (2.4.34) yields

$$\mathcal{E}(0,\nu,n) = \overline{S(\nu,n_1)}\, S(0,n_2)\, \overline{S(0,n_3)}\, \|n_2\|,$$

that, being substituted into (2.4.29), gives rise to Theorem 2.4.3. To prove Theorem 2.4.4 we have only to show that

$$\mathcal{E}(\mu,\nu,n) = \|n_2\|\, \overline{S(\mu n_1/n_2, n_3)}\, S(\mu, n_2, n_3^2)\, \overline{S(\nu, n_1, n_2^2 n_3)} \tag{2.4.35}$$

for $\mu n_1/n_3 \in q^*$, $\mu \neq 0$. Let us choose (see 2.2.7) the sets $\mathfrak{l}_1(n)$, $\mathfrak{l}_2(n, m_1, l_1)$ so that $\gcd(l_1, n_2) = \gcd(l_2, n_3) = 1$ for all $l_1 \in \mathfrak{l}_1(n)$, $l_2 \in \mathfrak{l}_2(n, m_1, l_1)$. This is possible for all n. Recall that the parameters l_j in (2.4.34) are determined $\bmod n_j$ by $m_j l_j \equiv -1 \pmod{n_j}$, $j = 1, 2$. Now from (2.4.34) we obtain

$$\mathcal{E}(\mu, \nu, n) = \sum_{m_1} \overline{\left(\frac{m_1}{n_1}\right)} e\Big(-\frac{\nu m_1}{n_1}\Big) \sum_{m_2} \left(\frac{m_2}{n_2}\right) e\Big(\mu \frac{l_1 m_2}{n_2}\Big) \nabla(\mu, l_2, n), \quad (2.4.36)$$

where

$$\nabla(\mu, l_2, n) = \gcd(2, n_2)^{-2} \sum_{m_3} \overline{\left(\frac{m_3}{n_3}\right)} e\Big(\mu \frac{n_1 l_2 m_3}{n_2 n_3}\Big).$$

Recall that $\gcd(l_2, n_3) = 1$. Obvious computation shows that

$$\nabla(\mu, l_2, n) = \|n_2\| \overline{S(\mu n_1/n_2, n_3)} \left(\frac{l_2}{n_3}\right) \qquad (2.4.37)$$

if $\mu n_1/n_2 \in q^*$, and $\nabla(\mu, l_2, n) = 0$ in opposite case. In what follows we assume (2.4.32). Substituting (2.4.37) into (2.4.36) we obtain

$$\mathcal{E}(\mu, \nu, n) = \|n_2\| \overline{S(\mu n_1/n_2, n_3)} \sum_{m_1} \overline{\left(\frac{m_1}{n_1}\right)} e\Big(-\nu \frac{m_1}{n_1}\Big) \widehat{\nabla}(\mu, l_1, n)$$

where

$$\widehat{\nabla}(\mu, l_1, n) = \sum_{m_2} \left(\frac{m_2}{n_2}\right) \left(\frac{l_2}{n_3}\right) e\Big(\mu \frac{l_1 m_2}{n_2}\Big).$$

Recall that $m_2 l_2 \equiv -1 \pmod{n_2}$. Under the conditions (2.4.32), we have $S(\mu n_1/n_2, n_3) \neq 0$ only if $\operatorname{ord}_p n_3 \equiv 0 \pmod 3$ for all prime $p \nmid n_2$. It follows from these two remarks that

$$\left(\frac{l_2}{n_3}\right) = \left(\frac{m_2}{n_3}\right)^{-1},$$

in all cases in which $S(\mu n_1/n_2, n_3) \neq 0$. Thus we have

$$\widehat{\nabla}(\mu, l_1, n) = \sum_{m_2} \left(\frac{m_2}{n_2}\right) \left(\frac{m_2}{n_3}\right)^{-1} e\Big(\mu \frac{l_1 m_2}{n_2}\Big) = \left(\frac{l_1}{n_3}\right) \left(\frac{l_1}{n_2}\right)^{-1} S(\mu, n_2, n_3^2)$$

and

$$\mathcal{E}(\mu, \nu, n) = \|n_2\| \overline{S(\mu n_1/n_2, n_3)}\, S(\mu, n_2, n_3^2) \cdot$$
$$\cdot \sum_{m_1} \overline{\left(\frac{m_1}{n_1}\right)} \left(\frac{l_1}{n_3}\right) \left(\frac{l_1}{n_2}\right)^{-1} e\Big(-\nu \frac{m_1}{n_1}\Big). \qquad (2.4.38)$$

Recall that $m_1 l_1 \equiv -1 \pmod{n_1}$. Under the conditions (2.4.32), the sum $S(\mu, n_2, n_3^2) \neq 0$ only if $\mathrm{ord}_p n_1 \geq 1$ or $\mathrm{ord}_p n_2 \equiv \mathrm{ord}_p n_3 \pmod 3$ for all prime p. These two remarks yield the interior sum (over m_1) in (2.4.38) is equal to $\overline{S(\nu, n_1, n_2^2 n_3)}$, and thus we have (2.4.35). ∎

2.4.5 Evaluation of integrals. Our aim in this subsection is to examine the functions $J_{\mu,\nu}$ that arose in Theorem 2.4.1. The formulae we give for $J_{\mu,\nu}$ with $\mu\nu = 0$, i. e., for a degenerate case are in agreement with those given in [33] in a more general context. The formula for $J_{\mu,\nu}$ with $\mu\nu \neq 0$ is new.

Recall that Γ and K mean the gamma-function and the Bessel–MacDonald function, see 0.1.6. We need one lemma. Let $z_2, z_3, z_4 \in \mathbf{C}$, $u, v \in \mathbf{R}_+^*$, and let

$$P = (v^2 + |z_3|^2)^{1/2}, \qquad Q = u^2 P^2 + |z_4|^2,$$
$$R = (v^2 P^{-4} Q^2 + |z_2|^2)^{1/2}. \tag{2.4.39}$$

Let p,$\tilde{p}$ and σ be the matrices defined in 2.1.3. Then for the matrix $\widehat{\sigma}$ in (2.4.13) one has $\widehat{\sigma} = \tilde{p}p\tilde{p}$.

Lemma 2.4.5 *With the notations* (2.4.39) *one has*

(a) *If* $\widehat{w} = (0, z_2, z_3, z_4, u, v)$, *then*

$$\tilde{p}\widehat{w} = (-z_4, z_2 - \bar{z}_3 z_4^2 P^{-2}, -\bar{z}_3 P^{-2}, \bar{z}_3 z_4 P^{-2}, uP, vP^{-2});$$

(b) *If* $w = (-z_4, z_2, -\bar{z}_3 P^{-2}, \bar{z}_3 z_4 P^{-2}, uP, vP^{-2})$, *then*

$$pw = (-\bar{z}_4 Q^{-1}, *, z_2, \bar{z}_3 z_4 P^{-2}, uPQ^{-1}, vP^{-2}Q),$$
$$\tilde{p}pw = (-z_2\bar{z}_4 Q^{-1} - \bar{z}_3 z_4 P^{-2}, *, *, *, uPQ^{-1}R, vP^{-2}QR^{-2}). \qquad \square$$

Proof. Apply successively parts (a) and (b) of Proposition 2.3.2. ∎

Now we can evaluate the integrals $J_{\mu,\nu}$ with $\mu = 0$.

Theorem 2.4.6 *For the integral* $J_{0,0}$ *from* (2.4.11) *we have*

$$J_{0,0}(u, v, s, t) = \frac{2^2 \pi^3 u^{6-s+t} v^{6-s}}{(s-t-2)(s+t-4)(s-3)}. \qquad \square$$

Theorem 2.4.7 *For the integrals* $J_{0,\nu}$ *from* (2.4.11) *we have*

$$J_{0,\nu}(u, v, s, t) = \frac{2^{l+2} \pi^{l+3} |\nu|^l u^{6-s+t} v^{6-s+l}}{(s-3)(s-t-2)\Gamma(l+1)} K_l(4\pi|\nu|v);$$

here $l = s/2 + t/2 - 2$ *and* $\nu \in \mathbf{C}$, $\nu \neq 0$. $\square$

Proof (of Theorem 2.4.6 and Theorem 2.4.7). We write $\widehat{\sigma}\widehat{w}$ in (2.4.11) as $(\tilde{p}p)(\tilde{p}\widehat{w})$, then we write $\tilde{p}\widehat{w}$ as in Lemma 2.4.5 (a), and then we change variable z_2 by $z_2 + \bar{z}_3 z_4^2 P^{-2}$ and apply Lemma 2.4.5 (b). After all we get

$$J_{0,\nu}(u,v,s,t) = u^{s+t}v^s \iiint\limits_{\mathbf{C}\mathbf{C}\mathbf{C}} Q^{-t}(PR)^{t-s}\overline{e(\nu z_3)}\, dz_2\, dz_3\, dz_4.$$

Notice that P and Q do not depend on z_2 and that

$$\int\limits_{\mathbf{C}} R^{t-s}\, dz_2 = \frac{2\pi}{s-t-2}(vP^{-2}Q)^{2-s+t},$$

by (0.1.33). Thus we have

$$J_{0,\nu}(u,v,s,t) = \frac{2\pi u^{s+t}v^{2+t}}{s-t-2}\iint\limits_{\mathbf{C}\mathbf{C}} Q^{2-s}P^{s-t-4}\overline{e(\nu z_3)}\, dz_3\, dz_4.$$

Now notice that P does not depend on z_4 and that

$$\int\limits_{\mathbf{C}} Q^{2-s}\, dz_4 = \frac{\pi}{s-3}(uP)^{6-2s},$$

again in view of (0.1.33). Thus we have

$$J_{0,\nu}(u,v,s,t) = \frac{2\pi^2 u^{6-s+t}v^{2+t}}{(s-t-2)(s-3)}\int\limits_{\mathbf{C}} P^{2-s-t}\overline{e(\nu z_3)}\, dz_3.$$

Using (0.1.33) we get Theorem 2.4.6 if $\nu = 0$ and Theorem 2.4.7 if $\nu \neq 0$. ∎

It is not so easy to evaluate the integrals $J_{\mu,\nu}$ with $\mu \neq 0$. Lemma 2.4.5 leads to the expression

$$J_{\mu,\nu}(u,v,s,t) = u^{s+1}v^s \cdot$$
$$\cdot \iiint\limits_{\mathbf{C}\mathbf{C}\mathbf{C}} K_{t-1}\Big(4\pi|\mu|u\frac{PR}{Q}\Big)e\Big(\mu\Big(\frac{z_2\bar{z}_4}{Q}+\frac{\bar{z}_3 z_4}{P^2}\Big)-\nu z_3\Big)\frac{dz_2\, dz_3\, dz_4}{Q(PR)^{s-1}}.$$

Now let us change variabels as $z_2 = vP^{-2}Qy_2$, $z_4 = uPy_4$, and then as $z_3 = vy_3$, and let us set $Y_j = (1+|y_j|^2)^{1/2}$, to get

$$J_{\mu,\nu}(u,v,s,t) = u^{7-s}v^{6-s}\int\limits_{\mathbf{C}} \tilde{J}(\alpha,\delta,s,y_3)\overline{e(\nu v y_3)}\frac{dy_3}{Y_3^{s-1}}, \tag{2.4.40}$$

where $\alpha = \mu u$, $\delta = t-1$, and

$$\tilde{J}(\alpha,\delta,s,y_3) = \iint\limits_{\mathbf{C}\,\mathbf{C}} K_\delta\Big(4\pi|\alpha|\frac{Y_2}{Y_3}\Big)e\Big(\alpha\frac{y_2\overline{y}_4+\overline{y}_3y_4}{Y_3}\Big)\frac{dy_2\,dy_4}{Y_2^{s-1}Y_4^{2s-4}}. \tag{2.4.41}$$

Theorem 2.4.8 *For the integrals* $J_{\mu,0}$ *from* (2.4.12) *we have*

$$J_{\mu,0}(u,v,s,t) = \frac{2^2\pi^3u^{7-s}v^{6-s}}{(s-t-2)(s-3)(s+t-4)}K_{t-1}(4\pi|\mu|u);$$

here $\mu \in \mathbf{C}$, $\mu \neq 0$. □

Theorem 2.4.9 *For the integrals* $J_{\mu,\nu}$ *from* (2.4.12) *we have*

$$J_{\mu,\nu}(u,v,s,t) = \frac{2^{2s-5}\pi^{2s-3}|\mu|^{s-7}|\nu|^{s-6}}{\Gamma\big(\frac{s-t}{2}\big)\Gamma\big(\frac{s+t}{2}-1\big)\Gamma(s-2)}R_{s-t,t}(|\mu|u,|\nu|v),$$

where $R_{s-t,t}(u,v) = u^{4+l}v^{3-l}$.

$$\cdot\int\limits_0^\infty\int\limits_0^\infty K_l\big(4\pi u\sqrt{1+x+y}\big)K_p\big(4\pi v\sqrt{(1+1/x)(1+1/y)}\big)\frac{dx\,dy}{Z(x,y)}$$

(is the radial part of Whittaker function attached to the characters $\chi_{s-t,t}$ *and* $\phi_{1,1}$*),* $l=(s-t)/2-1$, $p=(s+t)/2-2$,

$$Z(x,y) = x^{3-3s/4+t/4}y^{1-s/4+3t/4}(1+x+y)^{l/2}(1+x)^{-p/2}(1+y)^{p/2},$$

and $\mu,\nu \in \mathbf{C}$, $\mu\nu \neq 0$. □

Proof (of Theorem 2.4.8 and Theorem 2.4.9). We only have to evaluate the integrals (2.4.40), (2.4.41). We can and do assume $\Re(s)$ so large that these integrals and all the integrals in the computation below converge absolutely. Immediately from the definitions given in 0.1.6 we find that

$$K_\delta(ct) = \frac{c^\delta}{2}\int\limits_0^\infty \exp\Big(-(a+c^2a^{-1})\frac{t}{2}\Big)\frac{da}{a^{\delta+1}}, \tag{2.4.42}$$

$$c^{-\gamma+\delta/2} = \frac{1}{\Gamma(\gamma-\delta/2)}\int\limits_0^\infty \exp(-cb)b^{\gamma-\delta/2-1}\,db, \tag{2.4.43}$$

at least for $c,t \in \mathbf{R}_+^*$. Replacing $K_\delta(\ldots)$ in (2.4.41) by the right-hand side of (2.4.42) with $t = 4\pi|\alpha|Y_3^{-1}$, $c = Y_2$ and applying then (2.4.43) with $c = Y_2^2Y_3^{-1} = (1+|y_2|^2)Y_3^{-1}$, $\gamma = (s-1)/2$ we find:

$$2\Gamma(l+1)Y_3^{l+1}\tilde{J}(\alpha,\delta,s,y_3) \tag{2.4.44}$$

$$= \iiint\limits_{\mathbf{C}\,0\,0}^{\;\;\infty\,\infty} \nabla\exp\Big(-\frac{2\pi|\alpha|a+h}{Y_3}\Big)e\Big(\alpha\frac{\overline{y}_3y_4}{Y_3}\Big)\frac{b^l\,da\,db\,dy_4}{a^{\delta+1}Y_4^{2s-4}},$$

where $l = (s - \delta - 3)/2 = (s - t)/2 - 1$, $h = 2\pi|\alpha|a^{-1} + b$ and

$$\nabla = \int\limits_{\mathbf{C}} \exp\Big(- \frac{h|y_2|^2}{Y_3}\Big) e\Big(\alpha \frac{y_2 \overline{y}_4}{Y_3}\Big)\, dy_2 = \frac{\pi Y_3}{h} \exp\Big(- \frac{4\pi^2 |\alpha y_4|^2}{h Y_3}\Big),$$

by (0.1.32). Substituting the last expression for ∇ into (2.4.44) we obtain

$$2\pi^{-1}\Gamma(l+1) Y_3^l \widetilde{J}(\alpha, \delta, s, y_3) = \tag{2.4.45}$$

$$\iiint\limits_{\mathbf{C}\,0\,0}^{\ \ \infty\,\infty} \exp\Big(- \frac{2\pi|\alpha|a + h}{Y_3} - \frac{4\pi^2|\alpha y_4|^2}{hY_3}\Big) e\Big(\alpha \frac{\overline{y}_3 y_4}{Y_3}\Big) \frac{b^l\, da\, db\, dy_4}{Y_4^{2s-4} a^{\delta+1} h},$$

Let us now assume $h = 2\pi|\alpha|a^{-1} + b$ and $x = (2\pi|\alpha|)^{-1}ab$ as new variables in (2.4.45) instead of a and b. The old variables can be expressed as

$$a = 2\pi|\alpha|(1+x)h^{-1}, \quad b = hx(1+x)^{-1}.$$

The Jacobian equals $2\pi|\alpha|h^{-1}$. As a result, we have

$$2^{\delta+1}\pi^{\delta-1}|\alpha|^\delta \Gamma(l+1) Y_3^l \widetilde{J}(\alpha, \delta, s, y_3) \tag{2.4.46}$$

$$= \iint\limits_{0\,0}^{\infty\,\infty} \nabla \exp\Big(- \frac{4\pi^2|\alpha|^2(1+x)}{hY_3} - \frac{h}{Y_3}\Big) \frac{h^{(s+\delta-5)/2} x^l\, dx\, dh}{(1+x)^{(s+\delta-1)/2}}$$

with

$$\nabla = \int\limits_{\mathbf{C}} \exp\Big(- \frac{4\pi^2|\alpha|^2|y_4|^2}{hY_3}\Big) e\Big(\alpha \frac{\overline{y}_3 y_4}{Y_3}\Big) \frac{dy_4}{Y_4^{2s-4}}. \tag{2.4.47}$$

By means of the formula (2.4.43) with $c = Y_4^2 Y_3^{-1} = (1 + |y_4|^2) Y_3^{-1}$ and with $s - 2$ instead of $\gamma - \delta/2$ we get

$$Y_4^{4-2s} = \frac{1}{\Gamma(s-2)} Y_3^{2-s} \int\limits_0^\infty \exp\Big(- \frac{Y_4^2 \phi}{Y_3}\Big) \phi^{s-3}\, d\phi$$

and, substituting this into (2.4.47),

$$\nabla = \frac{1}{\Gamma(s-2)} Y_3^{2-s} \int\limits_0^\infty \widehat{\nabla} \exp\Big(- \frac{\phi}{Y_3}\Big) \phi^{s-3}\, d\phi, \tag{2.4.48}$$

where

$$\widehat{\nabla} = \int\limits_{\mathbf{C}} \exp\Big(- \frac{z|y_4|^2}{Y_3}\Big) e\Big(\alpha \frac{\overline{y}_3 y_4}{Y_3}\Big)\, dy_4, \quad z = 4\pi^2|\alpha|^2 h^{-1} + \phi.$$

According to (0.1.32), $\widehat{\nabla} = \pi Y_3 z^{-1} \exp(-4\pi^2|\alpha y_3|^2 z^{-1} Y_3^{-1})$, and this being substituted into (2.4.48) gives

$$\nabla = \frac{\pi}{\Gamma(s-2)} Y_3^{3-s} \int_0^\infty \exp\Big(-\frac{4\pi^2|\alpha y_3|^2}{zY_3} - \frac{\phi}{Y_3}\Big) z^{-1}\phi^{s-3}\, d\phi.$$

Substituting the last expression for ∇ into (2.4.46) we obtain the next equality

$$2^{\delta+1}\pi^{\delta-2}|\alpha|^\delta \Gamma(l+1)\Gamma(s-2) Y_3^{l+s-3} \widetilde{J}(\alpha,\delta,s,y_3) \tag{2.4.49}$$
$$= \int_0^\infty\int_0^\infty\int_0^\infty \exp\Big(-\frac{4\pi^2|\alpha|^2 x}{hY_3} - \frac{4\pi^2|\alpha y_3|^2}{zY_3} - \frac{h+z}{Y_3}\Big)\frac{h^{(s+\delta-5)/2}x^l\phi^{s-3}}{(1+x)^{(s+\delta-1)/2}z}\, dx\, dh\, d\phi,$$

where as above $z = 4\pi^2|\alpha|^2 h^{-1} + \phi$. Now let us change variabels in (2.4.49) by taking as new variabels $g = \big(4\pi^2|\alpha|^2 Y_3^2\big)^{-1} h\phi$, $n = \big(1+(2\pi|\alpha|)^{-2}h\phi\big)^{-1} x$ and z instead of h, ϕ and x. The old variabels can be expressed as

$$h = 4\pi^2|\alpha|^2(1+gY_3^2)z^{-1}, \quad \phi = zg(1+gY_3^2)^{-1}Y_3^2, \quad x = (1+gY_3^2)n.$$

The Jacobian equals $4\pi^2|\alpha|^2 Y_3^2(1+gY_3^2)z^{-1}$. As a result we obtain

$$\widetilde{J}(\alpha,\delta,s,y_3) = \frac{2^{s-4}\pi^{s-1}|\alpha|^{s-3}Y_3^{(s+\delta+1)/2}}{\Gamma(l+1)\Gamma(s-2)}\cdot \int_0^\infty\int_0^\infty \nabla \frac{n^l g^{s-3}\, dg\, dn}{(1+n+ngY_3^2)^{(s+\delta-1)/2}} \tag{2.4.50}$$

with

$$\nabla = \int_0^\infty \exp\Big(-\frac{4\pi^2|\alpha|^2(1+g)Y_3}{z} - \frac{(1+n)z}{Y_3}\Big) z^{(s-\delta-5)/2}\, dz.$$

To evaluate ∇ one can apply (2.4.42), and so to obtain

$$\nabla = 2\Big(2\pi|\alpha|\frac{\sqrt{1+g}}{\sqrt{1+n}}Y_3\Big)^l K_l\big(4\pi|\alpha|\sqrt{(1+g)(1+n)}\big).$$

Substituting this expression for ∇ into (2.4.50) we find that

$$\widetilde{J}(\alpha,\delta,s,y_3) = \frac{(2|\alpha|)^{l+s-3}\pi^{l+s-1}Y_3^{s-1}}{\Gamma(l+1)\Gamma(s-2)}\cdot \int_0^\infty\int_0^\infty K_l\big(4\pi|\alpha|\sqrt{(1+g)(1+n)}\big)\frac{n^l g^{s-3}(1+g)^{l/2}\, dg\, dn}{(1+n+ngY_3^2)^{(s+\delta-1)/2}(1+n)^{l/2}},$$

and then substituting the right-hand side of the last expression into (2.4.40) instead of $\widetilde{J}(\ldots)$ and replacing α by μu and δ by $t-1$ we obtain

$$J_{\mu,\nu}(u,v,s,t) = \frac{(2|\mu u|)^{l+s-3}\pi^{l+s-1}u^{7-s}v^{6-s}}{\Gamma(l+1)\Gamma(s-2)}\cdot \tag{2.4.51}$$
$$\cdot\int_0^\infty\int_0^\infty K_l\big(4\pi|\mu|u\sqrt{(1+g)(1+n)}\big)\frac{n^l g^{s-3}(1+g)^{l/2}}{(1+n)^{l/2}}\widetilde{\nabla}_\nu\, dg\, dn,$$

where $l=(s-t)/2-1$ and $\widetilde{\nabla}_\nu = \int\limits_{\mathbf{C}}(1+n+ngY_3^2)^{-(s+t-2)/2}e(-\nu v y_3)\,dy_3$.

By (0.1.33) we have $\widetilde{\nabla}_0 = \dfrac{2\pi}{(s+t-4)(1+n+ng)^p ng}$, and, for $\nu\neq 0$,

$$\widetilde{\nabla}_\nu = \frac{(2\pi)^{p+1}|\nu|^p v^p}{\Gamma(p+1)(1+n+ng)^{p/2}(ng)^{p/2+1}}K_p\big(4\pi|\nu|v\sqrt{1+1/g+1/(gn)}\big),$$

where $p=(s+t)/2-2$. Substituting these expressions for $\widetilde{\nabla}_\nu$ into (2.4.51) we obtain

$$J_{\mu,0}(u,v,s,t) = \frac{2^{l+s-2}\pi^{l+s}|\mu|^{l+s-3}u^{4+l}v^{6-s}}{\Gamma(l+1)\Gamma(s-2)(s+t-4)}I_{\mu,0} \tag{2.4.52}$$

with

$$I_{\mu,0} = \int_0^\infty\int_0^\infty K_l\big(4\pi|\mu|u\sqrt{(1+g)(1+n)}\big)\frac{n^{l-1}g^{s-4}(1+g)^{l/2}}{(1+n)^{l/2}(1+n+ng)^p}\,dn\,dg,$$

and also

$$J_{\mu,\nu}(u,v,s,t) = \frac{2^{2s-5}\pi^{2s-3}|\mu|^{l+s-3}|\nu|^p u^{4+l}v^{3-l}}{\Gamma(l+1)\Gamma(p+1)\Gamma(s-2)}I_{\mu,\nu} \tag{2.4.53}$$

with

$$I_{\mu,\nu} = \int_0^\infty\int_0^\infty K_l\big(4\pi|\mu|u\sqrt{(1+g)(1+n)}\big)K_p\big(4\pi|\nu|v\sqrt{1+1/g+1/(gn)}\big)\frac{dn\,dg}{\Omega(n,g)},$$

$$\Omega(n,g) = \frac{(1+n)^{l/2}(1+n+ng)^{p/2}}{n^{(s-3t)/4-1}g^{(l+s)/2-5/2}(1+g)^{l/2}}.$$

Notice that l and p in these formulae are the same as in Theorem 2.4.9, and that Theorem 2.4.9 follows from (2.4.53) if we change g by x and n by $(1+x)^{-1}y$ in the integral $I_{\mu,\nu}$. To prove Theorem 2.4.8 we have to evaluate integral $I_{\mu,0}$. Changing variable n by $y=(1+g)n$ we find that

$$I_{\mu,0} = \int_0^\infty \nabla\frac{y^{l-1}\,dy}{(1+y)^p}, \tag{2.4.54}$$

where

$$\nabla = \int_0^\infty K_l(4\pi|\mu|u\sqrt{1+g+y})\frac{g^{s-4}\,dg}{(1+g+y)^{l/2}}.$$

To evaluate ∇ we change g by x as $g=(1+y)x$ and then we apply (0.1.22) (with $c=4\pi|\mu|u\sqrt{1+y}$):

$$\nabla = (1+y)^{s-3-l/2}\int_0^\infty K_l(4\pi|\mu|u\sqrt{(1+y)(1+x)})(1+x)^{-l/2}x^{s-4}\,dx$$

$$= (2\pi|\mu|u)^{3-s}(1+y)^{(s-3-l)/2}K_{s-l-3}(4\pi|\mu|u\sqrt{1+y})\Gamma(s-3).$$

Substituting the last expression for ∇ into (2.4.54) and applying (0.1.22) once more we obtain

$$I_{\mu,0} = (2\pi|\mu|u)^{3-s}\Gamma(s-3)\int_0^\infty K_p(4\pi|\mu|u\sqrt{1+y})\frac{y^{l-1}\,dy}{(1+y)^{p/2}}$$

$$= (2\pi|\mu|u)^{3-s-l}\Gamma(s-3)\Gamma(l)K_{p-l}(4\pi|\mu|u),$$

and this after substituting into (2.4.52) gives Theorem 2.4.8. ∎

2.4.6 Mellin transform. We would like now to complete our exposition by deducing explicit formula for the Mellin transform of $J_{\mu,\nu}(\cdot,\cdot,s,t)$, i.e., for

$$M_{\mu,\nu}(a,b,s,t) = \int_0^\infty\int_0^\infty J_{\mu,\nu}(u,v,s,t)u^{a-1}v^{b-1}\,du\,dv, \tag{2.4.55}$$

where $\mu,\nu,s,t,a,b\in\mathbf{C}$, $\mu\nu\neq 0$, and we assume that $\Re(a)$ and $\Re(b)$ are sufficently large for absolute convergence of this integral.

Corollary 2.4.10 *For the integrals $M_{\mu,\nu}(a,b,s,t)$ defined by (2.4.55) with $\mu,\nu\in\mathbf{C}$, $\mu\nu\neq 0$, one has the expression*

$$M_{\mu,\nu}(a,b,s,t) = 2^{2s-16-a-b}\pi^{2s-10-a-b}|\mu|^{s-a-7}|\nu|^{s-b-6}C(a,b,s,t)\cdot$$

$$\cdot\frac{\Gamma\left(\frac{a+s-t}{2}+1\right)\Gamma\left(\frac{a}{2}+2\right)\Gamma\left(\frac{b+t}{2}+1\right)\Gamma\left(\frac{b-s}{2}+3\right)}{\Gamma\left(\frac{s-t}{2}\right)\Gamma\left(\frac{s+t}{2}-1\right)\Gamma(s-2)},$$

where $s,t\in\mathbf{C}$, and, if $\Re(a)$ and $\Re(b)$ are sufficiently large, $C(a,b,s,t)$

$$= \int_0^\infty\int_0^\infty \frac{x^{b/2+s/2-1}y^{b/2-t/2+1}\,dx\,dy}{(1+x)^{b/2-s/2+3}(1+y)^{b/2+t/2+1}(1+x+y)^{a/2+s/2-t/2+1}}. \tag{2.4.56}$$

Then we have $C(a,b,s,t) = G(a,b,s,t)\,X(a,b,s,t)$, *where one can take either*

$$G(a,b,s,t) = \frac{\Gamma\left(\frac{b-t}{2}+2\right)\Gamma\left(\frac{a+s+t}{2}\right)}{\Gamma\left(\frac{a+b+s}{2}+2\right)}, \tag{2.4.57}$$

$$X(a,b,s,t) = \int_0^\infty {}_2F_1\left(\frac{b+t}{2}+1, \frac{b-t}{2}+2; \frac{a+b+s}{2}+2; -x\right)\frac{x^{b/2+s/2-1}\,dx}{(1+x)^{a/2+2}},$$

or

$$G(a,b,s,t) = \frac{\Gamma\left(\frac{b+s}{2}\right)\Gamma\left(\frac{a-s-t}{2}+4\right)}{\Gamma\left(\frac{a+b-t}{2}+4\right)}, \tag{2.4.58}$$

$$X(a,b,s,t) = \int_0^\infty {}_2F_1\left(\frac{b-s}{2}+3, \frac{b+s}{2}; \frac{a+b-t}{2}+4; -y\right)\frac{y^{b/2-t/2+1}\,dy}{(1+y)^{a/2+2}}.$$

For X *in* (2.4.57) *one has the functional equation* $X(a,b,s,2-t) = X(a,b,s,t)$, *while* X *in* (2.4.58) *satisfies the functional equation* $X(a,b,6-s,t) = X(a,b,s,t)$. □

Proof. Let us substitute the expression given by Theorem 2.4.9 into (2.4.55) and change the order of integration to get $M_{\mu,\nu}(a,b,s,t) =$

$$\frac{2^{2s-5}\pi^{2s-3}|\mu|^{(3s-t)/2-4}|\nu|^{(s+t)/2-2}}{\Gamma\left(\frac{s-t}{2}\right)\Gamma\left(\frac{s+t}{2}-1\right)\Gamma(s-2)}\int_0^\infty\!\!\int_0^\infty A(x,y)B(x,y)\,\frac{dx\,dy}{Z_{s-t,t}(x,y)}, \tag{2.4.59}$$

where

$$A(x,y) = \int_0^\infty K_{(s-t)/2-1}\left(4\pi|\mu|u\sqrt{1+x+y}\right)u^{2+a+(s-t)/2}\,du,$$

$$B(x,y) = \int_0^\infty K_{(s+t)/2-2}\left(4\pi|\nu|v\sqrt{(1+1/x)(1+1/y)}\right)v^{3+b-(s-t)/2}\,dv$$

are the Mellin transforms of the Bessel–MacDonald functions. By means of (0.1.21) we find

$$A(x,y) = 2^{-5-a-s/2+t/2}\frac{\Gamma\left(\frac{a+s-t}{2}+1\right)\Gamma\left(\frac{a}{2}+2\right)}{\left(\pi|\mu|\sqrt{1+x+y}\right)^{3+a+s/2-t/2}},$$

$$B(x,y) = 2^{-6-b+s/2-t/2}\frac{\Gamma\left(\frac{b+t}{2}+1\right)\Gamma\left(\frac{b-s}{2}+3\right)}{\left(\pi|\nu|\sqrt{(1+1/x)(1+1/y)}\right)^{4+b-s/2+t/2}}.$$

Substituting the last expressions for $A(x,y)$ and $B(x,y)$ into (2.4.59) we obtain the desired expression for $M_{\mu,\nu}(a,b,s,t)$ with $C(a,b,s,t)$ as in (2.4.56). To get (2.4.57), let us rewrite (2.4.56) as

$$C(a,b,s,t) = \int_0^\infty I(x) \frac{x^{b/2+s/2-1}\,dx}{(1+x)^{b/2-s/2+3}} \tag{2.4.60}$$

with

$$I(x) = \int_0^\infty \frac{y^{b/2-t/2+1}\,dy}{(1+x+y)^{a/2+s/2-t/2+1}(1+y)^{b/2+t/2+1}}.$$

By means of (0.1.29) we find

$$I(x) = \frac{\Gamma\left(\frac{b-t}{2}+2\right)\Gamma\left(\frac{a+s+t}{2}\right)}{\Gamma\left(\frac{a+b+s}{2}+2\right)} \cdot {}_2F_1\left(\frac{a+s-t}{2}+1, \frac{b-t}{2}+2; \frac{a+b+s}{2}+2; \frac{x}{1+x}\right) \frac{1}{(1+x)^{a/2+s/2-t/2+1}}.$$

Substituting this expression into (2.4.60) and applying Euler's identity (0.1.30) we get (2.4.57). For (2.4.58) one should rewrite (2.4.56) as

$$C(a,b,s,t) = \int_0^\infty I(y) \frac{y^{b/2-t/2+1}\,dy}{(1+y)^{b/2+t/2+1}} \tag{2.4.61}$$

with

$$I(y) = \int_0^\infty \frac{x^{b/2+s/2-1}\,dx}{(1+x+y)^{a/2+s/2-t/2+1}(1+x)^{b/2-s/2+3}}.$$

Then one should apply (0.1.29) to get

$$I(y) = \frac{\Gamma\left(\frac{b+s}{2}-1\right)\Gamma\left(\frac{a-s-t}{2}+4\right)}{\Gamma\left(\frac{a+b-t}{2}+4\right)} \cdot {}_2F_1\left(\frac{a+s-t}{2}+1, \frac{b+s}{2}; \frac{a+b-t}{2}+4; \frac{x}{1+x}\right) \frac{1}{(1+x)^{a/2+s/2+t/2+1}},$$

and then apply (0.1.30) just like in the preciding case.

For the functional equations satisfied by $X(a,b,s,t)$ we only have to notice that the hypergeometric function ${}_2F_1$ is a simmetric function of the first two parameters, as it is clear just from the definition given in 0.1.6. ∎

Instead of using Euler's identity (0.1.30), to evaluate the integrals (2.4.60), (2.4.61) one can refer to the integral (0.1.28) so to obtain for $C(a,b,s,t)$ and for $M_{\mu,\nu}(a,b,s,t)$ expressions in terms of the gamma-function and the hypergeometric function ${}_3F_2(\ldots;1)$. The author prefer the formulae in Corollory 2.4.10

because of they give rise immediately to the functional equations satisfied by the radial parts of Whittaker functions (see Theorem 2.4.12 below), while the expressions with the hypergeometric function ${}_3F_2(\ldots;1)$ occur nonconvenient from this viewpoint. Nethertheless, the interested reader may try to find some other expressions for $M_{\mu,\nu}(a,b,s,t)$, or new relations for ${}_3F_2(\ldots;1)$ applying the known relations satisfied by ${}_3F_2(\ldots;1)$ (see, for example, Bailey [3]) and the functional equations satisfied by the radial parts of Whittaker functions.

2.4.7 More on Whittaker functions. As a consequence of our computation in 2.4.5, we got simple integral representation for radial parts of Whittaker functions. Let us state this result in a convenient form.

Theorem 2.4.11 *Given $s,t \in \mathbf{C}$, the function $R_{s,t}\colon \mathbf{R}_+^* \times \mathbf{R}_+^* \to \mathbf{C}$ defined as*

$$R_{s,t}(u,v) = u^{3+s/2}v^{4-s/2}\cdot \int_0^\infty\!\!\int_0^\infty K_l\big(4\pi u\sqrt{1+x+y}\big)K_p\big(4\pi v\sqrt{(1+1/x)(1+1/y)}\big)\frac{dx\,dy}{Z(x,y)}$$

with $l = s/2-1$, $p = s/2+t-2$ and

$$Z(x,y) = x^{3-3s/4-t/2}y^{1-s/4+t/2}(1+x)^{-p/2}(1+y)^{p/2}(1+x+y)^{l/2},$$

is the radial part of an Whittaker function attached to the characters $\phi_{1,1}$, $\chi_{s,t}$. □

Proof. According to Theorem 2.4.9 we have

$$J_{1,1}(u,v,s+t,t) = \frac{2^{2s+2t-5}\pi^{2s+2t-3}}{\Gamma(s/2)\Gamma(s/2+t-1)\Gamma(s+t-2)}R_{s,t}(u,v). \qquad (2.4.62)$$

Thus, $R_{s,t}$ is the radial part of Whittaker function attached to the same characters $\phi_{1,1}$, $\chi_{s,t}$ as $J_{1,1}(\cdot,\cdot,s+t,t)$ is. ■

The method we used to prove Theorem 2.4.9, and so to get the formula for the radial parts of Whittaker functions in Theorem 2.4.11, can be used to find similar integral representations for the radial parts of Whittaker functions for the group $\mathrm{Sp}(4,\mathbf{R})$. In the meantime, Niwa [66] and Oda [67] studied the Whittaker functions for $\mathrm{Sp}(4,\mathbf{R})$ by means of different methods, and with essentially different results. For better understood of both real and complex cases one needs further researh.

Now let us state in explicit form the functional equations mentioned in 2.3.7.

Theorem 2.4.12 *Given any $u,v \in \mathbf{R}_+^*$, the function $(s,t) \mapsto R_{s,t}(u,v)$, defined in Theorem 2.4.11 and in Theorem 2.4.9, is an entire function on $\mathbf{C}\times\mathbf{C}$ satisfying the functional equations*

$$\begin{aligned} R_{s,t} &= R_{s+2t-2,2-t} = R_{4-s,s+t-2} = R_{s+2t-2,4-s-t} \\ &= R_{6-s-2t,s+t-2} = R_{s,4-s-t} = R_{6-s-2t,t} = R_{4-s,2-t}. \end{aligned}$$ □

Proof. The fact that $(s,t) \mapsto R_{s,t}(u,v)$ is an entire function follows just from the integral expressions in Theorem 2.4.11, because of the Bessel–MacDonald function is an entire function of index and satisfies the conditions we need to differentiate under integral sign. To prove the functional equations let us look at the Mellin transform

$$M(a,b,s,t) = \int_0^\infty\int_0^\infty R_{s,t}(u,v)u^{a-1}v^{b-1}\,du\,dv.$$

In view of Corollary 2.4.10 and (2.4.62) we have

$$\begin{aligned} &2^{11+a+b}\pi^{7+a+b}\Gamma\Big(\frac{a}{2}+2\Big)^{-1}M(a,b,s,t) \\ &\quad = C(a,b,s+t,t)\Gamma\Big(\frac{a+s}{2}+1\Big)\Gamma\Big(\frac{b+t}{2}+1\Big)\Gamma\Big(\frac{b-s-t}{2}+3\Big), \end{aligned} \tag{2.4.63}$$

where $C(a,b,s+t,t) = G(a,b,s+t,t)\,X(a,b,s+t,t)$ and one can take either

$$G(a,b,s+t,t) = \frac{\Gamma\Big(\frac{b-t}{2}+2\Big)\Gamma\Big(\frac{a+s+2t}{2}\Big)}{\Gamma\Big(\frac{a+b+s+t}{2}+2\Big)} \tag{2.4.64}$$

with X satisfying the functional equation

$$X(a,b,s+t,t) = X(a,b,(s+2t-2)+(2-t),2-t), \tag{2.4.65}$$

(according to (2.4.57)), or

$$G(a,b,s+t,t) = \frac{\Gamma\Big(\frac{b+s+t}{2}\Big)\Gamma\Big(\frac{a-s-2t}{2}+4\Big)}{\Gamma\Big(\frac{a+b-t}{2}+4\Big)} \tag{2.4.66}$$

with X satisfying the functional equation

$$X(a,b,s+t,t) = X(a,b,(6-s-2t)+t,t) \tag{2.4.67}$$

(see (2.4.58)). It follows from (2.4.63), (2.4.64), (2.4.65) that $M(a,b,s,t)$ is invariant under the linear transformation $(s,t) \mapsto (s+2t-2, 2-t)$, just because of this transformation leaves invariant $X(a,b,s+t,t)$ and changes only the order of the factors in (2.4.63), (2.4.64). Taking inverse Mellin transform we find $R_{s,t}(u,v)$ is invariant under the linear transformation above, so we get the functional equation $R_{s,t} = R_{s+2t-2,2-t}$. In a similar way, it follows from (2.4.63), (2.4.66), (2.4.67) that $M(a,b,s,t)$ is invariant under the linear transformation $(s,t) \mapsto (6-s-2t,t)$, that yields the functional equation $R_{s,t} = R_{6-s-2t,t}$.

So we have proved $R_{s,t} = R_{s+2t-2,2-t} = R_{6-s-2t,t}$, and it remains to notice that the other functional equations follow immediately from these ones. ∎

Now let us look on some special cases.

Theorem 2.4.13 *If* (s,t) *is any of the eight pairs* $(8/3,4/3)$, $(10/3,2/3)$, $(4/3,2)$, $(10/3,0)$, $(2/3,2)$, $(8/3,0)$, $(2/3,4/3)$, $(4/3,2/3)$, *then one has*

$$R_{s,t}(u,v) = -2^{-7}3^{-6}\pi^{-5}\widetilde{R}((6\pi u)^{2/3},(6\pi v)^{2/3}),$$

where

$$\begin{aligned}\widetilde{R}(a,b) &= a^6b^3\int_0^\infty\int_0^\infty Ai(a(1+x+y)^{1/3})\,Ai'(b(1+1/x)^{1/3}(1+1/y)^{1/3})\frac{dx\,dy}{A(x,y)}\\ &= a^5b^4\int_0^\infty\int_0^\infty Ai(a(1+x+y)^{1/3})\,Ai'(b(1+1/x)^{1/3}(1+1/y)^{1/3})\frac{dx\,dy}{B(x,y)}\\ &= a^6b^3\int_0^\infty\int_0^\infty Ai'(a(1+x+y)^{1/3})\,Ai(b(1+1/x)^{1/3}(1+1/y)^{1/3})\frac{dx\,dy}{C(x,y)}\\ &= a^4b^5\int_0^\infty\int_0^\infty Ai'(a(1+x+y)^{1/3})\,Ai(b(1+1/x)^{1/3}(1+1/y)^{1/3})\frac{dx\,dy}{D(x,y)}\end{aligned}$$

with

$$A(x,y) = \frac{(1+y)^{2/3}(1+x+y)^{1/3}}{x^{2/3}y^{4/3}}, \qquad B(x,y) = \frac{(1+y)^{2/3}}{x^{4/3}y^2},$$

$$C(x,y) = \frac{(1+y)^{1/3}(1+x+y)^{2/3}}{x^{1/3}y^{2/3}}, \qquad D(x,y) = \frac{(1+y)^{1/3}}{x^{5/3}y^2}. \qquad \square$$

Proof. By the functional equations in Theorem 2.4.12 we have

$$\begin{aligned}R_{8/3,4/3} &= R_{10/3,2/3} = R_{4/3,2} = R_{10/3,0}\\ &= R_{2/3,2} = R_{8/3,0} = R_{2/3,4/3} = R_{4/3,2/3}.\end{aligned}$$

On the other hand, Theorem 2.4.11 gives us the expressions for $R_{s,t}$ with $l = \pm 1/3$ and $p = \pm 2/3$ for each of pairs (s,t) we deal with. Then we easily find that four of them are different, and that applying the formulae (0.1.26) to replace the Bessel–MacDonald function by the Airy function in these expressions we get the formulae stated in our theorem. ∎

Theorem 2.4.14 *If* (s,t) *is any of the four pairs* $(2,4/3)$, $(8/3,2/3)$, $(2,2/3)$, $(4/3,4/3)$, *then one has*

$$\begin{aligned}R_{s,t}(u,v) &= u^4v^3\int_0^\infty\int_0^\infty K_0(4\pi u\sqrt{1+x+y})\,K_{1/3}(4\pi v\sqrt{(1+1/x)(1+1/y)})\frac{dx\,dy}{A(x,y)}\\ &= u^{11/3}v^{10/3}\int_0^\infty\int_0^\infty K_{1/3}(4\pi u\sqrt{1+x+y})\,K_0(4\pi v\sqrt{(1+1/x)(1+1/y)})\frac{dx\,dy}{B(x,y)}\\ &= u^{13/3}v^{8/3}\int_0^\infty\int_0^\infty K_{1/3}(4\pi u\sqrt{1+x+y})\,K_0(4\pi v\sqrt{(1+1/x)(1+1/y)})\frac{dx\,dy}{C(x,y)},\end{aligned}$$

where $C(x,y) = (1+x+y)^{1/6} x^{2/3} y^{2/3}$ *and*

$$A(x,y) = \frac{(1+y)^{1/6} x^{5/6} y^{7/6}}{(1+x)^{1/6}}, \quad B(x,y) = \frac{x^{4/3} y^{4/3}}{(1+x+y)^{1/6}}. \quad \square$$

Proof. These formulae follow from Theorem 2.4.11 and the functional equations in Theorem 2.4.12, like in the proof of Theorem 2.4.13. ■

2.5 Eisenstein series $\mathrm{E}(\sigma(\cdot), s; \Theta_{\text{K-P}})$ and cubic theta functions

2.5.1 Preliminary remarks. The first goal of this section is to study the Fourier coefficients of the Eisenstein series

$$\mathrm{E}(\sigma(\cdot), s; \Theta_{\text{K-P}}) \colon \mathbf{X} \to \mathbf{C} \tag{2.5.1}$$

attached to the group $\Gamma(3)$ and to the Kubota–Patterson cubic theta function $\Theta_{\text{K-P}}$ defined in 0.3.2. Let us write $c_{\mu,\nu}$ for the Fourier coefficients of the Eisenstein series (2.5.1), see (2.4.2), (2.4.3). First of all, we shall find the expressions for $c_{\mu,\nu}$. Then it will be clear that the Eisenstein series (2.5.1) has simple poles at the points $s = 4$ and $s = 10/3$. We define the cubic theta functions as the residues of the Eisenstein series at these points and we give the expressions for their Fourier coefficients.

We shall use the results obtained in Section 2.4 taking $q = (3)$ and $f = \Theta_{\text{K-P}}$. With such a choice we have $q^* = (\sqrt{-3})^{-3}\mathcal{O}$, $\mathrm{vol}(\mathbf{C}/q) = 2^{-1}3^{5/2}$. Also, it follows from Theorem 0.3.8 (a), (c) that in the notations of Corollary 0.3.3, for $f = \Theta_{\text{K-P}}$ we have $\rho_+ = 0$, $\rho_- = 2^{-1}3^{-1/2}$, $r = 4/3$. Thus, in Theorem 2.4.1 for (2.5.1) we have

$$m_+ = 0, \quad m_- = 2^2/3^8, \quad m = 2^3/3^{15/2}, \quad r = 4/3. \tag{2.5.2}$$

Throughout this section

$$u, y \in \mathbf{R}_+^*, \quad \zeta_*(s) = (1-3^{-s})\zeta_{\mathbf{Q}(\sqrt{-3})}(s), \quad s \in \mathbf{C}, \quad \mu, \nu \in (\sqrt{-3})^{-3}\mathcal{O},$$

$$p, p' \equiv 1 \pmod 3 \quad \text{are prime}, \quad n_1, n_2 \in \mathcal{O}, \quad n_1, n_2 \equiv 1 \pmod 3.$$

2.5.2 Fourier coefficients $c_{0,\nu}$.

Theorem 2.5.1 *For the Eisenstein series* $\mathrm{E}(\sigma(\cdot), s; \Theta_{\text{K-P}})$ *Fourier coefficient* $c_{0,0}$ *we have*

$$c_{0,0} = h \frac{\zeta_*(3s-9)\zeta_*(3s/2-5)}{\zeta_*(3s-8)\zeta_*(3s/2-3)} I_{0,0}(u, v, s),$$

where $h = 2^4\pi^3/3^8$ *and*

$$I_{0,0}(u, v, s) = \frac{u^{20/3-s} v^{6-s}}{(3s-8)(3s-10)(s-3)}. \quad \square$$

Proof. From Theorem 2.4.1, (2.5.2) and Theorem 2.4.6, it follows that

$$c_{0,0} = 2^2 3^{-5} \mathcal{D}_{0,0}(s, 2/3) J_{0,0}(u, v, s, 2/3)$$

with

$$J_{0,0}(u, v, s, 2/3) = 2^2 3^2 \pi^3 I_{0,0}(u, v, s).$$

So, we have

$$c_{0,0} = h\mathcal{D}_{0,0}(s, 2/3) I_{0,0}(u, v, s),$$

and it only remains to apply Theorem 2.4.3 and the formulae (0.1.2), (0.1.3), (0.1.17) to find that

$$\mathcal{D}_{0,0}(s, 2/3) = \widetilde{\mathcal{D}}(s - 4/3)\widetilde{\mathcal{D}}(s - 2/3)\widetilde{\mathcal{D}}(2s - 4)$$

with

$$\widetilde{\mathcal{D}}(t) = \sum_{n \equiv 1(3)} \frac{S(0, n)}{|n|^t} = \frac{\zeta_*(3t/2 - 3)}{\zeta_*(3t/2 - 2)}. \quad \blacksquare$$

Theorem 2.5.2 *For the Eisenstein series* $\mathrm{E}(\sigma(\cdot), s; \Theta_{\text{K-P}})$ *Fourier coefficients* $c_{0,\nu}$, $\nu \neq 0$, *we have*

$$c_{0,\nu} = h\overline{\Psi}(\bar{s}/2 - 2/3, \nu)\frac{\zeta_*(3s-9)\zeta_*(3s/2-4)}{\zeta_*(3s-8)\zeta_*(3s/2-3)} I_{0,\nu}(u, v, s),$$

where $h = 2^4\pi^3/3^8$,

$$I_{0,\nu}(u, v, s) = \frac{u^{20/3-s} v^{6-s}}{(s-3)(s-8/3)\Gamma(s/2-2/3)} (2\pi|\nu|v)^{s/2-5/3} K_{s/2-5/3}(4\pi|\nu|v),$$

and Ψ *is the series* (0.3.11) (*with* $q = (3)$). $\square$

Proof. Just like in the proof of Theorem 2.5.1, we apply Theorem 2.4.1, Theorem 2.4.7 and (2.5.2) to get

$$c_{0,\nu} = h\mathcal{D}_{0,\nu}(s, 2/3) I_{0,\nu}(u, v, s).$$

Then, to receive the request form for $\mathcal{D}_{0,\nu}(s, 2/3)$, apply Theorem 2.4.3 and look at (0.1.17), (0.3.11):

$$\mathcal{D}_{0,\nu}(s, 2/3) = \overline{\Psi}(\bar{s}/2 - 2/3, \nu)\frac{\zeta_*(3s-9)\zeta_*(3s/2-4)}{\zeta_*(3s-8)\zeta_*(3s/2-3)}. \quad \blacksquare$$

2.5.3 Fourier coefficients $c_{\mu,0}$ with $\mu \neq 0$.

Theorem 2.5.3 *For the Eisenstein series* $\mathrm{E}(\sigma(\cdot), s; \Theta_{\text{K-P}})$ *Fourier coefficients* $c_{\mu,0}$, $\mu \neq 0$, *we have*

$$c_{\mu,0} = h\tau(\mu)\frac{\zeta_*(3s-9)\zeta_*(3s/2-5)}{\zeta_*(3s-8)\zeta_*(3s/2-3)} I_{\mu,0}(u,v,s),$$

where $h = 2^5\pi^3/3^{15/2}$, τ *is the function defined by* (0.3.15) *and*

$$I_{\mu,0}(u,v,s) = \frac{u^{7-s}v^{6-s}}{(s-3)(s-8/3)(s-10/3)} K_{1/3}(4\pi|\mu|u). \qquad \square$$

Proof. From Theorem 2.4.1, (2.5.2) and Theorem 2.4.8, it follows that

$$c_{\mu,0} = 2^3 3^{-15/2} \mathcal{D}_{\mu,0}(s) J_{\mu,0}(u,v,s,4/3)$$

and $J_{\mu,0}(u,v,s,4/3) = 2^2\pi^3 I_{\mu,0}(u,v,s)$. So, $c_{\mu,0} = h\mathcal{D}_{\mu,0}(s)I_{\mu,0}(u,v,s)$, and it remains only to show that

$$\mathcal{D}_{\mu,0}(s) = \tau(\mu)\frac{\zeta_*(3s-9)\zeta_*(3s/2-5)}{\zeta_*(3s-8)\zeta_*(3s/2-3)}. \tag{2.5.3}$$

For this we can apply Theorem 2.4.4. First, let us notice that

$$S(0, n_1, n_2^2 n_3) = \begin{cases} \tilde{\varphi}(n_1) & \text{if } n_1 n_2^2 n_3 \text{ is a cube,} \\ 0 & \text{otherwise.} \end{cases}$$

Then Theorem 2.4.4 yields

$$\mathcal{D}_{\mu,0}(s) = \nabla(\mu, 1), \tag{2.5.4}$$

where, for any $r \in q^*$ and non-zero $l \in \mathcal{O}$, we write

$$\nabla(r,l) = \sum_{n_1,n_2,n_3} \frac{\tau(rn_1/n_3)\overline{S(rn_1/n_2, n_3)}S(r, n_2, n_3^2)\tilde{\varphi}(n_1)}{|n_1 n_3|^{s-1}|n_2|^{2s-4}}, \tag{2.5.5}$$

where the summation is subject to the same conditions as in Theorem 2.4.4 and two supplementary conditions, on the whole

$$\begin{gathered} n_1, n_2, n_3 \equiv 1 \ (\mathrm{mod}\ 3), \quad n_1 n_2^2 n_3 \ \text{ is a cube,} \\ rn_1/n_2 \in q^*, \quad rn_1/n_3 \in q^*, \quad \gcd(l, n_1 n_2 n_3) = 1, \\ S(rn_1/n_2, n_3) \neq 0, \quad S(r, n_2, n_3^2) \neq 0. \end{gathered} \tag{2.5.6}$$

For prime $p \equiv 1 \ (\mathrm{mod}\ 3)$, let $\vartheta = \mathrm{ord}_p \mu$, $\mu' = \mu p^{-\vartheta}$. Express n_1, n_2, n_3 as the products

$$n_1 = m_1 p^\gamma, \quad n_2 = m_2 p^\delta, \quad n_3 = m_3 p^\varepsilon$$

with $m_1, m_2, m_3 \equiv 1 \pmod 3$, $p \nmid m_1 m_2 m_3$ and $\gamma, \delta, \varepsilon \in \mathbf{Z}$, $\gamma, \delta, \varepsilon \geq 0$. We have (see 0.1.4 and Proposition 0.3.9)

$$\tau(\mu n_1/n_3) = \left(\frac{\mu' m_1/m_3}{p}\right)^{\vartheta+\gamma-\varepsilon} \tau(\mu' m_1/m_3)\tau(p^{\vartheta+\gamma-\varepsilon}),$$

$$S(\mu n_1/n_2, n_3) = \left(\frac{\mu' m_1/m_2}{p}\right)^{-\varepsilon} \left(\frac{m_3}{p}\right)^{\delta-\vartheta-\gamma-\varepsilon} S(\mu' m_1/m_2, m_3) S(p^{\vartheta+\gamma-\delta}, p^{\varepsilon}),$$

$$S(\mu, n_2, n_3^2) = \left(\frac{\mu'}{p}\right)^{\varepsilon-\delta} \left(\frac{m_2}{p}\right)^{-\vartheta-\delta-\varepsilon} \left(\frac{m_3}{p}\right)^{\vartheta-\delta} S(\mu', m_2, m_3^2) S(p^{\vartheta}, p^{\delta}, p^{2\varepsilon}),$$

$$\tilde{\varphi}(n_1) = \tilde{\varphi}(m_1)\tilde{\varphi}(p^{\gamma}).$$

Substituting the right-hand sides into (2.5.4), (2.5.5) we find for $\mathcal{D}_{\mu,0}(s)$ the following expression

$$\mathcal{D}_{\mu,0}(s) = \left(\frac{\mu'}{p}\right)^{\vartheta} Q^{(p)}(\vartheta, s)\nabla(\mu', p), \tag{2.5.7}$$

where

$$Q^{(p)}(\vartheta, s) = \sum_{\substack{\gamma,\delta,\varepsilon\geq 0 \\ \vartheta+\gamma\geq\delta,\, \vartheta+\gamma\geq\varepsilon,\, \gamma+2\delta+\varepsilon\equiv 0(3)}} \frac{\tau(p^{\vartheta+\gamma-\varepsilon})\overline{S(p^{\vartheta+\gamma-\delta}, p^{\varepsilon})S(p^{\vartheta}, p^{\delta}, p^{2\varepsilon})\tilde{\varphi}(p^{\gamma})}}{|p|^{(s-1)(\gamma+\varepsilon)+2(s-2)\delta}}. \tag{2.5.8}$$

Now we can take some other prime $p' \equiv 1 \pmod 3$ and to repeat once more the above procedure to obtain

$$\mathcal{D}_{\mu,0}(s) = \left(\frac{\mu'}{p}\right)^{\vartheta} \left(\frac{\mu''}{p'}\right)^{\vartheta'} Q^{(p)}(\vartheta, s) Q^{(p')}(\vartheta', s)\nabla(\mu'', pp'),$$

where $\vartheta' = \operatorname{ord}_{p'}\mu' = \operatorname{ord}_{p'}\mu$, $\mu'' = \mu p^{-\vartheta} p'^{-\vartheta'}$. Applying this procedure to all prime of norm $\leq R$, where R is larger than the norm of all prime divisors of μ, we obtain

$$\mathcal{D}_{\mu,0}(s) = \tilde{k}(\mu)\nabla(\xi(\sqrt{-3})^t, l) \prod_{\substack{p\equiv 1(3) \\ p|l,\ p \text{ is prime}}} Q^{(p)}(\vartheta_p, s), \tag{2.5.9}$$

where

$$\mu = \xi(\sqrt{-3})^t \prod_{\substack{p\equiv 1(3) \\ p \text{ is prime}}} p^{\vartheta_p}, \quad \xi \in \mathcal{O}^*, \quad t \in \mathbf{Z}, \quad t \geq -3, \quad \vartheta_p = \operatorname{ord}_p \mu, \quad l = \prod_{\substack{p\equiv 1(3) \\ \|p\|\leq R,\ p \text{ is prime}}} p,$$

and

$$\tilde{k}(\mu) = \Omega(\mu) \prod_{\substack{p\equiv 1(3) \\ p \text{ is prime}}} \left(\frac{\xi(\sqrt{-3})^t}{p}\right)^{\vartheta_p}$$

with just the same Ω as in Corollary 0.3.10. It follows from Corollary 0.3.10 that

$$\tau(\mu) = \tilde{k}(\mu)\tau(\xi(\sqrt{-3})^t) \prod_{\substack{p\equiv 1(3)\\ p \text{ is prime}}} \tau(p^{\vartheta_p}). \tag{2.5.10}$$

Below we shall prove that

$$Q^{(p)}(\vartheta, s) = \tau(p^{\vartheta}) \frac{\left(1 - \frac{1}{\|p\|^{3s-8}}\right)\left(1 - \frac{1}{\|p\|^{3s/2-3}}\right)}{\left(1 - \frac{1}{\|p\|^{3s-9}}\right)\left(1 - \frac{1}{\|p\|^{3s/2-5}}\right)}. \tag{2.5.11}$$

One has, for any $r \in q^*$,

$$\nabla(r, l) \to \tau(r) \quad \text{as} \quad R \to \infty \tag{2.5.12}$$

because the term associated with $n_1 = n_2 = n_3 = 1$ in (2.5.5) is $\tau(r)$, and the sum of all other terms in (2.5.5) tends[†] to 0 as $R \to \infty$, due to the summation condition $\gcd(l, n_1 n_2 n_3) = 1$. Combining (2.5.9),..., (2.5.12) we get (2.5.3).

It remains only to prove (2.5.11). The computation is not very short. First, we have

$$\tau(p^{\vartheta+\gamma-\varepsilon}) = \begin{cases} \|p\|^{(\vartheta+\gamma-\varepsilon)/6} & \text{if } \vartheta+\gamma-\varepsilon \equiv 0(3), \\ \overline{S(1,p)}\|p\|^{(\vartheta+\gamma-\varepsilon-4)/6} & \text{if } \vartheta+\gamma-\varepsilon \equiv 1(3), \\ 0 & \text{if } \vartheta+\gamma-\varepsilon \equiv 2(3); \end{cases} \tag{2.5.13}$$

$$S(p^{\vartheta+\gamma-\delta}, p^{\varepsilon}) = \begin{cases} -\|p\|^{\varepsilon-1} & \text{if } \varepsilon = \vartheta+\gamma-\delta+1 \equiv 0(3), \\ S(1,p)\|p\|^{\varepsilon-1} & \text{if } \varepsilon = \vartheta+\gamma-\delta+1 \equiv 1(3), \\ \overline{S(1,p)}\|p\|^{\varepsilon-1} & \text{if } \varepsilon = \vartheta+\gamma-\delta+1 \equiv 2(3), \\ \tilde{\varphi}(p^{\varepsilon}) & \text{if } \varepsilon \le \vartheta+\gamma-\delta,\ \varepsilon \equiv 0(3), \\ 0 & \text{in all other cases}; \end{cases} \tag{2.5.14}$$

$$S(p^{\vartheta}, p^{\delta}, p^{2\varepsilon}) = \begin{cases} -\|p\|^{\vartheta} & \text{if } \delta = \vartheta+1,\ \delta+2\varepsilon \equiv 0(3), \\ S(1,p)\|p\|^{\vartheta} & \text{if } \delta = \vartheta+1,\ \delta+2\varepsilon \equiv 1(3), \\ \overline{S(1,p)}\|p\|^{\vartheta} & \text{if } \delta = \vartheta+1,\ \delta+2\varepsilon \equiv 2(3), \\ \tilde{\varphi}(p^{\delta}) & \text{if } \delta \le \vartheta,\ \delta+2\varepsilon \equiv 0(3), \\ 0 & \text{in all other cases}; \end{cases} \tag{2.5.15}$$

(see 0.1.4 and 0.3.8). To evaluate (2.5.8) let us express the set of terms in (2.5.8) as a disjnct union of several sets in such a way that

$$Q^{(p)}(\vartheta, s) = \mathcal{A}_0 + \mathcal{A}_1 + \mathcal{B}_0 + \mathcal{B}_1 + \mathcal{C}_0 + \mathcal{C}_1 + \mathcal{D}_0 + \mathcal{D}_1,$$

[†] If $\Re(s)$ is sufficiently large, that we suppose, certainly, in all our computations.

where the terms with $\vartheta+\gamma-\varepsilon \equiv t(3)$ are believed belonging

to $\mathcal{A}_t$ if $\varepsilon = \vartheta+\gamma-\delta+1$, $\delta = \vartheta+1$;
to $\mathcal{B}_t$ if $\varepsilon = \vartheta+\gamma-\delta+1$, $\delta \le \vartheta$, $\delta+2\varepsilon \equiv 0(3)$;
to $\mathcal{C}_t$ if $\varepsilon \le \vartheta+\gamma-\delta$, $\varepsilon \equiv 0(3)$, $\delta = \vartheta+1$;
to $\mathcal{D}_t$ if $\varepsilon \le \vartheta+\gamma-\delta$, $\varepsilon \equiv 0(3)$, $\delta \le \vartheta$, $\delta+2\varepsilon \equiv 0(3)$.

By means of (2.5.13), (2.5.14), (2.5.15) one can easily check that the terms in (2.5.8) which are not listed above are equal to 0. Moreover, it is easy to see that: for $\vartheta \equiv 0 \pmod 3$ the sums $\mathcal{A}_1, \mathcal{B}_0, \mathcal{C}_0, \mathcal{D}_1$ are empty; for $\vartheta \equiv 1 \pmod 3$ the sums $\mathcal{A}_0, \mathcal{B}_1, \mathcal{C}_1, \mathcal{D}_0$ are empty; for $\vartheta \equiv 2 \pmod 3$ all 8 sums are empty. Thus we have

$$Q^{(p)}(\vartheta, s) = \begin{cases} \mathcal{A}_0+\mathcal{B}_1+\mathcal{C}_1+\mathcal{D}_0 & \text{if } \vartheta \equiv 0(3), \\ \mathcal{A}_1+\mathcal{B}_0+\mathcal{C}_0+\mathcal{D}_1 & \text{if } \vartheta \equiv 1(3), \\ 0 & \text{if } \vartheta \equiv 2(3). \end{cases} \tag{2.5.16}$$

For $\vartheta \equiv 2 \pmod 3$ one has $\tau(p^\vartheta) = 0$, see (0.3.15). Hence, in this case (2.5.11) follows immediately from (2.5.16). Now let us consider successively the cases $\vartheta \equiv 0 \pmod 3$ and $\vartheta \equiv 1 \pmod 3$.

First, let $\vartheta \equiv 0 \pmod 3$. For this case we have

$$\mathcal{A}_0 = \tau(p^\vartheta)\frac{\tilde{\varphi}(p)}{\|p\|^{(\vartheta+3)s-3\vartheta-7}}\sum_{l\ge 0}\frac{1}{\|p\|^{(3s-9)l}},$$

$$\mathcal{B}_1 = \tau(p^\vartheta)\frac{\tilde{\varphi}(p)}{\|p\|^{(s/2-4/3)\vartheta+3s/2-3}}\sum_{l\ge 0}\frac{\tilde{\varphi}(p^l)}{\|p\|^{(3s-8)l}}\sum_{\vartheta/3-1\ge l\ge 0}\frac{1}{\|p\|^{(3s/2-5)l}},$$

$$\mathcal{C}_1 = \tau(p^\vartheta)\frac{\tilde{\varphi}(p)}{\|p\|^{(s-2)(\vartheta+1)+s/2-1-\vartheta}}\sum_{t\ge 0}\frac{1}{\|p\|^{(3s/2-5)t}}\sum_{t\ge l\ge 0}\frac{\tilde{\varphi}(p^l)}{\|p\|^{(3s/2-3)l}},$$

$$\mathcal{D}_0 = \tau(p^\vartheta)\sum_{t\ge 0}\frac{\tilde{\varphi}(p^t)}{\|p\|^{(3s/2-4)t}}\sum_{\vartheta/3\ge l\ge 0}\frac{\tilde{\varphi}(p^l)}{\|p\|^{(3s-8)l}}\sum_{\vartheta/3+t-l\ge k\ge 0}\frac{\tilde{\varphi}(p^k)}{\|p\|^{(3s/2-3)k}}.$$

To explain how to find these formulae, let us consider in detail $\mathcal{A}_0$. By definition, $\mathcal{A}_0$ is the sum of all terms in (2.5.8) with

$$\gamma,\delta,\varepsilon \ge 0, \quad \vartheta+\gamma \ge \delta, \quad \vartheta+\gamma\ge\varepsilon, \quad \gamma+2\delta+\varepsilon \equiv 0 \pmod 3,$$
$$\vartheta+\gamma-\varepsilon \equiv 0 \pmod 3, \quad \varepsilon = \vartheta+\gamma-\delta+1, \quad \delta = \vartheta+1.$$

Since $\vartheta \equiv 0 \pmod 3$, one can rewrite these conditions as

$$\gamma \ge 1, \quad \gamma \equiv 2 \pmod 3, \quad \varepsilon = \gamma, \quad \delta = \vartheta+1.$$

Expressing the Gauß sums in (2.5.8) according to the formulae (2.5.13), (2.5.14) and (2.5.15) and using $|S(1,p)|^2 = \|p\|$ we obtain

$$\mathcal{A}_0 = \frac{\tau(p^\vartheta)}{\|p\|^{(s-2)(\vartheta+1)}} \sum_{\substack{\gamma\geq 1\\ \gamma\equiv 2(3)}} \frac{\overline{S(p^{\gamma-1},p^\gamma)S(p^\vartheta,p^{\vartheta+1},p^{2\gamma})\tilde{\varphi}(p^\gamma)}}{\|p\|^{(s-1)\gamma}}$$
$$= \frac{\tau(p^\vartheta)}{\|p\|^{(s-2)(\vartheta+1)-\vartheta}} \sum_{\substack{\gamma\geq 1\\ \gamma\equiv 2(3)}} \frac{\tilde{\varphi}(p^\gamma)}{\|p\|^{(s-2)\gamma}}.$$

Expressing in the last sum $\tilde{\varphi}(p^\gamma)$ as $\tilde{\varphi}(p)\|p\|^{\gamma-1}$ and γ as $3l+2$ with $l \geq 0$, we obtain the desired formula for $\mathcal{A}_0$. The formulae for $\mathcal{B}_1$, $\mathcal{C}_1$, $\mathcal{D}_0$ may be obtained by the same manner and we omit the details for short. Then to evaluate $\mathcal{A}_0 + \mathcal{B}_1 + \mathcal{C}_1 + \mathcal{D}_0$ we have use the well-known formulae

$$\sum_{l\geq 0} \frac{1}{\|p\|^{tl}} = \left(1 - \frac{1}{\|p\|^t}\right)^{-1},$$
$$\sum_{l\geq 0} \frac{\tilde{\varphi}(p^l)}{\|p\|^{tl}} = \left(1 - \frac{1}{\|p\|^{t-1}}\right)^{-1}\left(1 - \frac{1}{\|p\|^t}\right),$$
$$\sum_{r\geq l\geq 0} \frac{1}{\|p\|^{tl}} = \left(1 - \frac{1}{\|p\|^t}\right)^{-1}\left(1 - \frac{1}{\|p\|^{(r+1)t}}\right),$$
$$\sum_{r\geq l\geq 0} \frac{\tilde{\varphi}(p^l)}{\|p\|^{tl}} = \left(1 - \frac{1}{\|p\|^{t-1}}\right)^{-1}\left(1 - \frac{1}{\|p\|^t} + \frac{1}{\|p\|^{rt+t-r}} - \frac{1}{\|p\|^{rt+t-r-1}}\right),$$

which have place for all $t \in \mathbf{C}$ with $\Re(t) > 0$ and $r \in \mathbf{Z}$, $r \geq 0$. By means of these formulae we can evaluate all the series in the formulae above for $\mathcal{A}_0$, $\mathcal{B}_1$, $\mathcal{C}_1$, $\mathcal{D}_0$ and then, by routine computation, we obtain (2.5.11) for $\vartheta \equiv 0 \pmod 3$.

The case $\vartheta \equiv 1 \pmod 3$ is quite similar. In this case we have

$$\mathcal{A}_1 = \tau(p^\vartheta)\frac{\tilde{\varphi}(p)}{\|p\|^{(\vartheta+2)s-3\vartheta-4}} \sum_{l\geq 0} \frac{1}{\|p\|^{(3s-9)l}},$$

$$\mathcal{B}_0 = \tau(p^\vartheta)\frac{\tilde{\varphi}(p)}{\|p\|^{(s/2-4/3)\vartheta+s-5/3}} \sum_{l\geq 0} \frac{\tilde{\varphi}(p^l)}{\|p\|^{(3s-8)l}} \sum_{(\vartheta-1)/3\geq l\geq 0} \frac{1}{\|p\|^{(3s/2-5)l}},$$

$$\mathcal{C}_0 = \tau(p^\vartheta)\frac{\tilde{\varphi}(p)}{\|p\|^{(s-2)(\vartheta+1)+s-3-\vartheta}} \sum_{t\geq 0} \frac{1}{\|p\|^{(3s/2-5)t}} \sum_{t\geq l\geq 0} \frac{\tilde{\varphi}(p^l)}{\|p\|^{(3s/2-3)l}},$$

$$\mathcal{D}_1 = \tau(p^\vartheta)\sum_{t\geq 0} \frac{\tilde{\varphi}(p^t)}{\|p\|^{(3s/2-4)t}} \sum_{(\vartheta-1)/3\geq l\geq 0} \frac{\tilde{\varphi}(p^l)}{\|p\|^{(3s-8)l}} \sum_{(\vartheta-1)/3+t-l\geq k\geq 0} \frac{\tilde{\varphi}(p^k)}{\|p\|^{(3s/2-3)k}}.$$

For short, let us restrict ourselves by evaluation $\mathcal{A}_1$. By definition $\mathcal{A}_1$ is the sum of all terms in (2.5.8) with

$$\gamma, \delta, \varepsilon \geq 0, \quad \vartheta + \gamma \geq \delta, \quad \vartheta + \gamma \geq \varepsilon, \quad \gamma + 2\delta + \varepsilon \equiv 0 \pmod 3,$$
$$\vartheta + \gamma - \varepsilon \equiv 1 \pmod 3, \quad \varepsilon = \vartheta + \gamma - \delta + 1, \quad \delta = \vartheta + 1.$$

Since $\vartheta \equiv 1 \pmod 3$, we can rewrite these conditions as

$$\gamma \geq 1, \quad \gamma \equiv 1 \pmod 3, \quad \varepsilon = \gamma, \quad \delta = \vartheta + 1.$$

Expressing the Gauß sums and ε and δ in (2.5.8) according to the formulae (2.5.13), (2.5.14) and (2.5.15) and using $|S(1,p)|^2 = \|p\|$ we obtain

$$\mathcal{A}_1 = \frac{\tau(p^\vartheta)}{\|p\|^{(s-2)(\vartheta+1)}} \sum_{\substack{\gamma \geq 1 \\ \gamma \equiv 1(3)}} \frac{\overline{S(p^{\gamma-1}, p^\gamma) S(p^\vartheta, p^{\vartheta+1}, p^{2\gamma})\tilde{\varphi}(p^\gamma)}}{\|p\|^{(s-1)\gamma}}$$

$$= \frac{\tau(p^\vartheta)}{\|p\|^{(s-2)(\vartheta+1)-\vartheta}} \sum_{\substack{\gamma \geq 1 \\ \gamma \equiv 1(3)}} \frac{\tilde{\varphi}(p^\gamma)}{\|p\|^{(s-2)\gamma}}.$$

Expressing in the last sum $\tilde{\varphi}(p^\gamma)$ as $\tilde{\varphi}(p)\|p\|^{\gamma-1}$ and γ as $3l+1$ with $l \geq 0$, we obtain the desired formula for $\mathcal{A}_1$. The formulae for $\mathcal{B}_0$, $\mathcal{C}_0$, $\mathcal{D}_1$ may be obtained by the same manner and we omit the details. Let us omit also all other computations which lead to (2.5.11). ■

2.5.4 Fourier coefficients $c_{\mu,\nu}$ with $\mu\nu \neq 0$. Let us begin with some notations. Recall that $\mu, \nu \in (\sqrt{-3})^{-3}\mathcal{O} \setminus \{0\}$, so $\mu\nu \in (\sqrt{-3})^{-6}\mathcal{O} \setminus \{0\}$. For prime $p \equiv 1 \pmod 3$, we write $p \mid \mu\nu$ if $\operatorname{ord}_p(\mu\nu) \geq 1$, and $p \nmid \mu\nu$ if $\operatorname{ord}_p(\mu\nu) = 0$. For $k \equiv 1 \pmod 3$, we write $k \mid (\mu\nu)^\infty$ if $\operatorname{ord}_p(\mu\nu) \geq 1$ for all prime $p \mid k$. Also, $\gcd(r,k) = 1$ with $r, k \in \mathbf{Q}(\sqrt{-3})$ means that there is no prime p (here $p = \sqrt{-3}$ is not excluded) such that both $\operatorname{ord}_p k \geq 1$ and $\operatorname{ord}_p r \geq 1$. We set

$$\Lambda(\mu\nu) = \{l \in \mathcal{O} \setminus \{0\} \mid \text{for each prime } p \equiv 1(3), \text{ if } p \mid l, \text{ then } p \mid \mu\nu\} \tag{2.5.17}$$

$$= \{\xi(\sqrt{-3})^n c \mid \xi \in \mathcal{O}^*,\ n \in \mathbf{Z},\ n \geq 0,\ c \in \mathcal{O},\ c \equiv 1(3),\ c \mid (\mu\nu)^\infty\}.$$

Given $l \in \Lambda(\mu\nu)$, we set

$$P_{\mu,\nu}(l;s) = \sum_{\substack{m \equiv 1(3) \\ \gcd(m,\mu\nu)=1}} \frac{\overline{S(1,m)^2}}{|m|^s} \left(\frac{l}{m}\right),$$

$$D_{\mu,\nu}(l;s) = \prod_{\substack{p \mid \mu\nu \\ p \text{ is prime, } p \equiv 1(3)}} \left(1 - \frac{1}{\|p\|^{3s/2-3}}\right)^{-1} \sum_{k_1,k_2,k_3} \frac{\tau(\mu k_1/k_3)\tilde{S}(\mu,\nu;k_1,k_2,k_3)}{|k_1k_3|^{s-1}|k_2|^{2s-4}}, \tag{2.5.18}$$

where $\widetilde{S}(\mu,\nu;k_1,k_2,k_3) = \overline{S(\mu k_1/k_2, k_3)} S(\mu, k_2, k_3^2) \overline{S(\nu, k_1, k_2^2 k_3)}$ and k_j are subject to the conditions

$$k_1, k_2, k_3 \equiv 1 \pmod 3, \quad \mu k_1/k_2 \in (\sqrt{-3})^{-3}\mathcal{O}, \quad \mu k_1/k_3 \in (\sqrt{-3})^{-3}\mathcal{O},$$

$$(k_1k_2k_3) \mid (\mu\nu)^\infty, \quad S(\mu k_1/k_2, k_3) \neq 0, \quad S(\mu, k_2, k_3^2) \neq 0, \tag{2.5.19}$$

$$\mu\nu k_1^2 k_2 k_3^2 l^{-1} \quad \text{is a cube in} \quad \mathbf{Q}(\sqrt{-3}).$$

The series $P_{\mu,\nu}(l;s)$ are nothing but the series $P_0(s/2;l,27\mu\nu)$ studied in 0.4.3. In view of Theorem 0.4.9, one can regard $P_{\mu,\nu}(l;\cdot)$ as meromorphic functions on $\mathbf{C}$, and as that they have no poles in $\{ s \in \mathbf{C} \mid \Re(s) \geq 10/3 \}$, except possible simple pole at the point 10/3. One can regard $D_{\mu,\nu}(l;\cdot)$ as meromorphic functions on $\mathbf{C}$ too. Namely, given prime $p \equiv 1 \pmod 3$, we have

$$\text{if } \operatorname{ord}_p k_1 \geq \operatorname{ord}_p \nu + 2, \text{ then } S(\nu, k_1, k_2^2 k_3) = 0;$$

$$\text{if } \operatorname{ord}_p k_2 \geq \operatorname{ord}_p \mu + 2, \text{ then } S(\mu, k_2, k_3^2) = 0;$$

$$\text{if } \operatorname{ord}_p k_3 > \operatorname{ord}_p(\mu k_1), \text{ then } \mu k_1/k_3 \notin (\sqrt{-3})^{-3}\mathcal{O}.$$

Hence, only a finite number of terms of the series defining $D_{\mu,\nu}(l;\cdot)$ are non-zero. So, we have that $D_{\mu,\nu}(l;\cdot)$ are meromorphic functions on $\mathbf{C}$ with only poles along the line $\Re(s) = 2$ produced by the Euler factors.

Theorem 2.5.4 *For the Eisenstein series* $\mathrm{E}(\sigma(\cdot), s; \Theta_{\text{K-P}})$ *Fourier coefficients* $c_{\mu,\nu}$, *with* $\mu\nu \neq 0$, *we have*

$$c_{\mu,\nu} = h\,\zeta_*(3s/2-3)^{-1} \Big\{ \sum_{\substack{l \in \Lambda(\mu\nu) \\ l \text{ is cube-free}}} P_{\mu,\nu}(l;s)\, D_{\mu,\nu}(l;s) \Big\}\, I_{\mu,\nu}(u,v,s),$$

where $h = 2^3/3^{15/2}$, $P_{\mu,\nu}$, $D_{\mu,\nu}$ *mean the functions defined in* (2.5.18), $\Lambda(\mu\nu)$ *as in* (2.5.17), *and* $I_{\mu,\nu}(u,v,s) = J_{\mu,\nu}(u,v,s,4/3)$, $J_{\mu,\nu}$ *being as in Theorem* 2.4.1 *and Theorem* 2.4.9. *In particular, if* $\operatorname{ord}_p \mu \equiv \operatorname{ord}_p \nu \equiv 2 \pmod 3$ *for at least one prime* $p \equiv 1 \pmod 3$, *then* $D_{\mu,\nu}(l;\cdot) = 0$ *for all* l *and, so,* $c_{\mu,\nu} = 0$. □

Proof. From Theorem 2.4.1 and (2.5.2), it follows that

$$c_{\mu,\nu} = h\mathcal{D}_{\mu,\nu}(s)\, J_{\mu,\nu}(u,v,s,4/3),$$

and thus we only have to prove

$$\zeta_*(3s/2-3)\mathcal{D}_{\mu,\nu}(s) = \sum_{\substack{l \in \Lambda(\mu\nu) \\ l \text{ is cube-free}}} P_{\mu,\nu}(l;s)\, D_{\mu,\nu}(l;s). \tag{2.5.20}$$

For this we apply Theorem 2.4.4 with $\rho = \tau$ and $q = (3)$. The series $\mathcal{D}_{\mu,\nu}(s)$ is very complicated to deal with, and we have to begin with some analysis. Let $p \equiv 1 \pmod 3$ be prime, $p \nmid \mu\nu$. Set $\gamma = \operatorname{ord}_p n_1$, $\delta = \operatorname{ord}_p n_2$ and $\varepsilon = \operatorname{ord}_p n_3$. The summation conditions (2.4.28) imply $\gamma \geq \max\{\delta, \varepsilon\} \geq 0$. On the other hand, we have (see 0.1.4)

$$S(\mu n_1/n_2, n_3) \neq 0 \quad \text{only if} \quad \varepsilon = \gamma - \delta + 1 \quad \text{or} \quad \varepsilon \leq \gamma - \delta, \quad \varepsilon \equiv 0 \pmod 3;$$

$$S(\mu, n_2, n_3^2) \neq 0 \quad \text{only if} \quad \delta = 1 \quad \text{or} \quad \delta = 0, \quad \varepsilon \equiv 0 \pmod 3;$$

$$S(\nu, n_1, n_2^2 n_3) \neq 0 \quad \text{only if} \quad \gamma = 1 \quad \text{or} \quad \gamma = 0, \quad 2\delta + \varepsilon \equiv 0 \pmod 3.$$

These remaks show that

$$\overline{S(\mu n_1/n_2, n_3)}\, S(\mu, n_2, n_3^2)\, \overline{S(\nu, n_1, n_2^2 n_3)} \neq 0 \tag{2.5.21}$$

only in the following four cases:

$$\gamma = \delta = \varepsilon = 0; \quad \gamma = 1, \quad \delta = \varepsilon = 0; \quad \gamma = \delta = 1, \quad \varepsilon = 0; \quad \gamma = \delta = \varepsilon = 1.$$

Since p here is arbitrary prime $\nmid \mu\nu$, we find that (2.5.21) has place only for n_1, n_2, n_3 of the form

$$n_1 = k_1 abc, \quad n_2 = k_2 ab, \quad n_3 = k_3 a, \tag{2.5.22}$$

where

$$\begin{gathered} k_1, k_2, k_3 \equiv 1 \pmod 3), \quad (k_1k_2k_3) \mid (\mu\nu)^\infty, \\ a, b, c \equiv 1 \pmod 3), \quad a, b, c \text{ are square-free}, \\ \gcd(\mu\nu, abc) = \gcd(a, b) = \gcd(a, c) = \gcd(b, c) = 1. \end{gathered} \tag{2.5.23}$$

Let us now express n_1, n_2, n_3 in Theorem 2.4.4 according to (2.5.22), (2.5.23) and replace the summation over n_1, n_2, n_3 by the summation over k_1, k_2, k_3 and a, b, c. By using the multiplicative properties of the Gauß sums (see 0.1.4, 0.3.8), one can rewrite the components of the terms as follows:

$$\tau(\mu n_1/n_3) = \left(\frac{\mu k_1/k_3}{bc}\right)\left(\frac{b}{c}\right)\tau(\mu k_1/k_3)\,\overline{S(1,b)}\,\overline{S(1,c)}\,|bc|^{-1},$$

$$S(\mu n_1/n_2, n_3) = \left(\frac{\mu c k_1/k_2}{a}\right)^2\left(\frac{k_3}{ac}\right)^2 S(\mu k_1/k_2, k_3)\, S(1,a),$$

$$S(\mu, n_2, n_3^2) = \left(\frac{\mu k_2 k_3}{b}\right)^2\left(\frac{k_2 k_3^2 b}{a}\right) S(\mu, k_2, k_3^2)\, C(1,a)\, S(1,b),$$

$$S(\nu, n_1, n_2^2 n_3) = \left(\frac{b}{ac}\right)\left(\frac{\nu k_1}{ac}\right)^2\left(\frac{k_1}{b}\right)\left(\frac{k_2^2 k_3}{abc}\right) S(\nu, k_1, k_2^2 k_3)\, C(1,b)\, S(1,ac).$$

Multiplying these and substituting into formula of Theorem 2.4.4 we obtain

$$\begin{aligned} &\tau(\mu n_1/n_3)\,\overline{S(\mu n_1/n_2, n_3)}\,S(\mu, n_2, n_3^2)\,\overline{S(\nu, n_1, n_2^2 n_3)} \\ &\quad = C(1,a)C(1,b)\overline{S(1,ac)}^2 \left|\frac{b}{c}\right| \left(\frac{\mu\nu k_1^2 k_2 k_3^2}{ac}\right) \tau(\mu k_1/k_3)\widetilde{S}(\mu, \nu; k_1, k_2, k_3), \end{aligned}$$

and then

$$\begin{aligned} \mathcal{D}_{\mu,\nu}(s) &= \sum_{k_1,k_2,k_3} \frac{\tau(\mu k_1/k_3)\widetilde{S}(\mu,\nu;k_1,k_2,k_3)}{|k_1k_3|^{s-1}|k_2|^{2s-4}} \nabla(k_1^2 k_2 k_3^2), \\ \nabla(k) &= \sum_{a,b,c} \frac{C(1,a)C(1,b)\overline{S(1,ac)}^2}{|a|^{4s-6}|b|^{3s-6}|c|^s}\left(\frac{\mu\nu k}{ac}\right); \end{aligned} \tag{2.5.24}$$

the summation is carried out over k_1, k_2, k_3 and a, b, c under the conditions

(2.5.23) and the complementary conditions

$$\mu k_1/k_2 \in (\sqrt{-3})^{-3}\mathcal{O}, \quad \mu k_1/k_3 \in (\sqrt{-3})^{-3}\mathcal{O},$$
$$S(\mu k_1/k_2, k_3) \neq 0, \quad S(\mu, k_2, k_3^2) \neq 0.$$

To examine the sum $\nabla(k)$ we introduce new variables $m = ac$ and $r = ab$, and by elementary transformations we get

$$\nabla(k) = \sum_{\substack{m\equiv 1(3)\\ \gcd(m,\mu\nu k)=1}} \frac{\overline{S(1,m)}^2}{|m|^s}\left(\frac{\mu\nu k}{m}\right)\tilde{\nabla}, \tag{2.5.25}$$

$$\tilde{\nabla} = \sum_{\substack{a,b\equiv 1(3)\\ \gcd(ab,\mu\nu k)=\gcd(m,b)=1\\ a|m,\ a,b \text{ are square-free}}} \frac{C(1,a)C(1,b)}{|ab|^{3s-6}} = \sum_{\substack{r\equiv 1(3)\\ r \text{ is square-free}\\ \gcd(r,\mu\nu k)=1}} \frac{C(1,r)}{|r|^{3s-6}} = \prod_{p\nmid \mu\nu k}\left(1-\frac{1}{|p|^{3s-6}}\right) \tag{2.5.26}$$

$$= \zeta_*(3s/2-3)^{-1}\prod_{p|\mu\nu k}\left(1-\frac{1}{\|p\|^{3s/2-3}}\right)^{-1};$$

the products here are over prime $p \equiv 1 \pmod 3$. Since $k = k_1^2k_2k_3^2 \mid \mu\nu$, we can, after substituting (2.5.25), (2.5.26) into (2.5.24), replace, in the product, the condition $p \mid \mu\nu k$ by $p \mid \mu\nu$ and, in the sum, the condition $\gcd(m, \mu\nu k) = 1$ by $\gcd(m, \mu\nu) = 1$. Thus we obtain

$$\zeta_*(3s/2-3)\prod_{p|\mu\nu}\left(1-\frac{1}{\|p\|^{3s/2-3}}\right)\mathcal{D}_{\mu,\nu}(s) \tag{2.5.27}$$

$$= \sum_{\substack{k_1,k_2,k_3\\ \text{as in (2.5.24)}}} \frac{\tau(\mu k_1/k_3)\widetilde{S}(\mu,\nu;k_1,k_2,k_3)}{|k_1k_3|^{s-1}|k_2|^{2s-4}} \sum_{\substack{m\equiv 1(3)\\ \gcd(m,\mu\nu)=1}} \frac{\overline{S(1,m)}^2}{|m|^s}\left(\frac{\mu\nu k_1^2k_2k_3^2}{m}\right).$$

Complementing the summation conditions for k_1, k_2, k_3 in (2.5.27) by the condition

$$\mu\nu k_1^2k_2k_3^2l^{-1} \quad \text{is a cube in} \quad \mathbf{Q}(\sqrt{-3})$$

and introducing into (2.5.27) the summation over all cube-free $l \in \Lambda(\mu\nu)$, we obtain (2.5.20).

To complete the proof it remains to show that $D_{\mu,\nu}(l;\cdot) = 0$ if there exist prime $p \equiv 1 \pmod 3$ such that $\operatorname{ord}_p \mu \equiv \operatorname{ord}_p \nu \equiv 2 \pmod 3$. Actually we can show now that, under this assumption, each term of the series defining $D_{\mu,\nu}(l;\cdot)$, see (2.5.18), is equal to 0. For this, let k_1, k_2, k_3 be as in (2.5.19). Given prime $p \equiv 1 \pmod 3$, we put

$$\theta = \operatorname{ord}_p \mu, \quad \zeta = \operatorname{ord}_p \nu, \quad \alpha = \operatorname{ord}_p k_1, \quad \beta = \operatorname{ord}_p k_2, \quad \gamma = \operatorname{ord}_p k_3.$$

If the term associated with k_1, k_2, k_3 in (2.5.18) is non-zero, then

$\tau(\mu k_1/k_3) \neq 0$, and hence $\theta + \alpha - \gamma \not\equiv 2 \pmod 3$;

$S(\mu k_1/k_2, k_3) \neq 0$, and hence either $\gamma = \theta + \alpha - \beta + 1$ or $\gamma \equiv 0 \pmod 3$;

$S(\mu, k_2, k_3^2) \neq 0$, and hence either $\beta = \theta + 1$ or $\beta + 2\gamma \equiv 0 \pmod 3$;

$S(\nu, k_1, k_2^2 k_3) \neq 0$, and hence either $\alpha = \zeta + 1$ or $\alpha + 2\beta + \gamma \equiv 0 \pmod 3$;

(see 0.1.4 and 0.3.8). If $\theta \equiv \zeta \equiv 2 \pmod 3$, these conditions are not satisfied by any α, β, γ, as claimed. ■

2.5.5 Poles of $c_{0,0}$. Let us turn to Theorem 1.5.1 and to look at the Eisenstein series $E(\sigma(\cdot), s; \Theta_{K\text{-}P})$ Fourier coefficient $c_{0,0}$. For fixed $u, v \in \mathbf{R}_+^*$), the coefficient $c_{0,0}$ is a meromorphic function on $\mathbf{C}$ and we would like to look at its singularities. With 0.1.5 and Theorem 2.5.1 we find that

$c_{0,0}$ has simple poles at the points 4 and 10/3,

just because of

ζ_* has a simple pole at the point 1,
ζ_* has first order zero at the point 0,
ζ_* is regular and non-zero at the points 2, 3, 4,
$I_{0,0}(u, v, \cdot)$ is regular and non-zero except simple poles at 8/3, 3, 10/3.

We find also that $c_{0,0}$ has no other poles in $\{ s \in \mathbf{C} \mid \Re(s) \geq 3 \}$.

2.5.6 Cubic theta functions. The general Eisenstein series theory and our consideration of the coefficient $c_{0,0}$ in 2.5.5 lead to the following theorem.

Theorem 2.5.5 *Eisenstein series $E(\sigma(\cdot), \cdot; \Theta_{K\text{-}P})$ has poles at the points 4 and 10/3. Its residues at these poles are cubic metaplectic forms on $\mathbf{X}$ under the group $\Gamma(3)$.* ■

It should be pointed out that, although the points 4 and 10/3 are simple poles of $c_{0,0}$, we can not say that these points are simple poles of the Eisenstein series. One need more detailed study of the Eisenstein series constant term to find the right orders of these poles. In any case, one has Laurent expansions of the Eisenstein series about the points 4 and 10/3, and the leading terms in these expansions are eigenfunctions of all the invariant differential operators, so satisfy (b) in the definition given in 2.3.2, while other terms satisfy (b′) only. (Compare with 1.5.5 and 0.3.8). In particular, in Theoerem 2.5.5 above we have in mind the residues of the Eisenstein series satisfy (b′).

Let us define cubic symplectic theta functions $\Theta_\natural : \mathbf{X} \to \mathbf{C}$ and $\Theta_\sharp : \mathbf{X} \to \mathbf{C}$ as the residues of the Eisenstein series —

$$\begin{aligned} \Theta_\natural &= 2^{-11/3} 3^{53/6} \pi^{-14/3} \zeta_*(4) \operatorname*{Res}_{s=4} E(\sigma(\cdot), s; \Theta_{K\text{-}P}), \\ \Theta_\sharp &= 2^{-3} 3^9 \pi^{-4} (\log 3)^{-1} \zeta_*(2)^2 \operatorname*{Res}_{s=10/3} E(\sigma(\cdot), s; \Theta_{K\text{-}P}). \end{aligned} \tag{2.5.28}$$

Up to the constants involved in the definition (2.5.28), the Fourier coefficients $c_{\mu,\nu}(u,v;\Theta_\natural)$, $c_{\mu,\nu}(u,v;\Theta_\sharp)$ of $\Theta_\natural$ and $\Theta_\sharp$ are the residues of the Fourier coefficients $c_{\mu,\nu}$ of the Eisenstein series. Let τ means Patterson's function (0.3.15), and let

$$\tilde{\tau}(\lambda) = \frac{\tau(\lambda)}{|\lambda|^{1/3}} \qquad \text{for} \quad \lambda \in (\sqrt{-3})^{-3}\mathcal{O}.$$

Theorem 2.5.6 *For the Fourier coefficients $c_{\mu,\nu}(u,v;\Theta_\natural)$, $\mu,\nu \in (\sqrt{-3})^{-3}\mathcal{O}$, of the cubic symplectic theta function $\Theta_\natural$ one has the following formulae.*

(a) $c_{0,0}(u,v;\Theta_\natural) = (6\pi)^{-2/3}u^{8/3}v^2$,

(b) $c_{\mu,0}(u,v;\Theta_\natural) = \tilde{\tau}(\mu)u^{8/3}v^2 Ai((6\pi|\mu|u)^{2/3})$ *if* $\mu \neq 0$,

(c) $c_{0,\nu}(u,v;\Theta_\natural) = \overline{\tilde{\tau}(\nu)}u^{8/3}v^2 Ai((6\pi|\nu|v)^{2/3})$ *if* $\nu \neq 0$,

(d) $c_{\mu,\nu}(u,v;\Theta_\natural) = 0$ *if* $\mu\nu \neq 0$. □

Proof. (a) Applying Theorem 2.5.1 (and taking into account the constant involved in the definition (2.5.28)) we find

$$c_{0,0}(u,v;\Theta_\natural) = 2^{-3}\pi^{-1}3^3u^{8/3}v^2 \operatorname*{Res}_{s=4} \zeta_*(3s/2-5).$$

This yields (a), since (see 0.1.5)

$$\operatorname*{Res}_{s=4} \zeta_*(3s/2-5) = 2^2\pi/3^{7/2}. \tag{2.5.29}$$

(b) From Theorem 2.5.3, we obtain

$$c_{\mu,0}(u,v;\Theta_\natural) = 2\pi^{-1}3^{3/2}\tau(\mu) \operatorname*{Res}_{s=4} \zeta_*(3s/2-5) I_{\mu,0}(u,v,4)$$

with $I_{\mu,0}(u,v,4) = 2^{-3}3^2u^3v^2K_{1/3}(4\pi|\mu|u)$, that being combined with (2.5.29) and (0.1.26) yields (b).

(c) From Theorem 2.5.2, it follows that

$$\begin{aligned} c_{0,\nu}(u,v;\Theta_\natural) &= 3\pi^{-1}\zeta_*(2) \operatorname*{Res}_{s=4} \overline{\Psi}(\bar{s}/2-2/3,\nu)I_{0,\nu}(u,v,4), \\ I_{0,\nu}(u,v,4) &= \frac{3\pi^{1/3}}{2^{5/3}\Gamma(4/3)}|\nu|^{1/3}u^{8/3}v^{7/3}K_{1/3}(4\pi|\nu|v). \end{aligned} \tag{2.5.30}$$

Then, according to Theorem 0.4.3 (b), we have

$$\begin{aligned} \operatorname*{Res}_{s=4} \overline{\Psi}(\bar{s}/2-2/3,\nu) &= 2\,\overline{\operatorname*{Res}_{s=4/3} \Psi(s,\nu)} \\ &= 2^{8/3}3^{-13/2}\pi^{5/3}\zeta_*(2)^{-1}\Gamma(2/3)^{-1}\frac{\overline{\tau(\nu)}}{|\nu|^{1/3}}. \end{aligned} \tag{2.5.31}$$

Taking into account that $\Gamma(2/3)\Gamma(4/3) = 2\pi/3^{3/2}$ (by triplication formula), we see that (c) follows from (2.5.30), (2.5.31), (0.1.26).

(d) As it was already noticed in 2.5.4, the functions $D_{\mu,\nu}(l;\cdot)$ and $P_{\mu,\nu}(l;\cdot)$ are regular at the point 4. The function $I_{\mu,\nu}(u,v,\cdot)$ also is regular at this point, as one can see from Theorem 2.4.9. This means that the point 4 is not a pole of the function under consideration, so we have (d). ∎

We see that the cubic symplectic theta function Θ_1 is singular, in the sence that its Fourier coefficients $c_{\mu,\nu}(u,v;\Theta_1)$ are zero for all non-zero μ,ν. The second cubic symplectic theta function Θ_2 is more complicated one.

Theorem 2.5.7 *For the Fourier coefficients $c_{\mu,\nu}(u,v;\Theta_2)$, $\mu,\nu \in (\sqrt{-3})^{-3}\mathcal{O}$, of the cubic symplectic theta function Θ_2 one has the following formulae.*

(a) $c_{0,0}(u,v;\Theta_2) = -2^{-1}3^{-1/2}u^{10/3}v^{8/3}$,

(b) $c_{\mu,0}(u,v;\Theta_2) = -\tau(\mu)u^{11/3}v^{8/3}K_{1/3}(4\pi|\mu|u)$ *if* $\mu \neq 0$,

(c) $c_{0,\nu}(u,v;\Theta_2) = \overline{\eta(\nu)}u^{10/3}v^{8/3}K_0(4\pi|\nu|v)$ *if* $\nu \neq 0$,

(d) $c_{\mu,\nu}(u,v;\Theta_2) = \lambda(\mu,\nu)u^{13/3}v^{8/3}R(4\pi|\mu|u, 4\pi|\nu|v)$ *if* $\mu\nu \neq 0$.

Here: $\tau(\mu)$, $\eta(\nu)$ are as in 0.3.8, 0.3.14; $\lambda(\mu,\nu)$ can be written as the finite sum

$$\lambda(\mu,\nu) = \frac{2^{5/3}3^{7/2}\zeta_*(2)|\mu|^{2/3}}{\pi^{1/3}\Gamma(1/3)^2\log 3} \sum_{\substack{l\in\Lambda(\mu\nu)\\ l \text{ is cube-free}}} D_{\mu,\nu}(l;10/3) \operatorname*{Res}_{s=10/3} P_{\mu,\nu}(l;s)$$

with $P_{\mu,\nu}$, $D_{\mu,\nu}$, $\Lambda(\mu\nu)$ as in (2.5.17), (2.5.18); by $R(u,v)$ we mean

$$\int_0^\infty\!\!\int_0^\infty K_{1/3}(u\sqrt{1+x+y})K_0(v\sqrt{(1+1/x)(1+1/y)})\frac{dx\,dy}{(1+x+y)^{1/6}x^{2/3}y^{2/3}}. \quad \square$$

Proof. (a) We have $\zeta_{\mathbf{Q}(\sqrt{-3})}(0) = -1/6$, and

$$\lim_{s\to 10/3}\frac{\zeta_*(3s/2-5)}{3s-10} = \zeta_{\mathbf{Q}(\sqrt{-3})}(0)\lim_{s\to 10/3}\frac{1-3^{5-3s/2}}{3s-10} = -\frac{\log 3}{12},$$
$$\operatorname*{Res}_{s=10/3}\zeta_*(3s-9) = 2\pi/3^{7/2}, \tag{2.5.32}$$

see 0.1.5. With these notices, Theorem 2.5.1 and (2.5.28) yield

$$c_{0,0}(u,v;\Theta_2) = -2^{-2}\pi^{-1}3^3u^{10/3}v^{8/3}\operatorname*{Res}_{s=10/3}\zeta_*(3s-9)$$
$$= -2^{-1}3^{-1/2}u^{10/3}v^{8/3}, \quad \text{as required.}$$

(b) Again, with (2.5.32) and (2.5.28) we deduce from Theorem 2.5.3:

$$c_{\mu,0}(u,v;\Theta_{\mathfrak{s}}) = 2\frac{3^{7/2}\tau(\mu)}{\pi\log 3}u^{11/3}v^{8/3}K_{1/3}(4\pi|\mu|u)\operatorname*{Res}_{s=10/3}\frac{\zeta_*(3s-9)\zeta_*(3s/2-5)}{s-10/3}$$
$$= -\tau(\mu)u^{11/3}v^{8/3}K_{1/3}(4\pi|\mu|u), \quad \text{as required.}$$

(c) From Theorem 2.5.2 and (2.5.28), it follows that

$$\begin{aligned} c_{0,\nu}(u,v;\Theta_{\mathfrak{s}}) &= \frac{6}{\pi\log 3}\operatorname*{Res}_{s=10/3}\nabla(s)\,I_{0,\nu}(u,v,10/3), \\ I_{0,\nu}(u,v,10/3) &= 2^{-1}3^2u^{10/3}v^{8/3}K_0(4\pi|\nu|v), \end{aligned} \tag{2.5.33}$$

where we write $\nabla(s)$ for $\overline{\Psi}(\overline{s}/2-2/3,\nu)\zeta_*(3s-9)\zeta_*(3s/2-4)$. By Theorem 0.4.3 (c) we have $\Psi(1,\nu)=0$, so $\Psi(t,\nu)=\Psi'(1,\nu)(t-1)+O((t-1)^2)$ as $t\to 1$. Consequently (see 0.1.5),

$$\operatorname*{Res}_{s=10/3}\nabla(s) = 3^{-2}\left\{\operatorname*{Res}_{t=1}\zeta_*(t)\right\}^2\overline{\Psi'(1,\nu)} = 2^23^{-7}\pi^2\overline{\Psi'(1,\nu)}.$$

Then, by Theorem 0.3.17 (c) we have $\Psi'(1,\nu)=2^{-2}3^4\pi^{-1}(\log 3)\eta(\nu)$, and so

$$\operatorname*{Res}_{s=10/3}\nabla(s) = 3^{-3}(\log 3)\pi\overline{\eta(\nu)}. \tag{2.5.34}$$

Combining (2.5.33) and (2.5.34) we obtain (c).
(d) Theorem 2.5.4 and (2.5.28) yield

$$\begin{aligned} c_{\mu,\nu}(u,v;\Theta_{\mathfrak{s}}) = \frac{3^{3/2}\zeta_*(2)}{\pi^4\log 3}&I_{\mu,\nu}(u,v,10/3)\cdot \\ &\cdot\sum_{\substack{l\in\Lambda(\mu\nu)\\ l\text{ is cube-free}}} D_{\mu,\nu}(l;10/3)\operatorname*{Res}_{s=10/3}P_{\mu,\nu}(l;s). \end{aligned} \tag{2.5.35}$$

(Recall, see the proof of Theorem 2.5.6 (d), that $D_{\mu,\nu}(l;\cdot)$ and $I_{\mu,\nu}(u,v,\cdot)$ are regular at 10/3.) From Theorem 2.4.9 and Theorem 2.4.14, we obtain

$$\begin{aligned} I_{\mu,\nu}(u,v,10/3) &= J_{\mu,\nu}(u,v,10/3,4/3) \\ &= \frac{2^{5/3}\pi^{11/3}}{\Gamma(4/3)^2|\mu|^{11/3}|\nu|^{8/3}}R_{2,4/3}(|\mu|u,|\nu|v) \\ &= \frac{2^{5/3}\pi^{11/3}}{\Gamma(4/3)^2}|\mu|^{2/3}u^{13/3}v^{8/3}R(4\pi|\mu|u,4\pi|\nu|v), \end{aligned}$$

that, after substituting into (2.5.35), gives (d). ■

According to Theorem 2.5.4, if $\mathrm{ord}_p\,\mu \equiv \mathrm{ord}_p\,\nu \equiv 2 \pmod 3$ for at least one prime $p \equiv 1 \pmod 3$, then $D_{\mu,\nu}(l; 10/3) = 0$. Thus, this case our formulae in Theorem 2.5.7 yield $c_{\mu,\nu}(u, v; \Theta_\sharp) = 0$. One can expect that $c_{\mu,\nu}(u, v; \Theta_\sharp) = 0$ also for some other μ, ν, but we have not full description of such μ, ν.

2.5.7 Commentary. The cubic symplectic metaplectic forms and theta functions were defined and studied by the author in [79], [80], [81]. Our present exposition in these notes is similar to that of [81], but includes study of the Whittaker functions and more detailed study of the cubic symplectic theta functions Fourier coefficients.

However, our formulae for the Fourier coefficients of $\Theta_\sharp$ in Theorem 2.5.7 (c), (d) are not conclusive, and we would like to give some additional commentary.

The first problem for further research is to determine η-function, i.e., to determine the Fourier coefficients $\eta(\nu)$ of the function $\Theta_!$ in 0.3.14. One may hope that methods similar to those used to study $\Theta_{K\text{-}P}$ can be applied to this problem.

The second problem concerns the coefficients $\lambda(\mu, \nu)$. One may expect that $\lambda(\mu, \nu)$ can be expressed somehow in terms of $\tau(\mu)$ and $\eta(\nu)$. To support such viewpoint we can refer to analoque with $SL(3, \mathbf{C})$-case (see 1.5.6) and to notice that $R(u, v)$ in part (d) of Theorem 2.5.7 is expressed in terms of the functions which occur in parts (b) and (c). It is not too difficult to evaluate the sums $D_{\mu,\nu}(l; 10/3)$, and thus to reduce the problem to evaluation of the residues

$$\operatorname*{Res}_{s=10/3} P_{\mu,\nu}(l; s). \tag{2.5.36}$$

As it is explained by Patterson (see remark in 0.4.2), there is a connection between (2.5.36) from one side, and the Fourier coefficients of six degree theta function on $\mathbf{H}$ from another side. Any progress in understanding of the six degree theta function on $\mathbf{H}$ will shed some light on the nature of (2.5.36), and visa versa.

One can introduce and apply Hecke operators acting on cubic metaplectic forms on $\mathbf{X}$. The Eisenstein series and so $\Theta_\sharp$ are eigenfunctions of the Hecke operators. This yields certain relations satisfied by the Fourier coefficients $\lambda(\mu, \nu)$ of $\Theta_\sharp$. However, it is clear these relations alone are not sufficient, and to determine $\lambda(\mu, \nu)$ one will need some more relations, maybe some analoque of the Kazhdan–Patterson periodicity theorem (see 1.5.7).

One more way to study the Fourier coefficients of $\Theta_\sharp$ is through studing the Eisenstein series $\widetilde{E}(\cdot, s; f)$ attached to the Jacobi maximal parabolic subgroup $\widetilde{P}$ instead of the Eisenstein series $E(\cdot, s; f)$ attached to the Siegel maximal parabolic subgroup P we dealt with. In view of the relations between the different type of Eisenstein series (see 2.3.6), one can define $\Theta_\sharp$ as the residue of the series $\widetilde{E}(\cdot, s; f)$ with $f = \Theta_!$. It seems reasonable to expect that formulae for the Fourier coefficients that one can obtain this way should be essentially different from those given in Theorem 2.5.7. If so, comparing the formulae one can obtain more information on $\lambda(\mu, \nu)$, $\eta(\nu)$ and the residues (2.5.36).

References

[1] Atkin A. O. L., Lehner J. *Hecke Operators on* $\Gamma_0(m)$. Mathematische Annalen, Band 185, 1970, 134–160.

[2] Baily W. L. *Introductory lectures on automorphic forms.* Publications of the Mathematical Society of Japan, Iwanami Shoten, Publishers and Princeton University Press, 1973.

[3] Bailey W. N. *Generalized hypergeometric series.* Cambridge at the University press, 1935.

[4] Bass H. *K-theory and stable algebra.* Institut des Hautes Études Scientifiques, Publ. Math., n° 22, 1964, 5–60.

[5] Bass H., Milnor J., Serre J.-P. *Solution of the congruence subgroup problem for* SL_n $(n \geq 3)$ *and* Sp_{2n} $(n \geq 2)$. Institut des Hautes Études Scientifiques, Publ. Math., n° 33, 1967, 59–137.

[6] Beltrami E. *Teoria fondamentale degli spazii di curvature costante.* Annali di Matematica pure ed applicata, ser. II, 2, 232–255, 1868. (*Opere Matematiche*, t. 1, Milano, 1902.)

[7] Berndt B. C., Evans R. J. *The determination of Gauss sums.* Bulletin of the Amer. Math. Soc., new series, Vol. 5, 1981, 107–129.

[8] Bianchi L. *Geometrische darstellung der Gruppen linearer substitutionen mit ganzen complexen coefficienten nebst anwendungen auf die zahlentheorie.* Mathematische Annalen, Band 38, 1891, 313–333. (*Opere*, Vol. I, Roma, 1952.)

[9] Bianchi L. *Sui gruppi de sostituzioni lineari con coefficienti appartenenti a corpi quadratici immaginari.* Mathematische Annalen, Band 40, 1892, 332–412. (*Opere*, Vol. I, Roma, 1952.)

[10] Borel A. *Commensurability classes and volumes of 3-manifolds.* Annali della Scuola Normale Superiore di Pisa, Cl. Sci., Serie IV, Vol. VIII, 1981, 1–33.

[11] Borevich Z. I., Shafarevich I. R. *Number Theory.* New York, 1966.

[12] Bourbaki N. *Éléments de Mathématique, Fonctions d'une variable réelle.* Paris.

[13] Bourbaki N. *Éléments de Mathématique, Groupes et algébres de Lie.* Paris.

[14] Bump D., Hoffstein J. *Cubic metaplectic forms on* $\mathrm{GL}(3)$. Inventiones Math., Vol. 84, 1986, 481–505.

[15] Bump D., Friedberg S. *On Mellin transform of unramified Whittaker functions on* $\mathrm{GL}(3,\mathbf{C})$. Journal of Math. analysis and applications, Vol. 139, 1989, 205–216.

[16] Bump D., Friedberg S., Hoffstein J. *The Kubota symbol for* $\mathrm{Sp}(4,\mathbf{Q}(i))$. Nagoya Math. J., Vol. 119, 1990, 173–188.

[17] Cassels J. W. S., Frohlich A. *Algebraic Number Theory.* London, 1967.

[18] Deligne P. *Sommes de Gauss cubiques et revetements de* SL(2), *d'apres S. J. Patterson.* Seminaire Bourbaki, Springer–Verlag Lecture Notes in Math. 770, 1980, 244–277.

[19] Deligne P., Serre J-P. *Formes modulaires de poids* 1. Annales scientifiques de l'École Normale Supérieure, 4e série, t. 7, fasc. 4, 1974, 507–530.

[20] Eckhardt C., Patterson S. J. *On the Fourier coefficients of biquadratic theta series.* Proceedings of the London Math. Soc. (3), Vol. 64, 1992, 225–264.

[21] Eisenstein G. *Beweis des Reciprocitätssatzes für die cubischen Reste*.... J. für die reine und angew. Math., Band 27, 1844, 289–310. (*Mathematische Werke*, Band I, 1975.)

[22] Eisenstein G. *Nachtrag zum cubischen Reciprocitätssatze*.... J. für die reine und angew. Math., Band 28, 1844, 28–35. (*Mathematische Werke*, Band I, 1975.)

[23] Eisenstein G. *Lois de réciprocité.* J. für die reine und angew. Math., Band 28, 1844, 53–67. (*Mathematische Werke*, Band I, 1975.)

[24] Elstrodt J., Grunewald F., Mennicke J. *Discontinuous groups on three-dimentional hyperbolic space: analytic theory and arithmetic applications.* Russian Mathematical Surveys, 38:1, 1983, 137–168.

[25] Elstrodt J., Grunewald F., Mennicke J. *Eisenstein series on three-dimentional hyperbolic space and imaginary quadratic number fields.* J. für die reine und angew. Math., Band 360, 1985, 160–213.

[26] Elstrodt J., Grunewald F., Mennicke J. *Arithmetic applications of the hyperbolic lattice point theorem.* Proceedings of the London Math. Soc. (3), Vol. 57, 1988, 239–283.

[27] Flicker Y. Z. *Automorphic forms on covering groups of* GL(2). Inventiones Math., Vol. 57, 1980, 119–182.

[28] Flicker Y. Z., Kazhdan D. A. *Metaplectic correspondence.* Institut des Hautes Études Scientifiques, Publ. Math., n° 64, 1986, 53–110.

[29] Gauss C. F. *Untersuchungen über höhere Arithmetik.* Springer–Verlag, 1889.

[30] Gradshteyn I. S., Ryzhik I. M. *Tables of integrals, series and products* Academic Press, New York, 1980.

[31] Grunewald F., Helling H., Mennicke J. SL_2 *over complex quadratic number fields.* Algebra i logika (Algebra and logic), Vol. 17, Number 5, 1978, 512–580.

[32] Harish-Chandra. *Automorphic Forms on Semi-Simple Lie Groups.* Springer-Verlag Lecture Notes in Math. 62, 1968.

[33] Hashizume M. *Whittaker functions on semisimple Lie groups.* Hiroshima Math. J., Vol. 12, 1982, 259–293.

[34] Hasse H. *Vorlesungen uber Zahlentheorie.* Berlin, 1950.

[35] Helgason S. *Differential geometry, Lie groups and symmetric spaces.* Academic Press, London, 1978.

[36] Helgason S. *Groups and Geometric Analysis.* Academic Press, London, 1984.

[37] Heath-Brown D. R., Patterson S. J. *The distribution of Kummer sums at prime arguments.* J. für die reine und angew. Math., Band 310, 1979, 111–130.

[38] Humbert G. *Sur la réduction des formes d'Hermite dans un corps quadratique imaginaire.* Comptes Rendus Acad. Sci. Paris, t. 161, 1915, 189–196.

[39] Humbert G. *Sur la mesure des classes d'Hermite de discriminant donné dans un corps quadratique imaginaire, et sur certains volumes non euclidiens.* Comptes Rendus Acad. Sci. Paris, t. 169, 1919, 448–454.

[40] Imamoglu Ö. *The Kubota symbol for* $\mathrm{Sp}(2n, \mathbf{Q}(i))$. Journal of Number Theory, Vol. 52, No. 1, 1995, 17–34.

[41] Ireland K., Rosen M. *A Classical Introduction to Modern Number Theory.* Graduate Texts in Mathematics 87, Springer–Verlag, New York, Heidelberg, Berlin, 1982.

[42] Iwasawa K. *On some type of topological groups.* Annals of Mathematics, second series, Vol. 50, 1949, 507–558.

[43] Jacquet H. *Functions de Whittaker associées aux groupes de Chevalley.* Bulletin de la Soc. Math. de France, Vol. 95, 1967, 243–309.

[44] Jacquet H., Piatetski-Shapiro I. I., Shalika J. *Automorphic forms on* $\mathrm{GL}(3)$. Annals of Mathematics, second series, Vol. 109, 1979, 169–212, 213–258.

[45] Kapitanski L. V., Proskurin N. V. (to appear).

[46] Kazhdan D. A., Patterson S. J. *Metaplectic forms.* Institut des Hautes Études Scientifiques, Publ. Math., n° 59, 1984, 35–142.

[47] Kazhdan D. A., Patterson S. J. *Towards a generalized Shimura correspondence.* Advances in Mathematics, Vol. 60, 1986, 161–234.

[48] Klein F. *Neue Beiträge zur Riemannschen Functionentheorie.* Mathematische Annalen, Band 21, 1883, 141–218.

[49] Klingen H. *Zum Darstellungssatz für Siegelsche Modulformen.* Mathem. Zeitschrift, Band 102, 1967, 30–43.

[50] Koblitz N. *Introduction to Elliptic Curves and Modular Forms.* Graduate Texts in Mathematics 97, Springer–Verlag, New York, Berlin, Heidelberg, Tokyo, 1984.

[51] Kubota T. *Ein arithmetischer Satz uber eine Matrizengruppe.* J. für die reine und angew. Math., Band 222, 1965, 55–57.

[52] Kubota T. *Topological covering of* $\mathrm{SL}(2)$ *over a local field.* J. Math. Soc. Japan, 19, 1967, 114–121.

[53] Kubota T. *Uber diskontinuierliche Gruppen Picardschen Typus und zigehorige Eisensteinsche Reihen.* Nagoya Math. J., Vol. 32, 1968, 259–271.

[54] Kubota T. *On automorphic functions and reciprocity law in a number theory.* Lecture in Math. Kyoto Univer., Vol. 2, 1969.

[55] Kubota T. *Two kinds of special functions related to the reciprocity law.* Proceedings of the summer institute on number theory, New York, Stony Brook, 1969.

[56] Kubota T. *Some number theoretical results on real analytic automorphic forms.* Springer–Verlag Lecture Notes in Math. 185, 1971, 87–96.

[57] Kubota T. *Some results concerning reciprocity law and real analytic automorphic forms.* Proceedings Symp. Pure Math. 20, 1971, 382–395.

[58] Kubota T. *Elementary theory of Eisenstein series.* Kodausha Ltd., Tokyo, New York, 1973.

[59] Langlands R. P. *On the functional equations satisfied by Eisenstein Series.* Springer–Verlag Lecture Notes in Math. 544, 1976.

[60] Luke Y. L. *The Special Functions and Their Approximations.* Academic Press, New York, 1969.

[61] Maaß H. *Über eine neue Art von nichtanalytischen automorphen Funktionen und die Bestimmung Dirichlet'sher Reihen durch Funktionalgleichungen.* Mathematische Annalen, Band 121, 1949, 141–183.

[62] Maaß H. *Siegel's modular forms and Dirichlet series.* Springer–Verlag Lecture Notes in Math. 216, 1971.

[63] Matsumoto H. *Sur les sous-groupes arithmétiques des groupes semi-simples déployés.* Annales scientifiques de l'École Normale Supérieure, 4e série, t. 2, 1969, 1–62.

[64] Milnor J. *Hyperbolic geometry: The first 150 years.* Bulletin of the Amer. Math. Soc., new series, Vol. 6, 1982, 9–24.

[65] Moore C. C. *Group extensions of p-adic and adelic linear groups.* Institut des Hautes Études Scientifiques, Publ. Math., n° 35, 1968, 157–222.

[66] Niwa S. *On generalized Whittaker functions on Siegel's upper half space of degree 2.* Nagoya Math. J. 121, 1991, 171–184.

[67] Oda T. *An explicit integral representation of Whittaker functions on* $\mathrm{Sp}(2,\mathbf{R})$ *for the large discrete series representations.* Tôhoku Math. J., Vol. 46, 1994, 261–279.

[68] Olver F. W. J. *Asymptotics and Special Functions.* Academic Press, New York and London, 1974.

[69] Patterson S. J. *A cubic analogue of the theta series* I, II. J. für die reine und angew. Math., Band 296, 1977, 125–161, 217–220.

[70] Patterson S. J. *On Dirichlet series associated with cubic Gauss sums.* J. für die reine und angew. Math., Band 303/304, 1978, 102–125.

[71] Patterson S. J. *On the distribution of Kummer sums.* J. für die reine und angew. Math., Band 303/304, 1978, 126–143.

[72] Patterson S. J. *Whittaker models of generalized theta series.* Sém. de th. des nombres, Paris 1982–1983, Birkhäuser, 1984.

[73] Patterson S. J. *Theta functions in finite characteristics.* Universität Hamburg Mathematisches seminar, International Symposium on number theory and analysis in Honour of Erich Hecke, 1987, 76–79.

[74] Patterson S. J., Piatetski-Shapiro I. I. *A cubic analogue of the cuspidal theta representations.* Journal de Mathématiques Pures et Appliquées, Vol. 63, n° 3, 1984, 333–375.

[75] Piatetski-Shapiro I. I. *Euler subgroups.* Lie Groups and their representations. Budapest, 1975. (Proceedings of the summer school on group representations, Bolyai János Mathematical Societ, 1971.)

[76] Picard É. *Sur un groupe de transformations des points de l'espace situés du même côté d'un plan.* Bulletin de la Soc. Math. de France, t. 12, 1884, 43–47. (*Œuvres*, t. I, Paris, 1978.)

[77] Poincaré H. *Mémoire sur les groupes kleinéens.* Acta Mathematica, Band 3, 1883, 49–92. (*Œuvres*, t. 2, Paris, 1916.)

[78] Proskurin N. V. *Automorphic functions and Bass–Milnor–Serre homomorphism*, I, II. Zap. Nauchn. Semin. LOMI, t. 123, 1983, 85–126, 127–163. (English translation: J. of Soviet Math., Vol. 29, Number 2, April 1985, 1160–1191, 1192–1219.)

[79] Proskurin N. V. *On cubic symplectic metaplectic forms.* Preprint LOMI, E-1-87, 1987.

[80] Proskurin N. V. *Cubic symplectic theta functions.* Zap. Nauchn. Semin. LOMI, t. 162, 1987, 186–188. (English translation: J. of Soviet Math., Vol. 46, Number 2, July 1989, 1841–1842.)

[81] Proskurin N. V. *On cubic symplectic metaplectic forms.* J. für die reine und angew. Math., Band 388, 1988, 158–188.

[82] Proskurin N. V. *Automorphic forms on the exceptional group* $G_2(\mathbf{C})$. Algebra and analysis, Vol. 1, Number 3, 1989, 196–226. (English translation: Leningrad Mathematical J., 1990, Vol. 1, Number 3, 765–797.)

[83] Proskurin N. V. *Cubic metaplectic forms on exceptional Lie group* G*-type.* Preprint LOMI, E-5-90, 1990.

[84] Proskurin N. V. *Cubic metaplectic forms on exceptional Lie group of type* G_2. Journal de Mathématiques Pures et Appliquées, Vol. 72, n° 3, 1993, 287–326.

[85] Proskurin N. V. *Integral addition formulae for Airy function.* Integral transforms and special functions (to appear).

[86] Sarnak P. *The arithmetic and geometry of some hyperbolic three manifolds.* Acta Mathematica, Band 151, 1983, 253–295.

[87] Selberg A. *Harmonic analysis and discontinuous groups in weakly symmetric Riemannian spaces with applications to Dirichlet series.* J. Indian Math. Soc., Vol. 20, 1956, 47–87.

[88] Selberg A. *Discontinuous groups and harmonic analysis.* Proc. ICM., Stocholm, 1962, Uppsala, Almqvist, Wikwells (1963).

[89] Serre J-P., Stark H. M. *Modular forms of weight* $1/2$. Springer–Verlag Lecture Notes in Math. 627, 1977, 27–67.

[90] Shalika J. *The multiplicity one theorem for* $\mathrm{GL}(n)$. Annals of Mathematics, second series, Vol. 100, 1974, 171–193.

[91] Shimura G. *Introduction to the arithmetic theory of automorphic functions.* Iwanami Shoten Publishers and Princeton University Press, 1971.

[92] Shimura G. *On modular forms of half-integral weight.* Annals of Mathematics, second series, Vol. 97, 1973, 440–481.

[93] Shimura G. *On the holomorphy of certain Dirichlet series.* Proceedings of the London Math. Soc. (3), Vol. 31, 1975, 79–98.

[94] Suzuki T. *Some results on the coefficients of the biquadratic theta series.* J. für die reine und angew. Math., Band 340, 1983, 70–117.

[95] Suzuki T. *Rankin–Selberg convolutions of generalized theta series.* J. für die reine und angew. Math., Band 414, 1991, 149–205.

[96] Swan R. G. *Generators and relations for certain special linear groups.* Advances in Mathematics, Vol. 6, 1971, 1–77.

[97] Vahlen R. *Über Bewegungen und complex Zahlen.* Mathematische Annalen, Band 55, 1902, 585–593.

[98] Varadarajan V. S. *An introduction to Harmonic analysis on semisimple Lie groups.* Cambridge University press, Cambridge, 1989.

[99] Venkov A. B. *Spectral theory of automorphic functions and its applications.* Kluwer Academic Publishers, Dordrecht, Boston, London, 1990.

[100] Vinogradov A. I., Tahtajan L. A. *Theory of Eisenstein series for the group* $\mathrm{SL}(3, \mathbf{R})$ *and its application to a binary problem.* Zap. Nauchn. Semin. LOMI, t. 76, 1978, 5–52. (English translation: J. of Soviet Math., Vol. 18, Number 3, February 1982, 293–324.)

[101] Watson G. N. *A treatise on the theory of Bessel functions.* Cambridge, 1966.

[102] Weil A. *Basic number theory.* Springer–Verlag, Berlin, Heidelberg, New York, 1967.

[103] Whittaker E. T., Watson G. N. *A course of modern analysis.* Cambridge, 1927.

Index